ADVANCES IN ELECTRONIC TESTING

CHALLENGES AND METHODOLOGIES

FRONTIERS IN ELECTRONIC TESTING

Consulting Editor
Vishwani D. Agrawal

Books in the series:

Introduction to Advanced System-on-Chip Test Design and Optimi...
Larsson, E., Vol. 29
ISBN: 1-4020-3207-2

Embedded Processor-Based Self-Test
Gizopoulos, D. (Ed.), Vol. 28
ISBN: 1-4020-2785-0

Advances in Electronic Testing : Challenges and Methodologies
Gizopoulos, D. (Ed.), Vol. 27
ISBN: 0-387-29408-2

Testing Static Random Access Memories
Hamdioui, S., Vol. 26,
ISBN: 1-4020-7752-1

Verification by Error Modeling
Radecka, K. and Zilic, Vol. 25
ISBN: 1-4020-7652-5

Elements of STIL: Principles and Applications of IEEE Std. 1450
Maston, G., Taylor, T. (et al.), Vol. 24
ISBN: 1-4020-7637-1

Fault Injection Techniques and Tools for Embedded systems Reliability ...
Benso, A., Prinetto, P. (Eds.), Vol. 23
ISBN: 1-4020-7589-8

Power-Constrained Testing of VLSI Circuits
Nicolici, N., Al-Hashimi, B.M., Vol. 22B
ISBN: 1-4020-7235-X

High Performance Memory Testing
Adams, R. Dean, Vol. 22A
ISBN: 1-4020-7255-4

SOC (System-on-a-Chip) Testing for Plug and Play Test Automation
Chakrabarty, K. (Ed.), Vol. 21
ISBN: 1-4020-7205-8

Test Resource Partitioning for System-on-a-Chip
Chakrabarty, K., Iyengar & Chandra(et al.), Vol. 20
ISBN: 1-4020-7119-1

A Designers' Guide to Built-in Self-Test
Stroud, C., Vol. 19
ISBN: 1-4020-7050-0

Boundary-Scan Interconnect Diagnosis
de Sousa, J., Cheung, P.Y.K., Vol. 18
ISBN: 0-7923-7314-6

Essentials of Electronic Testing for Digital, Memory, and Mixed Signal VLSI Circuits
Bushnell, M.L., Agrawal, V.D., Vol. 17
ISBN: 0-7923-7991-8

Analog and Mixed-Signal Boundary-Scan: A Guide to the IEEE 1149.4 Test ...
Osseiran, A. (Ed.), Vol. 16
ISBN: 0-7923-8686-8

ADVANCES IN ELECTRONIC TESTING

CHALLENGES AND METHODOLOGIES

Edited by

DIMITRIS GIZOPOULOS
University of Piraeus, Greece

Springer

A C.I.P. Catalogue record for this book is available from the Library of Congress.

ISBN-10 0-387-29408-2 (HB)
ISBN-13 978-0-387-29408-7 (HB)
ISBN-10 0-387-29409-0 (e-book)
ISBN-13 978-0-387-29409-4 (e-book)

Published by Springer,
P.O. Box 17, 3300 AA Dordrecht, The Netherlands.

www.springeronline.com

Printed on acid-free paper

Printed in the Netherlands.

Contents

Foreword

by Vishwani D. Agrawal

About five years ago, I was engaged in co-authoring a text-book on electronic testing for the *Frontiers in Electronic Testing* Book Series. We had to make some difficult decisions about what to include and what not to include. A text-book must contain all or most of the essentials of the established practices, and should not exceed a convenient, somewhat "standard," size. Those requirements do not leave room for what are the needs of the future and many of the groundbreaking developments. As a result, stuck-at tests win over delay tests, conventional analog tests are selected while radio frequency testing is ignored, and defect-oriented testing of nanometer devices is barely mentioned. So, no sooner the text-book was completed, I developed a feeling of discomfort about leaving out a vast amount of material on testing that could have been included.

What I have said about the text-book in the *Frontiers Series* applies to other text-books as well. As years go by the gap between those text-books and what is considered up-to-date has been widening. There is a definite need for documenting the advances in testing. It is for these reasons that I find the work of this edited volume by Dimitris Gizopoulos and his team of authors to be significant and timely.

The field of modern electronic testing can be regarded as about half a century old. Two things are obvious. First, the field has gained maturity. We have well established conferences and workshops all over the world organized by IEEE

Computer Society's Test Technology Technical Council. The attendance at these meeting remained quite steady even through down turns of the semiconductor industry. There is an over ten years old *Journal of Electronic Testing: Theory and Applications* that is entirely devoted to testing. The IEEE Computer Society has been publishing the *IEEE Design & Test of Computers* magazine for over two decades. Clearly, the field of electronic testing has developed a core but, and this is my second point, the field of testing now has a divergence of specializations. It is this divergence that an advanced book like this one captures.

While no one argues that the idea of an advanced book is good, there are problems with its implementation. Specialization demands experts and no single expert feels competent to write about all areas. Dimitris Gizopoulos has gathered a team of experts to write this book. Hence, the book provides, besides novel test methodologies, a collective insight into the emerging aspects of testing. This, I think, is beneficial to practicing engineers and researchers both of whom must stay at the forefront of technology.

Let me share a few of these insights with the reader. In Chapter 1, Rob Aiken states a theme, "*Defect-oriented tests for digital logic typically include comprehensive structural logic tests, a current test ... and at-speed tests. All these tests share the property that they measure some aspect of circuit behavior that is directly affected by defects,...*" before expanding on it.

In Chapter 2, Jaume Segura, Charles Hawkins and Jerry Soden give a motivation for statistical test methods by saying, "*Deep submicron structures don't affect test and diagnosis just because they are small. They primarily impact test because the manufacturing parameters are not tightly controlled as they were in the past.*"

Doug Josephson and Bob Gottlieb share their insights on silicon debug in Chapter 3. According to them, "*... test cases that are interesting for electrical validation are likely very different from those that are interesting for functional validation. For the ALU example, a CMOS dynamic circuit implementation may perform differently electrically if there are two add instructions executed in consecutive clock cycles than it would if there was a long period of inactivity between the add instructions ... from a functional validation point of view, these cases would be identical.*" Besides, I was fascinated with, "*An interesting example of "debugging" was in 1945 when a computer failure was traced down to a moth that was caught in a relay between contacts (Figure 3-1).*"

Discussing delay testing in Chapter 4, Adam Cron candidly admits, "*Much of this "information" about the prominent defect types is from informal discussions with engineers and researchers "in the trenches".* "

Continuing on the theme of high-speed test in Chapter 5, Wolfgang Maichen points out, "*... any chain is only as strong as its weakest link. In this case it means that even the best performing, highest bandwidth, most accurate tester will fail to reliably sort good devices from bad ones or give accurate characterization results if the connection between tester and device – i.e. the interface – does not perform equally well ...*"

Chapter 6 on low-cost testers, written by Al Crouch and Geir Eide, is particularly timely in view of the rising costs of testing and the test equipment. This subject is almost always found missing from the usual text-books.

Today, it is unthinkable that a VLSI chip will be designed without embedded cores. In Chapter 7, Rubin Parekhji points out the problems with applying the conventional test methodology to core-based System-on-Chip (SOC). He goes on to provide test solutions that use the conventional test methods and the IEEE standards like 1149.1 and 1500.

A majority of the embedded cores are memories. Dean Adams, the author of Chapter 8, describes the design for test structures and test methods in detail. The following sentence in that chapter very well represents the nature of the memory test problem and its solution: "*All of the testing and redundancy calculation must be performed by built-in self-test logic embedded on chip around the memory structures. The BIST must be implemented and integrated on the chip through the use of EDA tools which understand the memories, the process, and the physical constraints of the chip.*"

It is often said that testing of 10% analog circuitry of a mixed-signal device may contribute to 90% of the total test cost. Clearly, analog testing cannot be ignored. In Chapter 9, Stephen Sunter gives a complete coverage of analog test methodologies, fault modeling, design for testability including the IEEE 1149.4 test bus standard, and test tools.

For many digital test professionals radio frequency (RF) testing, mostly neglected during education, remains an unavoidable mystery. The wireless communication systems of today require SOCs that contain RF components. What I said above about the cost of analog testing is even more applicable to RF testing. Chapter 10 by Randy Wolf, Mustapha Slamani, John Ferrario and Jayendra Bhagat contains a comprehensive discussion on RF testing methods and tools that very few books on testing can boast of.

If we consider the varieties and the total number of printed circuit boards (PCB) manufactured in the world it will immediately become evident that the PCB test problem is no less important than the semiconductor device test problem. In Chapter 11, Kenneth Parker gives a detailed account of the PCB test methods oriented toward the board-specific defects, the conventional in-circuit testing (ICT), and the modern IEEE 1149.1 boundary-scan testing.

Considering that the eleven chapters of this book were written by different authors, the tasks of technical coordination and that of providing a uniform formatting and flow are not easy ones. I thank Dimitris Gizopoulos for his untiring effort on getting all chapters together and an excellent technical editing. This latest addition to the *Frontiers Series* is destined to serve an important role. However, "*Advances*" in the title of the book suggests that we keep track of the test technology as it advances. It is my hope that we will bring out future volumes of this type.

Vishwani D. Agrawal

Consulting Editor
Frontiers in Electronic Testing Book Series

September 2005

Preface

Electronic circuits testing has always been a very vibrant area of scientific research and development. The engineering adventure of discovering whether an integrated circuit has been properly manufactured and operates in accordance with its specifications advanced significantly over the last few decades. Every new manufacturing technology generation—already supporting feature sizes of a few tens of nanometers today—brings with it enhanced functionality, more transistors per unit area, elevated performance and reduced power consumption at lower costs per circuit unit. On the other hand, each shrinking manufacturing process also carries new types of failure mechanisms and defects which were either unknown or of less importance in previous generations of larger geometries and lower operating frequencies. Each integrated circuit generation packs more modules of improved or completely new functionality (digital logic, memories, analog and mixed-signal as well as radio frequency components) into half or even less of the space; as a consequence, improved or completely new testing techniques are necessary for them. To make matters worse, the rising instance count and shrinking pin-to-gate ratio exacerbate the difficulties of controlling and observing the internal nodes of the circuit.

The definitions of *test quality* and *test cost* have never been more complex than they are today. Electronic testing methodologies should be able to detect the new types of failure modes in modern manufacturing technologies. The population of integrated circuit physical defects that are not accurately modeled by traditional fault models is rapidly increasing. Moreover, the majority of defects can only be detected when the circuit operates at its regular, full-speed frequency. This can only be guaranteed if the performance and accuracy of test application and response capturing are extremely high. The volume of test data (stimuli and responses) that should be applied to each manufactured circuit to provide high levels of confidence of it being correctly implemented, along with the need of applying tests at very high performance, precision and accuracy have launched the costs of capable test equipment alarmingly upwards. Reducing the cost of automatic test equipment while maintaining the ability to perform high performance, precise and accurate testing has been and will continue to be a major concern: careful embedding of test-related mechanisms on the chip itself, and communicating this information to the tester has shown a promising path to reach this cost and quality goal.

If new defect types are not given special consideration, integrated circuits released to the market, mounted in boards, and connected in their final system have an increased probability of malfunctioning. Conversely, if testing the manufactured circuit takes too long and/or costs too much (in an effort to exhaustively deal with new defect types) the price of testing will severely affect product development costs and the resulting delayed entry to the market will jeopardize product success. If testing is not performed with accurate measurement techniques, the product development cost will also be severely affected due to yield loss: fault-free devices will be rejected only because of inaccuracies in high-speed test measurements.

The major *challenge* of test technology researchers and practitioners today is to define and apply electronic testing methodologies that keep a *balance* between quality and cost. This fundamental test technology challenge is an essential component of a key term of modern electronics: *manufacturability*—the extent to which a new product can be easily and effectively manufactured at minimum cost and with maximum quality and reliability meeting customer expectations. Production of high quality electronic circuits at profitable yield levels requires carefully implemented testing strategies.

For the majority of electronic circuits today, it is crucial that the appropriate test budget (in terms of time or allocated expenses—these are usually directly related) must be utilized, no more, no less. All the *advances* of the last decades in electronic circuits test technology have led us to a maturity point that can make this happen. *Methodologies* and *practices* of the near future should take advantage of this knowledge base to effectively answer today's *challenges* of test cost and test quality; all that needs to be done is to understand and focus on these challenges. This is the motivation and inspiration behind this book: to provide a comprehensive text that focuses on the advances of the research and development community in key test technology topics, records today's industrial practices and new needs, elaborates on the challenges that emerging testing methodologies have to deal with, and provides a vision for the near future of this amazing journey. Hopefully, the book provides the necessary information to understand and assess the tradeoffs to achieve the ideal balance of meeting the appropriate test budget.

PURPOSE AND CONTENT OF THIS BOOK

This edited volume is a unique compilation of chapters on many electronic testing topics of importance today and in the foreseeable future. The topics discussed are those on which the vast majority of the research and development community in test technology works today; topics where the electronic circuits industry needs effective answers, methodologies and practices that can be applied in the short term.

Every chapter of the book includes the following pieces of information for the reader:

- *Insight* about the *importance* of the chapter topic today. Unless improved or new methodologies and solutions are devised in the topic, electronic testing will either lead to poor test quality or unacceptable test costs. This part of the chapters gives the *motivation* for further research and development in each topic.
- Detailed snapshot of the *state-of-the-art* in the topic and recent *advances* and *industry practices* related to it. The chapter authors allocated a significant portion of their efforts in distilling the literature and providing a comprehensive set of references that represent significant recent research in the topic and can be used as a compass for further in-depth study of each area.
- Identification of the *challenges* in the topic today. Challenges in a topic are either due to the topic being in its infancy and the lack of effective methodologies providing solutions, or due to new problems that emerging manufacturing technologies or product needs have introduced to mature topics. Both types of challenges are discussed in this book along with *vision* and *forecasting* about the near future as well as *guidelines* for the focus of emerging testing methodologies.

Chapter authors provide all this information based on their long experience in the corresponding topic, lots of industrial success and failure cases, supported by a deep understanding of what test quality and test cost mean today for the electronics market. The entire book has a strong industrial and practical orientation. Each chapter is written in a unique way corresponding to the specifics of the topic and representing the authors' background, experience, and way of addressing challenges, problems and solutions. There are several interconnecting relationships among the chapters of the book: chapters touch on the topics of each other, and the reader of one piece can refer to other locations in the book where more specialized elaboration can be found. The matched pieces of this puzzle give the entire picture of *Advances in Electronic Testing: Challenges and Methodologies.*

This book serves a different and unique purpose compared to the comprehensive list of test technology books in the *Frontiers in Electronic Testing* series—a series that continues to support the education of the international test technology community and has done so for the past ten years. This book is neither an introductory book in test technology nor a detailed and specialized study of a single research topic. These two purposes are very successfully served by the other books of the series. *Advances in Electronic Testing: Challenges and Methodologies* enriches the series with a new type of edited volume on *recent advances* in modern electronic circuits testing. The book focuses on a carefully selected and broad set of topics among those in which intensive research and development takes place today and is expected to continue attracting the interest of researchers and practitioners in the near future.

The intention of this edited volume is to be an advanced textbook and valuable reference point for senior undergraduate students, graduate students in MSc or PhD tracks, researchers and professors conducting research in the electronic testing domain; this book can support orientation of their research plans. The book is also for industry engineers and managers seeking a global view and understanding of test technology and a dense elaboration on test technology issues they deal with in their development projects.

The book chapters are organized in a coherent sequence covering several aspects of electronic testing: failures, defects, bugs, fault models, test interfaces, tester-related considerations, circuit-specific aspects (cores, Systems-on-Chips, memories, mixed-signal and radio frequency circuits) as well as loaded boards testing. The list of chapters and contributing authors is as follows.

Chapter	**Author(s)**
Defect-Oriented Testing	Robert C. Aitken
Failure Mechanisms and Testing in Nanometer Technologies	Jaume Segura Charles Hawkins Jerry Soden
Silicon Debug	Doug Josephson Bob Gottlieb
Delay Testing	Adam Cron
High-Speed Digital Test Interfaces	Wolfgang Maichen
DFT-Oriented, Low-Cost Testers	Al Crouch Geir Eide
Embedded Cores and System-on-Chip Testing	Rubin Parekhji
Embedded Memory Testing	R. Dean Adams
Mixed-Signal Testing and DfT	Stephen K. Sunter
RF Testing	Randy Wolf Mustapha Slamani John Ferrario Jayendra Bhagat
Loaded Board Testing	Kenneth P. Parker

ACKNOWLEDGEMENTS

The authors of the book chapters have devoted several hours in putting together the material of this volume and presenting it in an as useful as possible way to the reader sharing their knowledge and experience in test technology. The most difficult task we faced during the writing of this book was to squeeze so much technical information in a topic in just a few pages; this took several revisions, lots of disk space and hundreds of emails. I want to warmly thank all authors for their contribution to this volume and for having the patience to revise the chapters several times. Rob, Jaume, Chuck, Jerry, Doug, Bob, Adam, Wolfgang, Al, Geir, Rubin, Dean, Steve, Randy, Mustapha, John, Jayendra, Ken, it has been a pleasure working with you for this book.

I would like to acknowledge the continuous support and useful suggestions of Prof. Vishwani D. Agrawal, the Consulting Editor of *Frontiers in Electronic Testing* book series. It is my honor to have the Foreword of this volume written by him.

I also want to thank Mark de Jongh, Springer's Senior Publishing Editor, Cindy Zitter, Senior Assistant as well as Springer's production staff for the excellent collaboration.

Several corrections and improvements are due to the help of Mihalis Psarakis who did a thorough review of the entire book. Dimitris Kostakis saved many hours of work by providing valuable hints and answers for the formatting of the book. Ken Parker provided a great help since the beginning with his advises and hints on the structure and layout of the book. Many people have provided inputs and reviews to the authors of the chapters and are separately acknowledged in each chapter.

Dimitris Gizopoulos
September 2005

Contributing Authors

Chapter 1
Robert C. Aitken ARM, Inc. [Physical IP, Sunnyvale, CA, USA]

Chapter 2
Jaume Segura Univ. Illes Balears [Dept. Fisica, Palma de Mallorca, Spain]
Charles F. Hawkins Univ. New Mexico [EECE Dept., Albuquerque, NM, USA]
Jerry M. Soden Sandia Natl. Labs [Failure Analysis Dept., Albuquerque, NM, USA]

Chapter 3
Doug Josephson Intel Corp. [Digital Enterprise Group, Fort Collins, CO, USA]
Bob Gottlieb Intel Corp. [Digital Enterprise Group, Santa Clara, CA, USA]

Chapter 4
Adam Cron Synopsys, Inc. [Test Automation, Mountain View, CA, USA]

Chapter 5
Wolfgang Maichen Teradyne, Inc. [IC Enabling Technology, Agoura Hills, CA, USA]

Chapter 6
Alfred L. Crouch Inovys Corp. [Research and Development, Austin, TX, USA]
Geir Eide Mentor Graphics Corp. [Design Verif. & Test Div., Wilsonville, OR, USA]

Chapter 7
Rubin Parekhji Texas Instruments Pvt. Ltd. [SOC Design Techn., Bangalore, India]

Chapter 8
R. Dean Adams Magma Design Automation [DFT, Santa Clara, CA, USA]

Chapter 9
Stephen Sunter LogicVision (Canada), Inc. [Ottawa, Ontario, Canada]

Chapter 10
Randy Wolf, Mustapha Slamani, John Ferrario, Jayendra Bhagat
IBM [RF & Analog Test Group, Essex Junction, VT, USA]

Chapter 11
Kenneth P. Parker Agilent Techn. [Manufacturing Systems Div., Loveland, CO, USA]

Dedication

This book is dedicated to the moments when
one becomes two, two become three and ten become eleven.

Chapter 1

Defect-Oriented Testing

Robert C. Aitken

The integrated circuit manufacturing process is imperfect, and as a result defects are introduced into some of the fabricated chips. Defects take a wide variety of forms, from localized spot defects, typically extra or missing material caused by contamination, to defects affecting much larger areas, such as transistor changes caused by implantation variation.

Some defects affect circuit behavior. Testing is used to find these before products are shipped, in order to ensure high quality. Defect-oriented testing is a way to improve the efficiency of testing by targeting tests directly at the defects that cause incorrect circuit operation.

Defects occur in random places and can have unpredictable effects. The processes that cause them are continuous over a wide range of variables. In order to simplify the problem of identifying defective circuits, this infinite defect space is approximated by a finite set of faults. A fault is a deterministic, discrete change in circuit behavior. Faults are often thought of as being localized within a circuit (e.g. a particular gate is broken), but they may also be modeled mathematically as transformations that change the Boolean function implemented by a circuit. Many fault models are time-independent; some use an arbitrary form of time progression,

D. Gizopoulos (ed.), Advances in Electronic Testing: Challenges and Methodologies, 1-42.

while a few include time behavior explicitly. The fault effects associated with fault models can be as simple as replacing a subcircuit function with a constant value or be so complex as to require SPICE simulation to evaluate. The choice of fault model depends on its intended use (e.g. test generation, manufacturing quality prediction, defect diagnosis, characterization for defect tolerance, etc.)

This chapter provides an overview of the Defect-Oriented test approach, beginning with a brief history of the subject, an overview of defect mechanisms, with special emphasis on advanced technologies. This is followed by a discussion of how these change the behavior of circuits (fault models). Next is a catalog of the types of tests used in Defect-Oriented test, and a survey of relevant published experimental results on the effectiveness of Defect-Oriented test approaches. Finally, some thoughts on future directions are given as part of the conclusions.

1.1 HISTORY OF DEFECT-ORIENTED TESTING

Historically, all testing was functional, and asked the question "Does the device do what it is supposed to?" Functional tests are primarily defined logically (outputs are a function of the inputs). For digital logic, functional tests became too expensive to develop, due partly to the amount of manual effort required to write the tests, but more to the complexity required to translate verification testbenches and tests from a simulation or characterization environment into an ATE environment. As a result, functional tests in production have largely (but not completely) been replaced by structural tests. Structural tests changed the basic questions being asked by test, and expanded the "Does it work?" question into a new question and a syllogism: "Are all circuit elements present and working? If so, and the design is correct, then it must work". This approach is the basis of scan testing. *Defect-Oriented testing* takes a step beyond structural testing to ask, "What could go wrong with this design, how would the design's behavior change if this happened, and how can that be measured?" Any measurable circuit property could be affected: logical values, timing, current consumption, etc., whether part of the specification or not. Defect-Oriented tests for digital logic typically include comprehensive structural logic tests, a current test (typically a variant of I_{DDQ} test[1]), and at-speed tests. All these tests share the property that they measure some aspect of circuit behavior that is directly affected by defects, regardless of their direct applicability to normal circuit functionality.

A key aspect of Defect-Oriented Testing is measurability, and measurability involves overcoming multiple sources of error or variability, including:

1. Defect-dependent variation, since defects can change circuit behavior in a variety of ways.
2. Circuit-dependent variation, including manufacturing process related variation (e.g. gate length, oxide thickness, etc.).

---

[1] I_{DDQ} testing measures the quiescent (Q) current (I_{DD}) consumed by a device. In standard CMOS circuits, the defect-free current is mainly transistor leakage. Many defects raise the level of this current enough to be measured in production test.

3. Environment-dependent variation, since many circuit behaviors change depending on temperature and voltage.
4. Equipment-dependent variation, due to limitations in resolution, repeatability of measurement equipment, as well as drift over time and other machine-dependent variations.

A defect detection method is not useful unless it consistently separates defect-dependent variation from circuit- and environment-dependent variation in the context of equipment variation. We will concentrate mainly on the first two sources of variation in this chapter.

An interesting question is whether we can separate defect behavior from circuit variation. A given flaw, such as a resistive via, can have a range of resistance values. Some of these will cause the circuit to fail logically, others will cause timing failures under some operating conditions, and others will never fail. Defect-Oriented Testing needs to work with the design margin process in order to effectively distinguish between these cases.

Defect-Oriented test requires information exchange between design, test development, process R&D, wafer manufacturing and manufacturing test to be successful. Historically, all these elements have been present inside vertically organized companies, and this is still true today in many cases. In these cases, information exchange between various entities is relatively straightforward, since all are motivated to action by the ultimate financial success of the company. Most published success stories in Defect-Oriented Testing originate within vertically integrated companies.

However, the semiconductor industry has substantially disaggregated, allowing chip designers the freedom to select from a number of IP providers, wafer foundries, packaging and test houses, among others. Some of these complex relationships are shown in Figure 1-1.

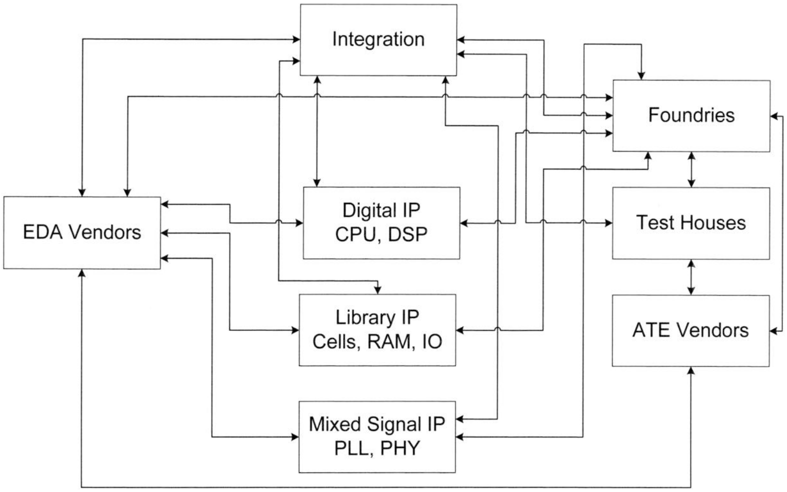

Figure 1-1: Value chain relationships in semiconductor industry.

In this situation, information exchange is not guaranteed, and in some cases may be hindered, based on economic interests of the various parties involved. To succeed in this context, a Defect-Oriented test methodology must provide some form of value to all companies involved. This constrains the problem significantly from the original vertically integrated case, but successful partnership approaches are possible, since all parties benefit from successful silicon.

This new format of the semiconductor industry should still be able to implement Defect-Oriented Testing, although in a modified form. For instance, in a single company, process engineers, library design engineers, and test engineers could easily collaborate on the best approach to designing a defect-robust scan flip-flop. In the disaggregated model, each of these three could belong to a different organization, different even from the design integrator and end customer. In such a model, the library design engineers need to develop flip-flop architectures that were robust across a variety of foundries, when used with standard EDA tools in standard ways. Similarly, the test house would need to be prepared for a variety of approaches. In each case, an effective flow can be established through standardization, through a robust methodology that can account for a variety of implementation techniques, or through a specific cooperative effort between organizations.

1.2 CLASSIC DEFECT MECHANISMS

The causes and manifestations of CMOS failures are many, but they have historically been lumped into two broad categories: Shorts, where conduction occurs when none is desired, and opens, where desired conduction does not occur. In aluminum processes, shorts have been more common and more problematic than opens, and so most research has focused on them [1]. Both shorts and opens have standard electrical properties. Of these, the most commonly studied has been resistance. Failure mechanisms will be discussed in detail in Chapter 2 of this book, but are introduced in brief here since they are key to understanding Defect-Oriented test.

1.2.1 Shorts

Shorts can be caused both by extra conducting material and by missing insulating material. Examples include:

- Photolithographic printing error
- Conductive particle contamination
- Incomplete etch
- Incomplete metal polish
- Crack in the insulator
- Gate oxide defect causing pinhole

For a comprehensive list, see Chapter 2 of this book. An example metal short by a conductive particle and gate oxide short are shown in Figure 1-2.

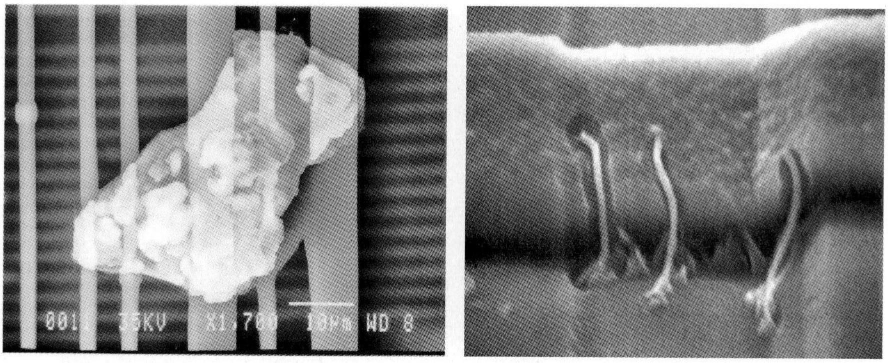

Figure 1-2: Particle and gate oxide shorts.

The electrical behavior of a short is determined by where it lies in a circuit. Shorts at the diffusion level often involve only the terminals of a single transistor. Shorts in poly or metal 1 affect the internals of one or more standard cells. Shorts in higher levels of metal interconnect typically involve gate outputs, power, and /or ground.

Shorts are active when the two nodes involved are driven to opposite values. A conducting path is formed from the power supply through active P transistors through the short (modeled as a resistance) through active N transistors to ground. The shorter (less resistive) the short, the more likely this conducting path will disrupt circuit operation. This basic idea can be extended to the concept of a "critical resistance" [2] below which the short "wins" and the circuit operates incorrectly, and above which the circuit "wins" and continues to operate correctly. There are actually multiple critical resistances for any short, as shown by Table 1-1 below, which lists delays associated with a bridge between an inverter output and ground in 0.13um technology. The waveforms associated with this Table can be seen in Figure 1-3. There is a logical critical resistance, where the circuit will fail under all circumstances (below about 1700 ohms), a set of timing critical resistances (e.g. at 1800 ohms, a delay of about 150ps results), where the critical resistance depends on required timing and also on operating environment, and finally, a set of I_{DDQ} critical resistance, where the defect will cause a significant enough increase in I_{DDQ} to be observed. An I_{DDQ} technique with 100uA resolution will be able to identify this short at 1.0V for resistances below 10kohms.

Resistance (Ω)	1700	1720	1730	1750	1800	2000	3000
Delay	SA0	600ps	400ps	250ps	150ps	70ps	<10ps

Table 1-1: Resistance and delay for short in 0.13um technology [3].

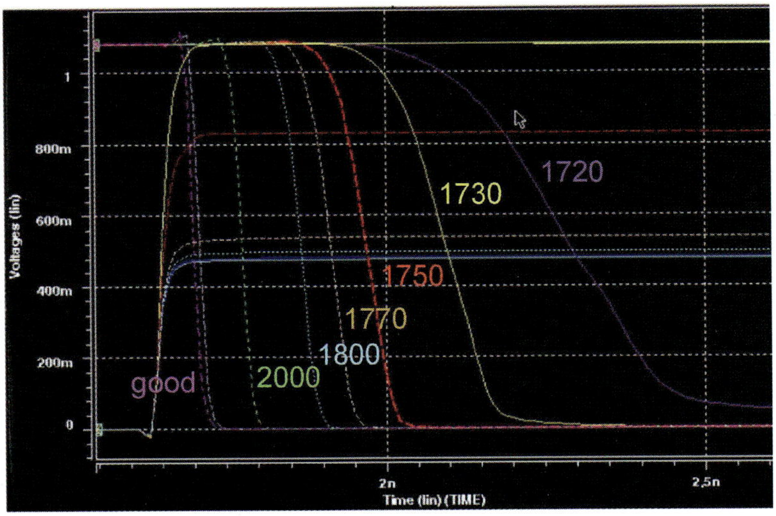

Figure 1-3: Resistance delay curves for a short in 0.13um technology.

Actual bridge resistance has been studied experimentally, and a wide distribution has been reported [4]. As technology advances, critical resistance is tending to rise, because timing requirements are tighter. Slack times – the difference between the required arrival of a signal and its actual arrival – are shrinking to the order of 100ps or less. At the same time, better synthesis algorithms are making more timing paths critical or near critical. So with more paths having less slack, even high resistance shorts are able to disturb circuit behavior enough to cause a fault.

1.2.2 Opens

Opens are caused by missing conducting material or extra insulating material. Examples of these include:

- Photolithographic printing error
- Step coverage
- Incompletely filled via
- Electromigration
- Silicide agglomeration
- Incomplete via etch or via foreign material
- Insulating particle contamination

As with shorts, the behavior of an open is determined by where it is located, whether in the transistor structure (diffusion, poly or metal 1) or in the interconnect between transistors (higher level metal). Logically, opens can be within a cell, or between cells. In many cases, a complete open results in a node that is electrically isolated from its surroundings. Charge stored on this node during fabrication can affect its subsequent operation. For small opens, Fowler-Nordheim tunneling[2] can occur, resulting in a circuit that operates more slowly than expected. Many complex open behaviors have been postulated (see Chapter 2 of this book for details), but some of these have proven difficult to identify in practice. High levels of leakage and mutual capacitance in modern processes mean that open behavior will likely be more deterministic in future, in that stored charge will bleed away, or nodes will follow their neighbors.

In practice, many opens are partial or "almost opens". Such opens are challenging to detect and will be addressed later in this chapter. An experimental analysis of resistances was made in [5] and is given in Figure 1-4.

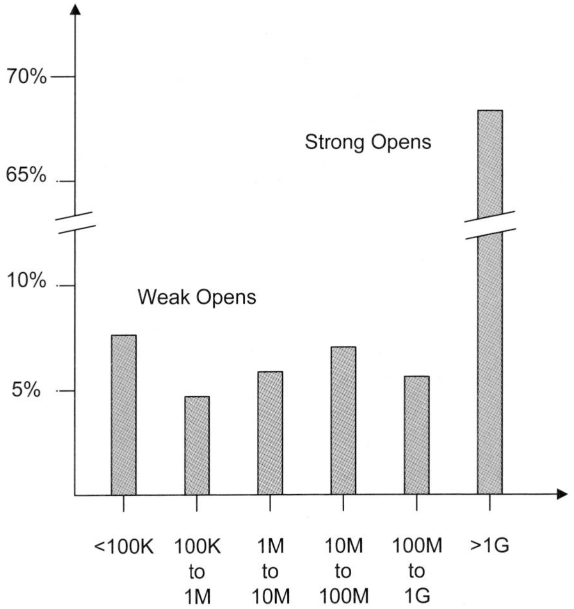

Figure 1-4: Resistance distribution for opens [5].

1.2.3 Parametric Changes

Defective behavior is not always caused by a single isolated problem such as a short or an open. Sometimes a circuit parameter is out of specification across a wide area, and this can cause a failure, or an increased susceptibility to other problems (e.g. temperature effects, crosstalk, etc.). These problems will also be discussed later in the chapter. Parametric variation begins with a physical change (e.g. variation in printed transistor gate length) and affects the circuit via electrical change (e.g. transistors that are faster and leakier than expected). In addition to gate length, other

---

2 Fowler-Nordheim tunneling is the process by which an electron is able to "tunnel" under the potential barrier represented by the physical break in conducting material. As a result, current is still able to flow across a physically small open.

parameters of interest include dopant concentration (device mobility, capacitance), metal thickness (resistance, capacitance), oxide thickness (leakage, performance). Some variation is expected, and design must accommodate it, but at some point variation will exceed the tolerance or margin in the design, and it becomes a defect.

1.3 DEFECT MECHANISMS IN ADVANCED TECHNOLOGIES

This section covers some defect mechanisms that are becoming increasingly common in manufacturing processes at the 130nm node and below, and which need to provide the basis for ongoing efforts in Defect-Oriented test.

1.3.1 Copper-related Defects

For most of the history of CMOS, metal has meant aluminum. Since 130nm, however, copper has become the metal of choice, and this means that changes must be made in the way metal defects are considered. Aluminum metallization is a subtractive process: an entire layer of metal is deposited, a mask is applied and unwanted metal is etched away. This metal etch is inherently "dirty" and results in many particles being present, some of which lead to shorts. The etching process has led to shorts being far more common faults than opens in CMOS processes for many years.

Copper, on the other hand, uses a dual *damascene*[3] process. A layer of insulator is applied to the wafer. Next troughs for wires are etched (first damascene step). Next additional troughs for vias are etched. A layer of copper is electroplated into the trenches (on top of a small tantalum barrier/seed layer). Finally, excess copper is removed via *chemical/mechanical polishing* (*CMP*). There is no metal etch to provide a slurry of particles. As a result, additional defect mechanisms become important. Figure 1-5 summarizes the processing differences. Note that the process is highly simplified when copper is used.

In addition to a metal etch, aluminum processing also features two via/wire interfaces (tungsten to aluminum). Copper has only a single via/wire interface, since both vias and wires are deposited together. The absence of the metal etch and the reduced number of interfaces can result in a lower defect level for copper metalization than aluminum. However, there are some specific new defect mechanisms associated with copper. These were particularly troublesome in the early days of copper processing, but are now at more acceptable levels.

---

[3] The term "damascene" comes from metal inlaid with gold or silver, an intricate craft associated with the city of Damascus.

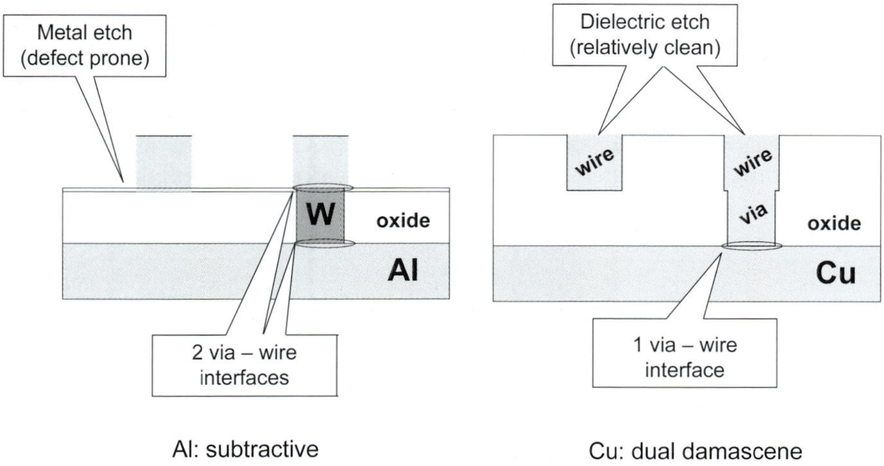

Figure 1-5: Comparison of Al and Cu processing.

The polishing process for copper leads to four major defect mechanisms: *under polishing*, which leaves copper or slurry remaining where it is not wanted, which can lead to shorts, *dishing*, where over polishing has occurred, and *erosion*, where insulator has been polished as well as copper (see Figure 1-6). In each case, the defect leaves a surface that is poorly planarized and susceptible to defects at higher levels (e.g. wide metal dishing on one layer can cause shorts in the layer above as wires fold in on each other). The fourth mechanism, *scratches*, result in combinations of shorts and opens, spread across relatively large areas [6].

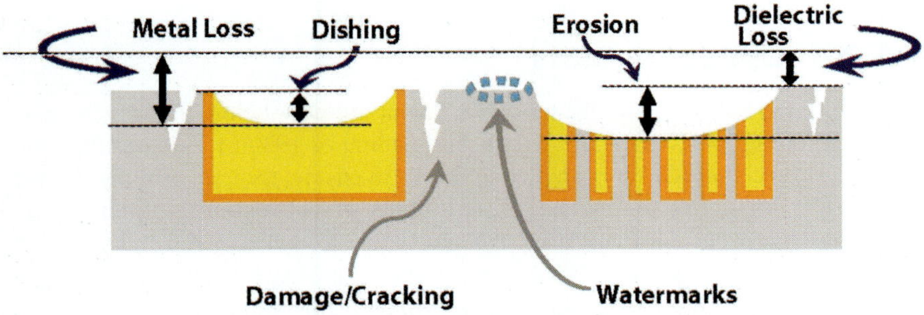

Figure 1-6: Dishing and Erosion (courtesy Applied Materials).

In addition to CMP-induced defects, *voids*[4] (also known as partial opens) are another major defect mechanism. These defects often have an easily recognizable signature: delay faults at low temperatures. This happens because of thermal expansion of the copper into the void. As temperature increases, the void shrinks and the two metals make better contact, as shown in Figure 1-7. Cold temperature delay

---

[4] Voids are empty spaces in the metal left by incomplete copper deposition.

mechanisms exist in aluminum processes as well (voids, via contamination, and silicide cracks, for example), but they are more common in copper processes.

The copper void defect mechanism shows that while high temperatures are the worst operating condition for defect-free circuits, defective circuits may have worst-case behavior at room temperature, or even lower. If possible, delay tests should be applied at two temperatures to account for void defects and process variation at tolerance imits.

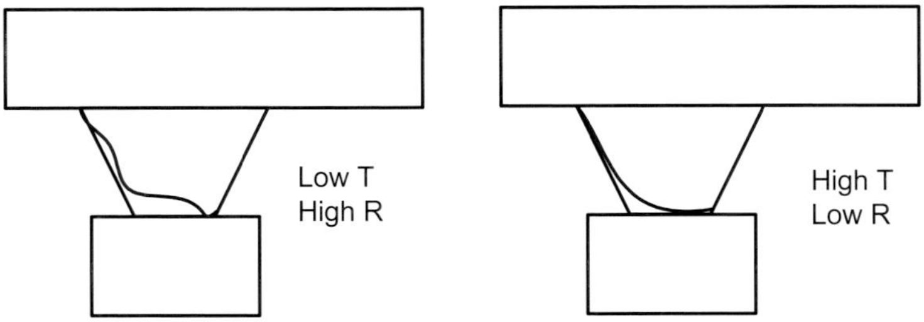

Figure 1-7: Temperature dependence for copper voids (T=temperature, R=resistance).

Automatic Test Pattern Generation (ATPG) for void-related defects should emphasize via crossings, especially those where only a single via has been used (redundant vias are the standard design-for-manufacturability – DFM – approach to via voids, since the spare vias can carry the required current if voids are present).

1.3.2 Optical Defects

At 130nm and below, many features of a layout are now below the wavelength of light used during photolithography (typically 248nm or 193nm). As a result, *optical proximity correction* (*OPC*) is applied to masks to ensure accurate printing. In many ways, OPC is similar to the use of serif fonts in conventional printing, which developed to ensure adequate ink coverage (e.g. "X" versus "**X**"). OPC can add additional features (e.g. hammerheads or outriggers) or shrink features.

OPC leads to two major classes of defect: *mask errors* and *Design Rule Checking* (*DRC*) *escapes*. In a mask error[5] (see Figure 1-8), OPC fixes result in a short or open when features interfere with one another. In general, mask defects are catastrophic and result in 100% yield loss simply because all chips manufactured using the bad masks will be faulty. Reticle inspection helps reduce the number of these, but inspection programs have been confused by OPC features that look like defects but are legitimately needed for the design [7].

---

[5] A mask error is present on the masks used for photolithography. This error is then printed instead of the desired circuit.

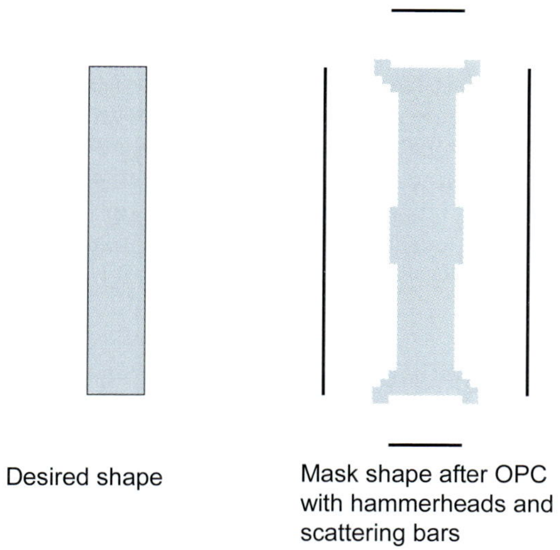

Figure 1-8: Mask error after OPC.

In a DRC escape, OPC is meant to fix a particular feature, but can't, due to neighboring circuitry. Consider the layout of Figure 1-9 where hammerheads should be added to each gate, but nearby circuitry prevents this from being possible on the leftmost gate. As a result, the gate is foreshortened. In some cases, catastrophic failure will result, but often only weaker than usual gates will be formed. The resulting channel is shorter in on portion and narrower as well. This defect will typically behave as a delay fault, and can often be detected by transition fault tests.

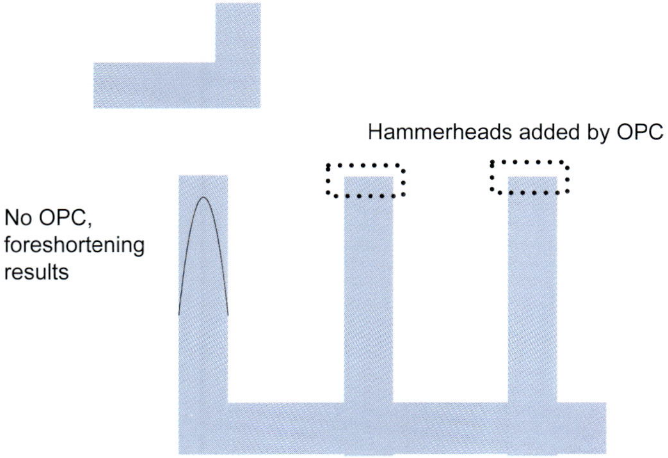

Figure 1-9: OPC limited by nearby layout.

1.3.3 Design-related Defects

While process technology is the main driving force between new defect types, the impact of design methodology should not be ignored. At the most obvious level, decisions made by routing tools determine which wires are likely to short and which aren't. Even adjacent wire spacing can be controlled. This section explores two more subtle mechanisms, both relating to low power design and thus appearing in advanced manufacturing technologies. Additional discussion of design related defects is available in [3], [8].

Multi-V_T Libraries

Transistor off current and drive current both depend on threshold voltage, as shown in Figure 1-10. A low threshold device has somewhat higher saturation current and thus higher performance, but at the cost of substantially higher off current. By using both high and low V_T transistors in a design, leakage and performance can be traded off more finely than by using exclusively one or the other. For consistency, we will use "HV_T" for the high threshold voltage devices and "RV_T" for the regular threshold voltage (V_T) devices. A common approach to get the reduced leakage of HV_T devices with the performance of RV_T devices is to mix them at the gate level – some gates will use each type [9]. Typically, the critical path and near critical paths are implemented with RV_T gates, and the rest with HV_T. Making this work easily in a tool path requires that the HV_T and RV_T gates be artwork compatible (same cell height, routing pitch, etc.). Ideally, they should be the same artwork with different implants, in order to allow drop-in substitution after placement. Using these techniques can save 70-80% of leakage power, even for high performance designs [10]. Two-variant processes are discussed here, but processes supporting 4 or 5 variants are currently available at 90nm technology in foundries such as TSMC and IBM.

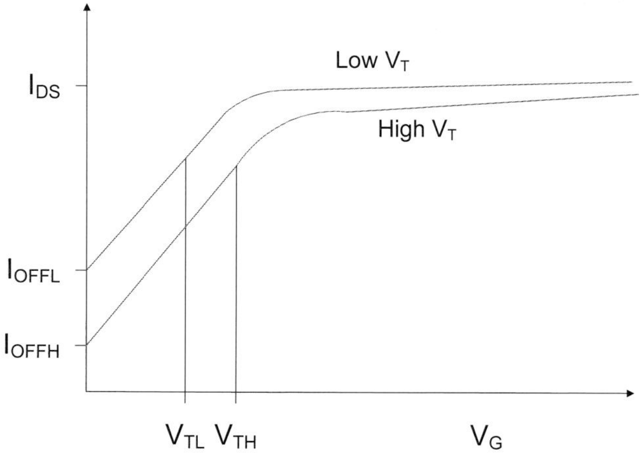

Figure 1-10: Relationship between off current, drive current and threshold voltage [11].

Multi-V_T cells can increase the chances of "Byzantine Generals" behavior (see section 1.4.3) of bridging faults because of varying thresholds among gates and varying drive currents. HV_T devices tend to have higher variability, and their weaker drive currents give them higher critical resistances than RV_T devices.

Multiple Voltage Domains

Power consumption varies as the square of voltage. Low power design approaches, especially for battery-powered applications, are increasingly using variable power supply voltages. Two basic approaches are possible: open loop, where voltage levels are predetermined based on characterization and expected power consumption, and closed loop, where the system dynamically adjusts its voltage to the lowest level able to sustain its present level of activity (i.e. embedded CPU voltage will be higher during peak activity times, such as displaying an MPEG video, and will be reduced during lower activity times such as voice transmission), and can be reduced even further or even shut off during sleep modes [12].

Keeping voltage domains separate is best accomplished by using physical isolation together with level shifting circuitry, but when they interact new defect behavior is possible. For example, bridging fault models usually make the simplifying assumption that a bridging fault is not activated if the involved nodes are at the same logic value. However, consider the circuit in Figure 1-11. The P transistor in the transmission gate has both its gate and source at logic 1 in the low voltage domain (1_L), so the gate is off in the fault free case. A short (or even capacitive coupling) with the higher voltage domain leads to a voltage on the source above the V_{DD} value of 1_L, denoted by 1_H. This causes the difference between the gate and source voltages of the transistor to be greater than its threshold value ($V_{sg}>V_{th}$), turning the transistor on. The faulty value is now free to propagate as any with other fault. Physical isolation of level shifters can help reduce this problem, but will not totally eliminate it. These shorts will be more of a concern for diagnosis than for bridging ATPG, which should try to sensitize them the conventional way (opposite values).

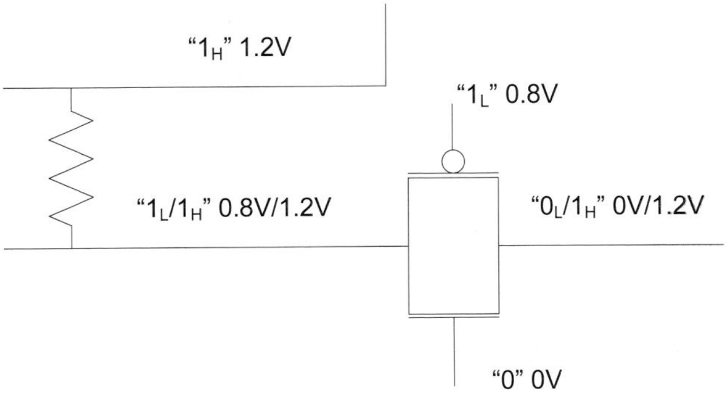

Figure 1-11: Faulty behavior with multiple V_{DD} values bridged.

Defects in the level shifters themselves cannot be avoided, but careful attention needs to be paid in layout in order to ensure that they result in catastrophic (easily measurable) failures, rather than subtle problems that might escape detection during manufacturing test.

1.4 DEFECTS AND FAULTS

Defects manifest themselves in a variety of ways. At any given circuit node X, some will be detectable only by test method 1, others by test method 2, and others by both (see Figure 1-12). A fourth category will not be detectable by either 1 or 2, but could be detected by another approach. Ideally, it would be desirable to model defects directly and independently from the methods used to detect them, but realistically defect behavior must be extrapolated and simplified into *fault models*[6]. The factors affecting fault modeling in nanometer technologies can be divided into three broad categories: increasing circuit size, increasing defect subtlety, and decreasing numbers of manufacturing processes. Each of these has major implications on the nature of faults that will be targeted in future designs.

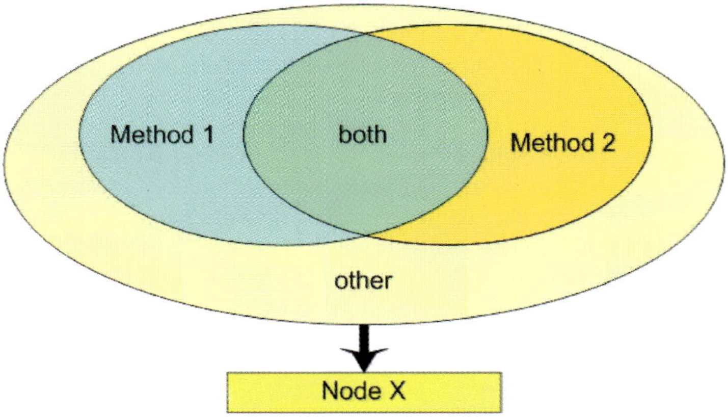

Figure 1-12: Relationship between defects and test methods.

As transistor manufacturing cost has declined over time, chip producers have largely responded by including more functionality on chip, rather than producing cheaper chips. This explosive growth in transistor counts means that fault models and defect analysis must be tractable on ever-increasing circuit size. At the same time, increasing defect subtlety is being observed. The primary concern of manufacturing testing in the past has been the identification of permanent, deterministic, defective behavior, so-called "hard faults". The fault models and associated tools developed reflect this. Test generation tools strive for completeness: generating a test which detects a fault, or proving that the fault is untestable. There is no middle ground. Defects, of course, follow no such rules. In order to account for

[6] A fault model is an abstraction of defect behavior. To be useful, a fault model must help in the development or evaluation of tests, or in yield improvement.

their behavior correctly, fault modeling will need to move from the discrete to the continuous domain.

As the cost of developing CMOS manufacturing technology skyrockets, the number of independent efforts is declining. Bringing a new process technology to market already costs multiple billions of dollars, and this cost is rising with every process generation. With these sorts of up-front costs, more and more companies are forming joint ventures and working to co-develop technology, or are simply dropping out of the manufacturing business altogether and contracting wafer manufacturing to others. As this happens, the number of *different* technologies will drop with every process generation; perhaps until only a handful remain[7]. Fewer processes means greater opportunity to amortize high model development cost, and thus has several beneficial side effects.

In order to see more concretely the effects described above, it is worthwhile to consider some specific fault models in common use today and see where they are likely to be headed. First, recall the main uses of fault models.

1.4.1 Uses of Fault Models

Test Generation

A fault model accomplishes two purposes during test generation. First, it provides a measure of completeness (how much work remains to be done). In an explicit fault model[8], keeping track of the faults for which test generation has been attempted will suffice for this task. The second role is test effectiveness. By calculating both the fraction of faults detected (fault coverage), and the fraction of faults which cannot be detected (untestable faults), a test generator can demonstrate that it has achieved complete fault efficiency; i.e. it has developed a test for all the faults it can develop a test for and has proven the rest untestable.

Fault Simulation

Fault simulation requires a fault model, primarily for coverage measurement. Except for a few trivial cases (such as stuck-at coverage in combinational circuits with 20 or fewer inputs), fault simulation cannot prove that a fault is untestable, but it can evaluate tests and determine their coverage. A simulator may work in conjunction with a test generator, as a check for a test generator, or as a means of evaluating tests from some other source.

Quality Prediction

The ultimate purpose of fault coverage metrics is to provide a means of ensuring that tests developed for a chip will allow it to meet its necessary quality goals. Of course,

---

[7] Note that this does not involve the number of fabs producing wafers, or the number of transistor types available within these fabs' processes, only the number of different manufacturing processes used by these fabs.

[8] An explicit fault model is one where the faults can be enumerated.

quality is always measured in terms of defects, while coverage can only be measured in terms of faults. But how do defects and faults relate to one another? It is sometimes desirable to use a simple model as a "surrogate" for a more complex model in order to quantify this relationship [13].

Fault Diagnosis

Fault models can be used as part of fault diagnosis and thus help find the root cause of defective chips. In its most basic form, algorithmic fault diagnosis consists of using a fault model to predict the behavior of faulty circuits, comparing these predictions to the actual observed behavior of defective chips, and identifying the predicted behavior(s) which most closely match the observations. The goal of the process is to enable failure analysis by identifying promising locations for further study. The process is successful if the actual defect is contained in the list of possible locations, and if that list is sufficiently small to permit a failure analysis engineer to investigate each possibility.

1.4.2 Single Stuck-at Faults

The earliest fault model, and still the most common, is the single stuck-at fault model, originally published by Eldred [14] for circuits consisting of resistors and vacuum tubes. In the single stuck-at fault model, a single node somewhere in the circuit is assumed to have taken a fixed logic value (and is thus "stuck-at" either 0 or 1). A stuck-at fault generally represents a direct connection between a node and power (for stuck-at 1) or ground (for stuck-at 0). The stuck-at fault model has several advantages:

Simplicity—There are exactly two faults for every circuit node. The single stuck-at fault model is usually applied at the gate level, and each logic gate has two faults for each of its inputs and two for its output.

Logical behavior—Each fault is confined to a single location in the circuit and its effect is readily modeled as a boolean equation. As a result equivalence relationships can be developed: A stuck-at 0 fault on the output of an AND gate is equivalent to a stuck-at 0 fault on any of its inputs.

Tractability—The number of faults is directly proportional to the size of the circuit, so very large circuits can be analyzed. Stuck-at fault based test generation and simulation can be applied to designs containing millions of logic gates in a single pass.

Measurability—Because the number of faults can be counted and fault behavior is so precise, it is possible to precisely determine whether or not a given set of circuit inputs detects a fault. By collecting these measurements, it is possible to quantify the total "fault coverage" of a circuit inputs sequence (known as test set).

The single stuck-at fault model has remained the most commonly used through multiple technologies and numerous process shrinks, and has the most mature tools associated with it. Will its dominance disappear with the advent of nanometer technologies? In all probability, no, although new fault types will continue to become more common. The advantages described earlier for the stuck-at model will continue to hold, even with the emergence of exotic defect types. While high stuck-at coverage is not enough to guarantee high quality, it remains a baseline needed in virtually all designs.

One promising line of research in the area is to make explicit the advantages and limitations of stuck-at faults by attempting to control so-called *peripheral coverage* where tests generated for stuck-at faults will detect other types of defects simply by activating logic in a particular area and making it observable at circuit primary outputs. The work of Grimaila et al [15] shows that stuck-at tests generated to force repeated observation of all circuit nodes are more effective than standard tests in finding defective manufactured parts. By recognizing the capabilities of the stuck-at model as a facilitator of defect detection, rather than a perfect representation of defect behavior, we can be assured that the stuck-at model will be with us for a long time to come.

1.4.3 Bridging Faults

Since stuck-at faults may be thought of intuitively as shorts to power and ground, it is easy to conceive of an extension to the stuck-at fault model where nodes are permitted to be shorted to one another. This model was dubbed the bridging fault model in [16]. Shorts have generally been found to be the most common type of defect in CMOS circuits, so the model is appealing. Of course, it is simple enough to declare that two signals are shorted together, but much more difficult to say precisely what that means in terms of circuit behavior. Early bridging fault models postulated that the additional connection would form an implicit (i.e. "wired") logic gate, such as an AND or OR (see Figure 1-13a), and some logic processes behave this way in practice. Unfortunately, CMOS, in general, does not. In some cases, CMOS bridge behavior is fairly straightforward: A wire driven by large powerful transistors will always overpower one driven by smaller transistors (Figure 1-13b).

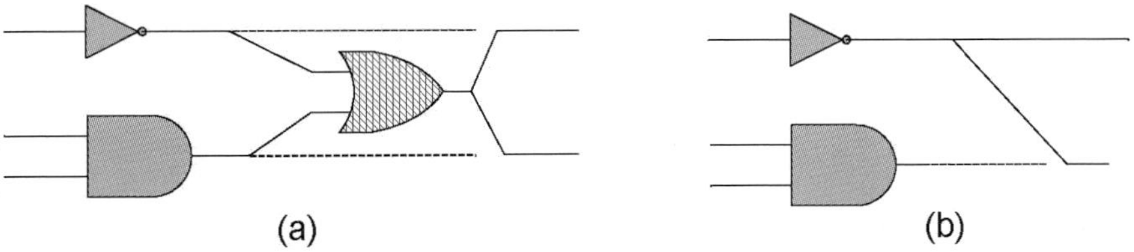

Figure 1-13: Simple bridging fault models (a) wired-OR bridging fault and (b) dominance bridging fault.

When strengths are similar, the result is less clear. An example is the "Byzantine Generals" behavior shown in Figure 1-14, where a bridged line is interpreted differently by different downstream logic gates with different gate thresholds. A short in a CMOS circuit results in an intermediate voltage level somewhere between supply and ground. Timing and logic design needs cause different CMOS gates to switch anywhere between 25% and 75% of nominal supply voltage. As a result, an intermediate voltage level can often be interpreted as a 0 by one gate and a 1 by another.

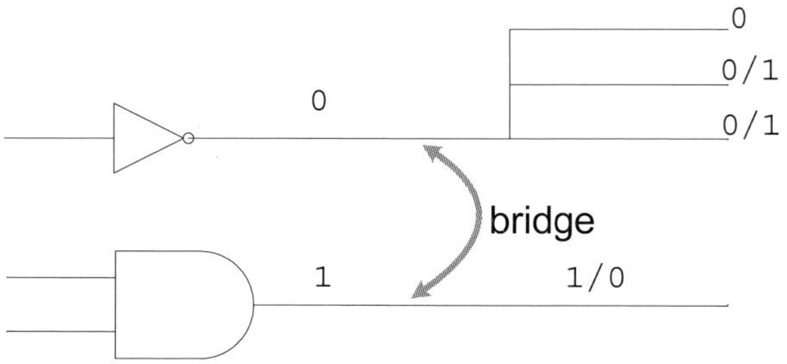

Figure 1-14: Byzantine generals behavior for bridging faults.

Randomly-placed bridging defects are and will continue to be a serious problem in CMOS processes. As technology advances to smaller geometries, more metal layers, reduced design margin and higher frequencies, the effects of these defects will grow in complexity and the variety of bridging behavior that needs to be detected will increase. Nonetheless, it is and will remain impractical to develop complete models for bridging defects. Instead, simplifying assumptions will need to be made to produce tractable fault models. Some of the issues are explained in detail below.

Bridging fault models arise from a desire to predict the behavior of bridging defects. An example of a single layer bridge defect is shown in Figure 1-15. Two metal lines AB and CD are shorted by a conducting particle. The defect is characterized by at least four parameters: its center point (x and y), its radius (r), and resistance (R). The resistance parameter may be some function of the other three, related in some way to the material forming the defect. There are other parameters which could potentially be considered (the defect might not be round, for instance), but these four will serve as a starting point.

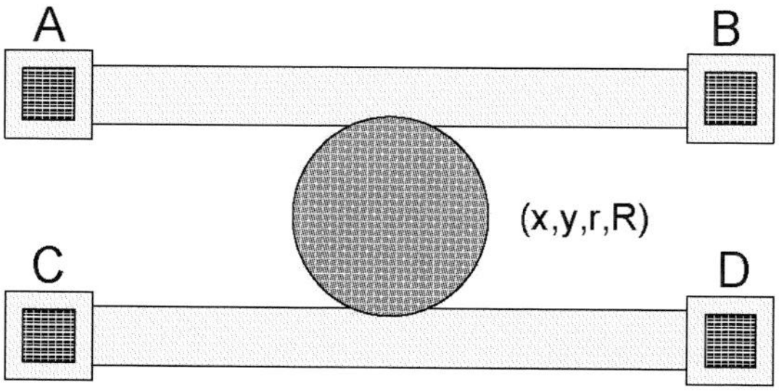

Figure 1-15: Example defect.

Each of the four parameters is continuously variable over some range, giving an infinite variety of potential bridges between just these two nodes. However, fault simulation is a discrete activity, so the range of possible defects is always simplified into some fault model.

A stuck-at fault assumes that either AB or CD is a power or ground rail, and that the defect resistance is such that the other line is pulled to that rail value. A bridging fault obtained by inductive fault analysis [1] assumes only that the lines are connected somehow. Specific simulation algorithms make additional assumptions about bridge resistance and logic behavior. Virtually all bridging approaches assume that the entire line AB is at the same voltage. However, circuit wires are in general complicated RLC structures, and there is no guarantee that the constant voltage assumption will hold true, particularly at high frequencies. The "critical resistance" of the defect (the resistance above which the defect cannot be detected) will depend on the maximum frequency that can be measured by a given test. Many tests are run at substantially less than the system clock frequency.

To further complicate matters, bridging defects can change the sequential behavior of a circuit if the bridge creates logical feedback in the circuit. Also, a bridge may affect more than two lines, and a particle-induced defect can span multiple layers. See [17] for a description of the current challenges.

Future Directions

Shorts, even in copper technology, remain a major source of defects. Some of these are due to classic effects, such as particle contamination, while others result from CMP scratches, optical effects, and more [6]. Shorts involving many lines are typically catastrophic, so while they are vital to yield calculation, they are less important for fault modeling. Many models have been proposed over the years, but we will consider the dominance model (strongest node always wins), the voting model [18] (winner depends on relative drive strengths of opposing values), and analog models (including Byzantine generals behavior).

Dominance and Voting Model

The voting model is most appropriate in cases where drive strengths are near equal, as shown in Table 1-2 (vote values, which relate to W/L ratios, but are scaled down for PFETs, are shown for each drive strength for a sample 130nm technology). A standard cell library contains a variety of drive strengths, built as multiples of a base value (x1 is the base, x2 has transistors twice as powerful as the base, x4 transistors are four times as powerful, and so on). Each entry in the Table is either D0 (0 dominates, so N transistors always win), D1 (1 dominates, so P transistors always win), or V (sometimes P wins, sometimes N wins) – voting model must be used for each specific case. The discussion below looks at some of these values.

	P drive	x1	x2	x4	x6	x8	x12
N drive	Vote	2	4	8	12	16	24
x1	4	V	V	V	V	D1	D1
x2	8	V	V	V	V	V	D1
x4	16	D0	V	V	V	V	V
x6	24	D0	D0	V	V	V	V
x8	32	D0	D0	D0	V	V	V
x12	48	D0	D0	D0	D0	V	V

Table 1-2: Dominance and Voting behavior for shorts of different drive strengths.

In this technology, an x1 P drive is virtually always dominated by an x2 N drive, but there are some exceptions (e.g. a NAND3x1 with all three PFETs active versus an INVx2), so the voting model is still necessary. This is not the case with the x4 drive strength, which dominates all x1 drives. CMOS PFETs are generally weaker, so voting is still needed even for an x4 PFET fighting an x1 NFET. Methods for applying voting to complex gate structures are given in [18] and [19]. Notice that for the most part, NFETs dominate, with a 0 result for a bridge.

As an example, a bridging fault between two small NAND gates (NAND2x1) is considered in a particular 0.13um technology. The behavior depends on the resistance of the defect (this is a slight variant of the well-known critical resistance concept) as shown in Table 1-3.

Bridge resistance	Behavior (1.2V, 25° C)
< 3000 ohms	Voting model
3000-3700 ohms	Analog/Byzantine
> 3700 ohms	Delay fault

Table 1-3: Example bridge behavior at 1.2V, 25° C.

The Byzantine behavior (as shown in Figure 1-16), while real, is highly dependent on defect resistance and supply voltage. Running the same test at multiple voltages can eliminate the problem, as shown in Table 1-4. At 1.2V, defects in the range of 3K–3.7K ohms will produce Byzantine behavior, while at 1.3V the range changes to 2.5K–3K ohms. Defect resistance will remain roughly constant, so the Byzantine behavior will be absent at one voltage or the other.

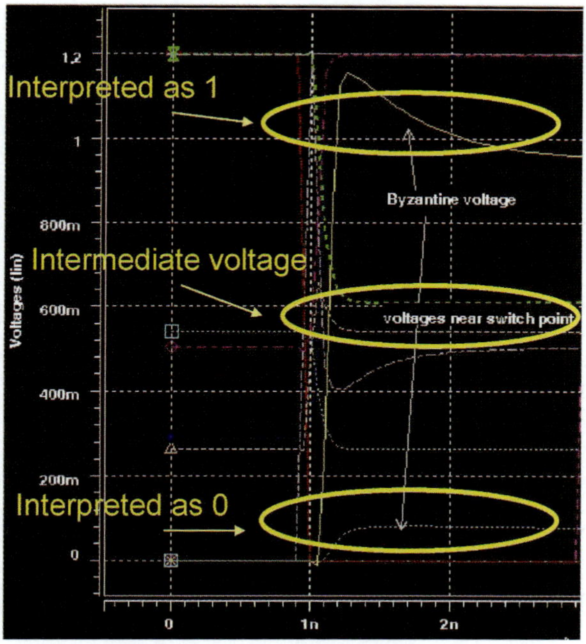

Figure 1-16: Byzantine bridging behavior at 130nm.

Bridge resistance	Behavior (1.3V, 25° C)
<2500 ohms	Voting model
2500-3000 ohms	Analog/Byzantine
>3000 ohms	Delay fault

Table 1-4: Example bridge behavior at 1.32V, 25°C.

Does this mean that the Byzantine Generals have been demoted and are no longer of interest for determining bridging fault behavior? Yes and no. Multiple voltage tests will eliminate delay-independent Byzantine behavior. However, in addition to gate switching voltages, Byzantine behavior can be caused by gate switching times. Long settling times for bridging voltages lead to large skews on gate inputs, which in turn result in different switching times on gate outputs. If path slack is low enough that values are latched before switching is complete, Byzantine behavior can result, as shown in Figure 1-17. In theory, increasing clock speed could eliminate this problem, by latching data at the "latch 0" point rather than the "latch 1" point, but in practice running such tests is difficult. As a result, diagnosis of short-induced delay faults should consider Byzantine behavior a strong possibility.

Also, as was seen with the copper voids, resistance is not a fixed property for defects. Voltage changes result in localized temperature changes around a defect (in general, higher voltages lead to higher currents, and these lead to higher local temperatures). The results are not always predictable. Bit line to ground shorts in memory, for example, have been observed to cause both operating voltage floors and operating voltage ceilings, in some cases on the same RAM instance (but different die).

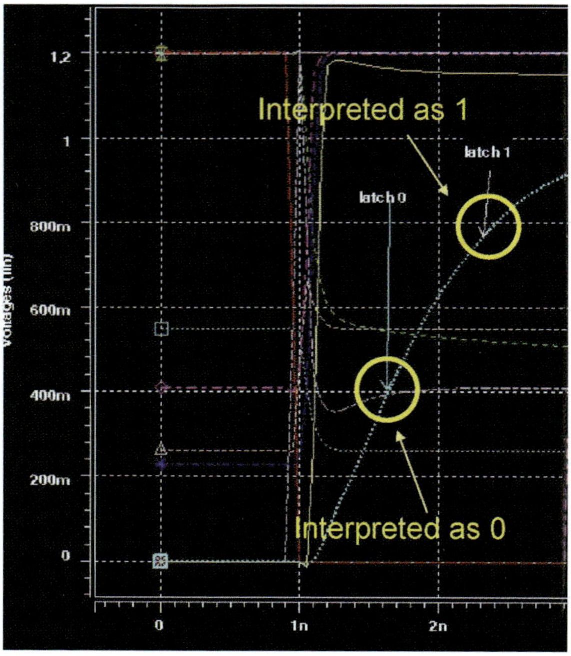

Figure 1-17: Byzantine delay behavior.

1.4.4 Open Fault Models

While shorts remain the most common type of defect in most CMOS processes, open faults are also a cause for concern. As the number of wiring levels in circuits increases, the number of vias proliferates, now running into the tens to hundreds of millions. The effects of missing vias, partial vias, and resistive vias on circuit operation are not easily modeled. In some cases the circuit will still function correctly, but at a lower speed. In others, the orphaned lines will float to a stable voltage. If the voltage is close to a rail value, then a stuck-at fault test will be able to detect the fault. If it floats to an intermediate value, a logic test may not catch the defect, but a current test likely will, since a logical path from power to ground may be active.

The best-known open model is the stuck-open fault model, introduced by Wadsack [20]. In this model, the gate to a given transistor is fixed at the "open" or off value. As a result, the transistor will be unable to pull the cell output to its voltage. Consider the NOR gate in Figure 1-18. If the n-transistor connected to input A is permanently off, then when input A is 1, the output Y will remain unchanged. If B had been 1 for the previous vector, then the output will be correct (logic 0). A stuck-open fault requires a two-pattern test. The first sets up the fault by pulling the circuit output to a known value (applying A=B=0 sets Y to logic 1). The second pattern activates the fault (applying A=1 will not cause Y to change, hence a 1 will be observed rather than the expected 0).

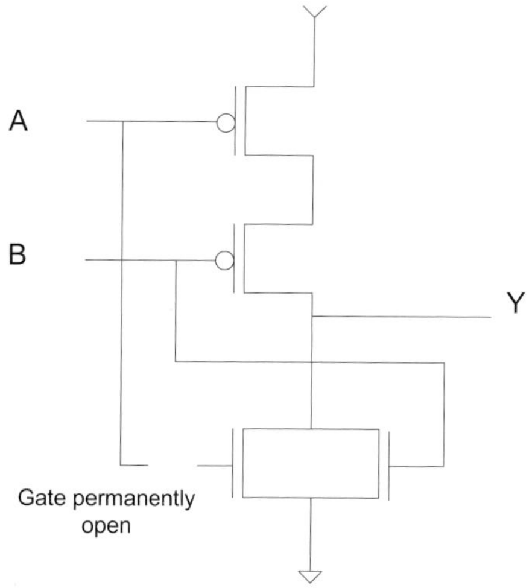

Figure 1-18: N-transistor stuck-open.

The stuck-open model is one example of a fault model targeted at open defects, but it has not found to be representative of many actual CMOS defects, even in current geometries. Because predicting the behavior of opens is so difficult and error-prone, it is likely that test methods that target opens implicitly, such as delay tests or current tests, will be more successful than explicit open models.

Future Directions

As mentioned earlier, partial opens due to copper voiding are a common defect mechanism at 130nm. The two Shmoo[9] plots shown in Figure1-19 are characteristic of this type of failure. Figure 1-19a (a) shows a typical pattern of voltage versus delay failure (at 85°C), where the circuit operates slightly beyond its rated frequency and voltage (red areas are failures, while white areas are passes). The X axis is frequency (increases left to right) and the Y axis is voltage (increases bottom to top). Figure 1-19(b) (25°C) shows the same chip failing tests well into its operating region. Room temperature delay testing is clearly needed here, and is a good idea for any circuit using copper.

Other open failures at 130nm and below behave similarly to those in previous generations, and similar detection methods apply, with the exception of CMP scratches, which produce complex patterns of shorts and opens, although these are usually readily detectable in logic and do not require specific ATPG.

---

[9] The "Shmoo" was originally a blob-like character in the comic strip "Li'l Abner" The patterns of passing and failing tests in a Shmoo plot were reminiscent of its shape.

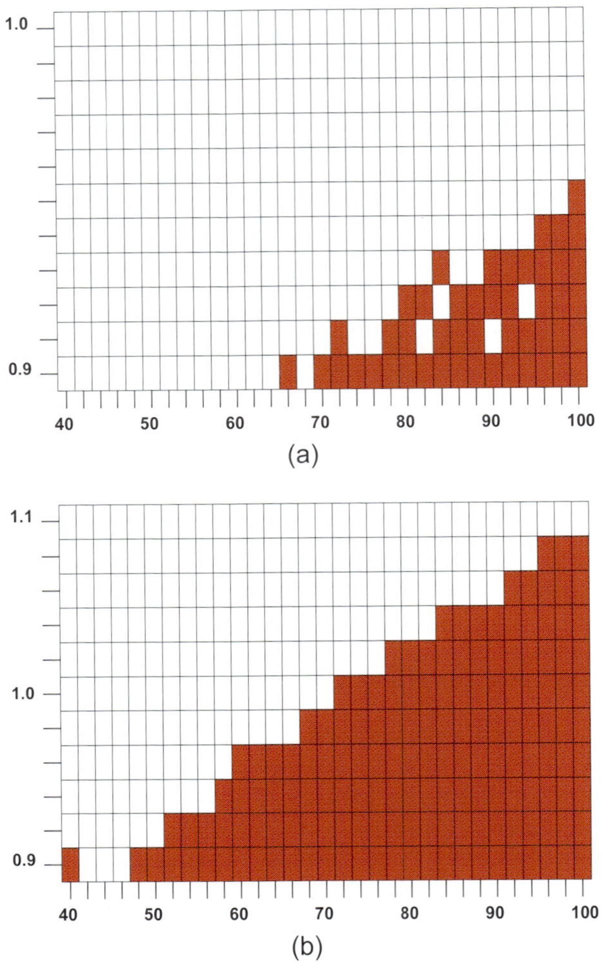

Figure 1-19: Voltage versus timing shmoos at (a) 85°C and (b) 25°C showing copper void behavior.

1.4.5 Timing-related or Delay Faults

Not all defects permanently alter circuit behavior, the way the stuck-at fault model requires. Instead, some change the timing behavior of the circuit, causing incorrect operation only at certain frequencies. These are called *timing-related defects* and the fault models associated with them are *delay fault models*. There are two broad classes of delay faults: *gate delay faults*, which model defects local to individual circuit elements, and *path delay faults*, which model defects resulting from parametric shifts across larger portions of the circuit.

Because delays refer to differences in behavior over time, delay fault models are concerned with *transitions* in logic values, rather than the values themselves. A test for a delay fault consists of an initialization pattern, which sets the stage for a transition, and a propagation pattern, which creates a desired transition and propagates it to an observable point.

Gate Delay Faults

A gate delay fault occurs when a gate operates slower than it is expected to. In the OR gate shown in Figure 1-20, the rising transition on the input is expected to be followed soon after by a rising transition on the output. Instead, the signal rises, but much later. This is known as a *slow-to-rise* fault. A stuck-at-0 fault can be thought of as an extreme case of a slow-to-rise fault. A similar relationship exists between *slow-to-fall* faults and stuck-at-1 faults. *Transition faults* can also be thought of as a type of gate delay fault, where the rise or fall time becomes infinite.

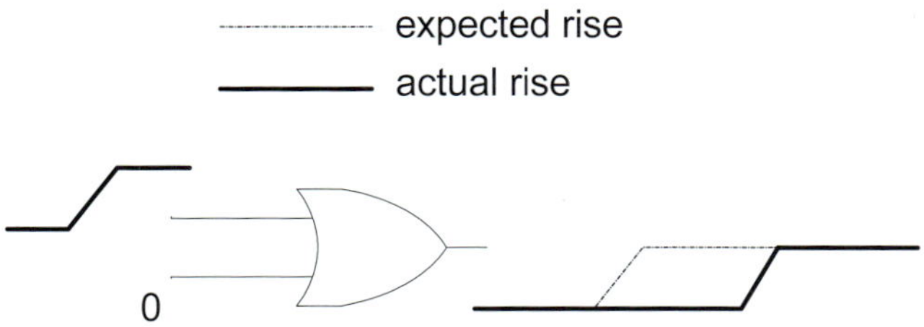

Figure 1-20: Slow-to-rise fault.

The transition used to detect a gate delay fault needs to be propagated to a circuit output, much as a stuck-at fault needs to be propagated, but the actual path of propagation is unimportant (which is not the case in path delay fault testing).

Path Delay Faults

A path delay fault assumes that a logic transition is delayed along an entire path. A gate delay fault can be thought of as a path delay fault through a single gate. Consider the path from A through D, H and I to J in Figure 1-21.

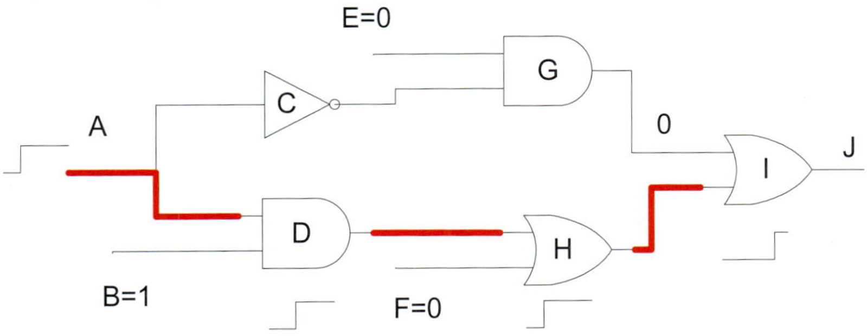

Figure 1-21: Example circuit for path delay faults.

As the rising transition moves through the circuit, it will be delayed. If the signal is delayed beyond the sampling time at J (say a clock edge at a flip-flop), then a delay fault has occurred along that path. In order to detect the fault, the transition must be propagated along the entire path (thus, the inverted transition through gate C must be blocked at G by setting E to 0). For the fault to be detected *robustly*, the existence of other delay faults in the circuit must not prevent detection. As an example of this, input F to gate H must be a fixed 0 value throughout the test – no hazards are allowed on F.

A detailed discussion of delay fault models and testing is provided in a separate chapter of the book.

Capacitive Defects

It has long been known that logic synthesis will result in additional near critical paths and hence additional delay faults [21]. As designs shrink, two additional issues come to the fore: capacitive defects and signal integrity effects. Let us look in more detail at the former. These have the property that they result in timing related failures at low operating frequency, but not high frequency.

Originally, bridging faults were considered to have zero resistance. Later resistance models were added. Now, defect capacitance must also be considered. Intuitively, the increasing importance of wire capacitance in delay calculations is due to the increasing relative capacitance of wiring versus gates as a result of CMOS scaling. In order to measure timing behavior accurately, parasitic capacitances between wires must now be extracted for almost all designs. Increasing frequency and decreasing slack means that the capacitive properties of defects can significantly affect their behavior (RC time constants of defects are well within the range of standard parasitics, and strongly influence timing).

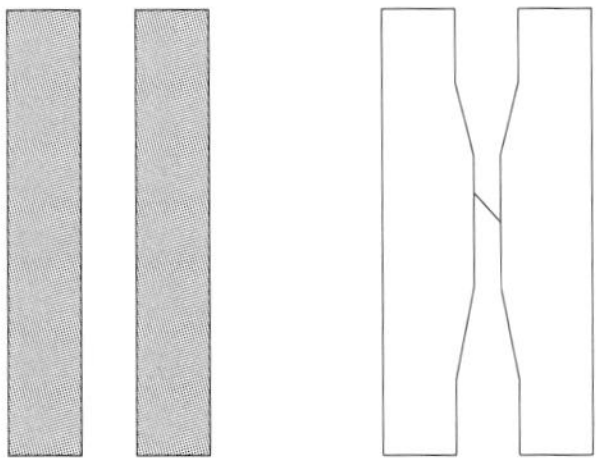

Figure 1-22: Proposed mechanism for capacitive faults (reduced separation gives higher C).

The leakage failure discussed in the next section is an example of a delay failure that operates correctly at full speed but fails at lower speeds (similar in behavior, but not mechanism, to the well-known retention faults in SRAM). Another failure with this behavior is a high resistance, high capacitance bridge. The physical motivation for these is shown in Figure 1-22. This type of behavior, which has been observed in silicon, although not in silicon that is within process specifications (at least to the author's knowledge), shows that low speed tests are needed, in addition to full speed tests, to ensure high quality.

1.4.6 I_{DDQ} Models

In addition to conventional logic testing, where inputs are applied to a circuit and output responses are measured, circuits can be tested parametrically. One example of such a test is an I_{DDQ} test (I_{DD} is the drain current in a CMOS circuit, the "Q" refers to the quiescent or static state). In an I_{DDQ} test, a set of inputs is applied, then the circuit is allowed to settle and the current flowing through the power supply is measured. A circuit that draws more current than some threshold value is identified as defective. Although the test method is quite different than logic testing, similar fault models can still be used (see [22] for an overview).

As CMOS technologies scale down, background leakages are rising inexorably, primarily due to device subthreshold leakage[10] [23]. As a result, conventional single threshold pass/fail I_{DDQ} testing is not viewed as economical below 0.25 micron. Various alternatives have been proposed to extend the lifetime of the technique. Some will work for a single generation (e.g. splitting circuit power supplies), others may last longer (measurement analysis techniques which treat finding defect current within an I_{DDQ} measurement as signal extraction from a noisy reading). Other approaches include dynamic current measurement, energy consumption, and thermal testing. In each of these cases defect observation is implicit and activation is explicit, so fault models developed for I_{DDQ} testing will continue to work.

Leakage Induced Faults

In 130nm technology and below, leakage is a significant problem, especially in the fast process corner and at high temperatures. Leakage is mainly dominated by subthreshold leakage, although gate leakage[11] is a significant factor at lower temperatures. Leakage is not directly controlled by many foundries, but instead is a byproduct of other parameters (e.g. channel length). Because of this, it is possible for leakage on individual transistors to be well beyond specified process limits. As an example, a latching circuit that uses a weak pull-up transistor as a holding device has been observed to fail in silicon due to leakage on its input, as shown in Figure 1-23. The transistors connected to A and B are comparatively large, and off-state leakage on A was more powerful than the pull-up drive of the PFET. This eventually pulled the input of the inverter to zero, which changed its output to a 1.

The future of I_{DDQ} testing is twofold. First, there is a large class of chips where low leakage power is critical to correct operation and is part of the product specification (this includes virtually anything that operates from a battery). I_{DDQ} will continue to be used for these circuits for the foreseeable future. For some other chips, usually high performance devices that are "plugged into a wall", I_{DDQ} is no longer

---

[10] Subthreshold leakage occurs across the channel of a transistor even though the gate voltage is too low to turn on the transistor. It increases as channel length shrinks and also strongly varies with increasing temperature.

[11] Gate leakage is conduction across the oxide of a transistor between the gate and substrate. It increases as oxide thickness decreases, and has much less temperature dependence than subthreshold leakage.

practical as a production test, but even for these chips it remains an effective diagnosis tool.

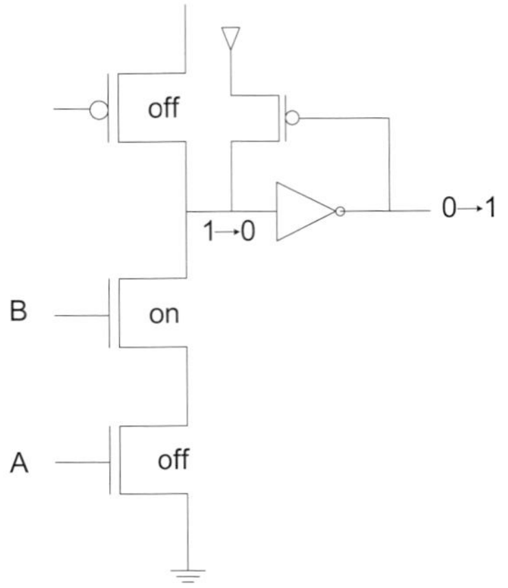

Figure 1-23: Leakage induced faulty latch behavior.

1.5 DEFECT-ORIENTED TEST TYPES

Now that we have looked at both defects and fault models, we are ready to apply them into a Defect-Oriented test strategy. This section describes the required elements of such a strategy; that is, a comprehensive approach to chip testing beginning with a defect analysis and comprising multiple test types or methods.

1.5.1 Logic Tests

Logic testing, most typically using scan vectors, remains the most common and useful technique for quickly identifying bad parts. The vast majority of these tests are developed using the single stuck-at fault model, and most commercial ATPG tools are based on it. Since few defects manifest themselves as single stuck-at faults[12] [24], what should be done for logic testing in a Defect-Oriented test strategy[13]?

---

[12] Note that this is not the same thing as saying that few defects are detected by stuck-at fault tests. Most defects can be detected by test vectors generated for single stuck-at faults, even though they do not behave exactly as postulated by the model. As a simple example, 01, 10, and 11 constitute a complete single stuck-at test set for a two input NAND gate, with each of three basic stuck-at faults failing one of the patterns. Most actual defects will fail for more than one pattern, or will fail a pattern intermittently; e.g. an NFET transistor gate shorted to the cell output can potentially fail all 4 input patterns.

[13] A Defect-Oriented test strategy begins with a decision to use Defect-Oriented methods as the basis for production test. This leads to decisions about which tests to include, and this is the subject of this section.

In general, a threefold approach is best:

- First, logic tests are a key part of any test methodology.
- Second, the vectors may be generated with the stuck-at model, but they do not have to be *graded* that way[14].
- Third, test application conditions are as important as the vectors themselves.

Given that logic tests are effective, and that few organizations have the resources needed to develop non-stuck-at test generators (e.g. bridging faults), then it is important to use stuck-at based logic tests effectively. If an "n-detect"[15] ATPG is available, it should be used. The use of n-detect patterns forces logic errors to be propagated from the fault site when it has been made active (sensitized) in at least n different ways. If a given defect has some probability of being activated whenever the fault site is sensitized, then multiple independent sensitizations will increase its probability of being detected. If not, then a combination of DFT and stuck-at ATPG should be used to obtain as high a fault coverage as practicable. Note, though, that splitting a fixed test effort across multiple test types is more effective at improving quality than focusing only on a single type. Tests generated with stuck-at ATPG can be graded for other models, such as bridging faults, or open faults. This will provide some added value, especially in achieving very high quality tests.

Lastly, test application is a key. Tests should always be applied at operating voltage extremes, both high and low, and also at multiple temperatures, as shown in Table 1-5.

Method	Wafer	Package
Slow speed scan	25C, V_{DD} +/-10%	85C, V_{DD} +/-10%
At speed scan	25C, V_{DD} +/-10%	85C, V_{DD} +/-10%
I_{DDQ}	25C, V_{DD} +10%	85C, V_{DD} +10%
Min- V_{DD}	25C	85C

Table 1-5: Sample defect oriented production test suite.

1.5.2 Current-based Tests

Current-based tests have several inherent advantages over voltage-based tests: The power supply is ubiquitous, making observation straightforward; many defects draw extra supply current; and the inherent parallelism of the tests can result in reduced vector sets[16]. From the early 1980s to the mid 1990s, these advantages led to the widespread adoption of I_{DDQ} testing (measurement of the static power supply current – meant to be zero in CMOS). As technology advanced to the 0.35u level and below, however, it became clear that I_{DDQ} testing was becoming increasingly difficult, and

---

[14] Fault grading is the process of simulating faults to determine fault coverage.

[15] An n-detect test set forces each fault to be detected n different ways rather than just once.

[16] Current-based testing looks for defects in larger areas than a node which voltage-based tests examine.

that the line between good and bad parts was increasingly difficult to draw (see the histogram of Figure 1-24 [25]). The consequences are clear: setting a limit too high results in shipping bad parts, while setting it too low results in scrapping good parts.

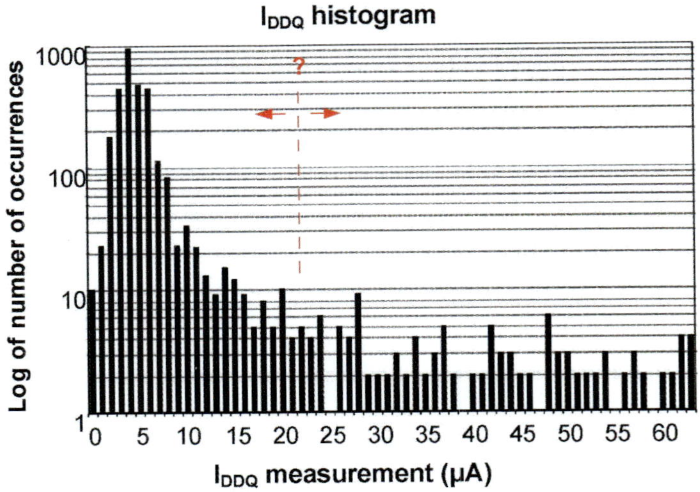

Figure 1-24: Histogram of I_{DDQ} values [25].

One observation that has helped develop production worthy current-based tests in the deep submicron regime is that defects that can affect circuit operation will draw different amounts of current: more when they are activated than when they are not. Other defects are "benign" or not defects at all (e.g. transistor length reduction will simultaneously increase leakage). As a result, a variety of "delta I_{DDQ}" techniques have been proposed. The basic rationale is shown in Figure 1-25, which is taken from a SEMATECH experiment [26] that will be described in detail in Section 1.6. Background leakage is about 2mA, but a defective behavior can be clearly seen raising this value to 5 and even 8mA.

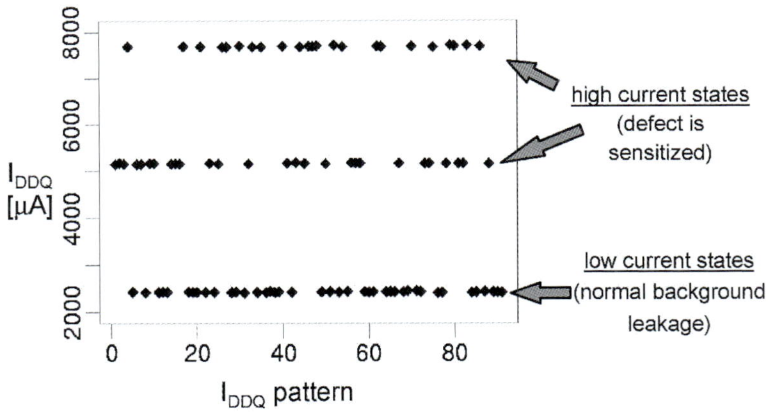

Figure 1-25: Multiple observed current levels [26].

The current ratios technique [27] sets a maximum delta for subsequent measurements based on an initial reading. Currents above a maximum or below a minimum are cause for rejecting a part. In one experiment [27], 14 die were rejected with ratios. Each of these parts exhibits significant vector-dependent deviation in I_{DDQ} from expected values and so each is clearly a defective part. A single threshold measurement of 330uA applied to the same sample also rejected 14 parts, but only 3 of these were in the original set. Reducing the threshold to 100uA finds 11 of the 14 parts, but rejects 44 other presumably good die. At 50uA, 13 of 14 are found, but at the expense of 56 good die. Table 1-6 shows the results.

Approach	Total rejected die	"Bad" die rejected	Other die rejected
Current ratios	14	14	0
330uA threshold	14	3	11
100uA threshold	55	11	44
50uA threshold	69	13	56

Table 1-6: Rejected die for various I_{DDQ} approaches [27].

Other delta methods include pre and post stress I_{DDQ} measurements, measurements at different temperatures, delta across wafer, and vector-to-vector deltas (the latter allows for on-chip temperature changes during test). New emphasis on leakage reduction techniques as part of low-power design will serve to extend the life of I_{DDQ} testing in many applications.

Several techniques for measuring dynamic current have been proposed (e.g. monitoring charge consumption, average supply current, the current waveform, and a single sample point), and some even used in production [28], but by and large these remain an active research area. The basic approach is apply vectors to circuit, sample current at one or multiple points, potentially convert to frequency or other spectrum, and compare with expected values. The challenge comes in making repeatable measurements, in light of process variations, as well as difficulties with the equipment itself.

1.5.3 Delay Test

Considerable discussion of timing testing occurs in other chapters of the book, but for this section we will focus on the need for delay testing and how it should be performed. Most of the defect types described earlier have an effect on circuit operation over a limited frequency range. A slow logic test simply will not detect them. In addition, most delay defects manifest themselves mainly at high temperatures. A leakage-related problem, for example, may not be observable at room temperature. On the other hand, as noted earlier, a copper void may only be detectable at room temperature or below. Thus Defect-Oriented delay tests need to be run at multiple temperatures.

The specific type of test run is also important. A transition fault test gives high coverage of circuit nodes, but will mainly target short paths [29]. A path delay test may be needed to isolate very small timing defects (on the order of picoseconds), but it is impossible to apply these tests to all circuit nodes. Efforts such as [30] on path distribution are valuable in this respect.

Internal versus external application of tests (e.g. BIST versus AC scan, PLL versus ATE) are of less importance to defect detection than the type of vectors and the manner that they are applied. In general, it is better to have good accuracy on waveforms, especially clock, and comparable IR drop[17] conditions to normal operation (e.g. fraction of signals switching near the observed path). If full circuit operating speed cannot be obtained, a reduced speed is often very effective (Maxwell et al found that 13 of 15 at-speed functional test fails would fail the same test at half speed [31]).

1.5.4 Very Low Voltage

Another approach to finding delay problems that has found some success is *very low voltage (VLV) testing*. The basic idea is that delay increases substantially as V_{DD} lowers, and that this effect is more pronounced for faulty circuits. Thus, a test at low V_{DD} can more easily detect defects difficult to detect at nominal V_{DD}. Much of the research on the subject has come from Stanford (e.g. [32]). The authors show that VLV testing can detect many defects, including:

- Shorts (gate oxide, metal, etc.)
- Weakly driven gates
- V_T shifts
- Transmission gate opens
- Defective interconnect buffers
- Tunneling opens

Determining the test voltage has proven to be difficult in practice, and this is aggravated by variations in threshold voltage and lower operating V_{DD} with scaling (e.g. Chang et al recommend setting the VLV limit at 2 to 2.5 times V_T, but for many circuits in the 90nm generation, this is the operating voltage). However, one method that works in practice is to look for "Min V_{DD}" or the lowest voltage where a particular chip will function (as defined by passing a given test at a given frequency). Like I_{DDQ}, setting a specific threshold for Min V_{DD} is difficult, for two reasons: first, process variation will result in substantial variation in the minimum voltage at which a circuit will operate, and second, SPICE models are significantly less accurate at low voltages far away from nominal supply values. Instead, it may be more useful to use statistical analysis and look for outliers [33].

[17] *IR drop* refers to the drop in supply voltage caused by sudden increases in current consumption across a resistive power supply network. *Ground bounce* is the analogous increase in ground voltage.

1.5.5 Stress Testing

Voltage stress testing is useful for detecting a variety of gate oxide defects, ranging from pinholes to contamination to excessive thinning. Other circuit elements, such as vias and bulk, are also stressed, but oxide defects are the customary targets. In earlier generations of CMOS technology, gate oxides were not significantly stressed and power supply voltage was set primarily for TTL compatibility. Over the past few generations, however, shrinking geometries and higher performance requirements have mandated lower voltages, while I/O compatibility still favors higher voltages. Dual gate oxide processes[18] attempt to cover both contingencies, while still maintaining reasonable electric fields across the oxides. As competitive pressures mount, process development engineers push oxide fields further, approaching the inherent limits of the materials over expected device lifetimes.

The acceleration factor, AF, of a voltage stress test can be written as [34]:

$$AF = He^{\gamma(V_{stress} - V_{udr})}$$

In this equation, H is a constant, and γ is a field acceleration parameter that depends on oxide thickness. Stress voltage (V_{stress}) must be kept high for long enough time to damage bad (thin) oxides, but not so long that it destroys good oxides. The acceleration factor in the equation above depends on the difference between the use voltage (V_{udr}) and the stress voltage (V_{stress}). Righter [35] showed good results with 8V, less good with 7V on a 5V process, implying that stress voltages closest to the absolute tolerance of the device are most effective. For current generation 1.0V CMOS the differential stress exponent is reduced from 3V to 0.4V or less. This reduction in exponent means that the acceleration factor will be lower for current technology than it was for previous generations, and that as a result longer stress times will produce less effective results.[19]

The future of stress testing is an open question. Burn-in[20] and voltage stress continue to be widely used, but it is clear that the acceleration available is diminished. Interested readers are referred to [34] or the proceedings of the International Reliability Physics Symposium for recent discussions and results.

---

[18] It is desirable to scale gate oxide thickness to attain high performance, and this is what leads to scaled power supply voltages. I/O voltages are often driven by standards that do not scale, but transistor oxides that are able to withstand these high voltages are too thick to obtain the required performance away from the I/O. Instead, two oxide thicknesses are used, one for the I/O and the other for the logic in the chip's core.

[19] Assuming gamma is roughly constant, a change in voltage delta from 3 to 0.4 reduces the acceleration factor by about 13X (from 20 to 1.5).

[20] Burn in is accelerated life testing accomplished by raising the ambient temperature and often operating voltage. Its acceleration behavior is similar to voltage stressing. Subclasses of burn-in include "bake" (no voltage applied), "static" (chip is powered up, but not operating), and "dynamic" (chip is operating during burn-in). Some modern chips generate enough heat (especially at accelerated voltages), that burn-in ovens require significant cooling to prevent thermal runaway – a positive feedback between increasing ambient temperature and increasing heat generation from the chip.

1.6 EXPERIMENTAL RESULTS

A number of results have been published in Defect-Oriented testing over the years. Some of these have been pivotal and will be reviewed here.

In the early 1990s, structural test was just beginning to achieve wide adoption. Some companies, such as IBM, had been using scan design for many years, but for most of the industry, scan was something new and untried. At HP, scan had first been adopted in the late 1980s, and a key question was how effective were scan tests: How many bad parts did they find? How many good parts did they reject? Around 1990, HP also began investigating I_{DDQ} tests in production. These tests tended to fail numerous parts, so effectiveness questions were equally important. Many of the results of HP's studies have been published. Some of the key ones are [31], [36] [37] and [38].

1.6.1 Fault Coverage, Scan vs. Functional

Fault coverage has always been used as an approximation of defect coverage. An interesting series of studies was carried out at HP to determine the relationship between different types of fault coverage and defect coverage, as measured by both test escapes and fallout. The results clearly showed that while defect coverage and fault coverage were related, the type of test was more influential in determining quality level than the coverage number. Thus a scan test with 92% single stuck-at fault coverage had lower overall defect coverage than a combined scan and functional test with only 82% single stuck-at fault coverage (see Figure 1-26).

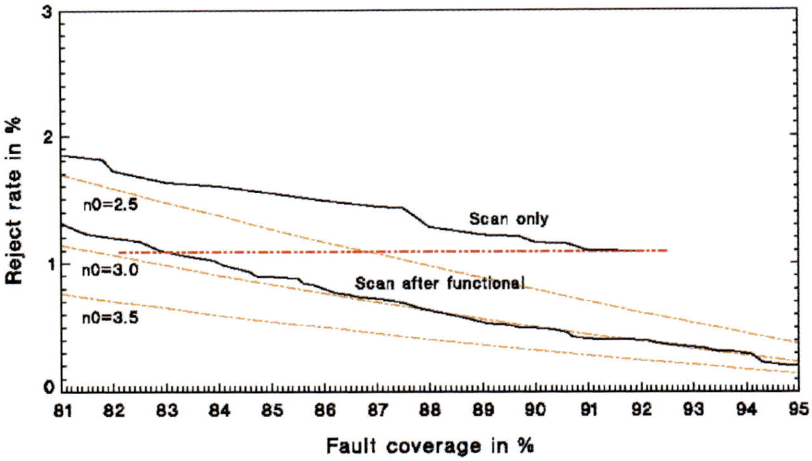

Figure 1-26: Effectiveness of tests versus fault [36].

1.6.2 Effectiveness of I_{DDQ}, Scan, At-speed Tests

This experiment was reported at ITC in 1996 [38]. Five test types were applied to a 25K gates synthesized standard cell design (a small to medium sized ASIC at the time): slow speed scan, AC scan (same vectors, clocked faster), I_{DDQ}, slow functional and at-speed functional. Details of the chip are below:

- Fully static, 5V operation
- 0.8 um process, 3 metal layers
- 1497 flip-flops, full scan design
- Single clock domain, 33 MHz operating frequency

These features made the design very amenable to analysis, and allowed a variety of tests to be developed, including:

- Functional vectors with 83.7% single stuck-at fault coverage
- 284 stuck-at scan vectors with 99.0% single stuck-at fault coverage (100% detectable coverage)
- Combined coverage of functional + stuck-at scan 99.3%
- 369 I_{DDQ} vectors with 98.6% pseudo-stuck-at coverage [21]
- 554 transition tests with 99.4% transition coverage
- 148 path delay tests (fault coverage here is not meaningful, since only a miniscule subset of paths are tested, but these are the longest sensitizable paths reported by the timing analysis tool).

Data were gathered for parts from 3 wafer lots, in order to reduce lot-specific effects on the results. As with any such experiment, there are many ways of looking at the results, but two approaches are key. First, effectiveness of at-speed scan testing versus functional testing, as shown by the Venn diagram in Figure 1-27.

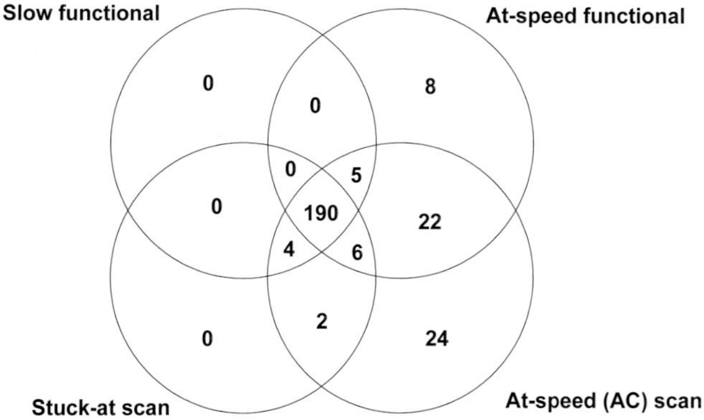

Figure 1-27: Effectiveness of various test types [38].

Three results are immediately clear: 1) at-speed tests are required to detect significant numbers of failing parts, 2) at-speed scan tests find most of the same failing parts as at-speed functional tests, plus some others, and 3) at-speed functional tests still find parts missed by other tests.

---

[21] Pseudo-stuck-at fault coverage is calculated by applying a full stuck-at fault set within a standard cell, and requiring that these faults be propagated to the output of the cell.

The second aspect of the results related to the effectiveness of I_{DDQ} testing. Two Venn diagrams are shown in Figure 1-28. Again, several results are clear: (1) Many die fail I_{DDQ} testing uniquely. At a 50uA threshold (considered reasonable at the time), more parts fail I_{DDQ} than all other tests combined. (2) The number of I_{DDQ} fails is strongly tied to the threshold chosen. (3) A small number of I_{DDQ} fails also fail at-speed tests (note how 25 AC scan and functional fails becomes 22 when the threshold is reduced from 200uA to 50uA). Again, these results were used to justify multiple, sometimes conflicting, actions. At HP, they spurred the effort that eventually lead to the current ratios technique, while for others they showed that I_{DDQ} was a wasteful test that should be removed from production.

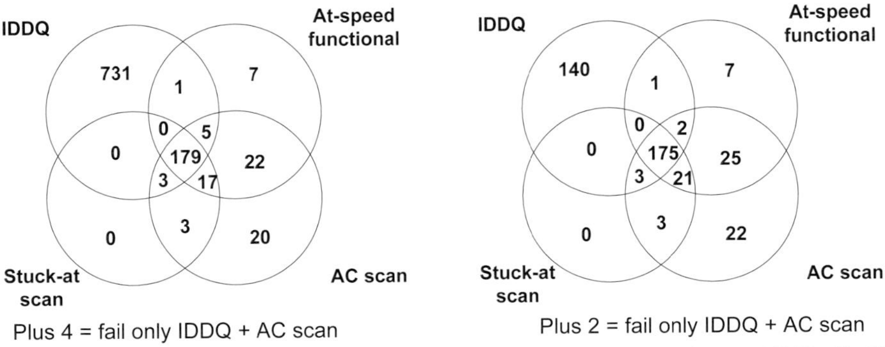

Figure 1-28: Effectiveness of I_{DDQ} versus other tests [38] (left diagram for 50uA, right diagram for 200uA).

The nature of the parts used for these tests, and the confusion left over in some of the results, meant that additional experiments were called for. SEMATECH sponsored several projects, including one intended to reproduce the test types examined in the HP experiment. This experiment used a larger chip, and also included burn-in on a number of the parts to address reliability issues.

This experiment was reported at VTS and ITC in 1997 and 1998 [39], [40]. Details of the chip are below:

- Fully static, 3.3V operation
- 0.8 um process, 3 metal layers
- 116K gates
- 5280 LSSD latches, full scan design (2640 equivalent flip-flops)
- Two clock domains, 40 and 50 MHz operating frequency

Tests applied included:

- Functional vectors with 52% single stuck-at fault coverage (532K cycles, the best vectors available – their low coverage contributed significant questions to the overall results)
- 8023 stuck-at scan vectors with 99.79% single stuck-at fault coverage (100% detectable coverage)

- Combined coverage of functional + stuck-at 99.3 (note that this is less than the reported single stuck-at coverage because some areas of the circuit were inaccessible to the scan tests and not included in the model)%
- 195 I_{DDQ} vectors with >95.7% pseudo-stuck-at coverage
- 5232 transition tests with 91% transition coverage
- Scan flush test[22] (tests entire scan path for delays)

Numerous significant results came from this experiment, and work from the data was reported as late as 2002 [41]. It is instructive to look at the data in the same Venn diagram format as the HP data (see Figure 1-29).

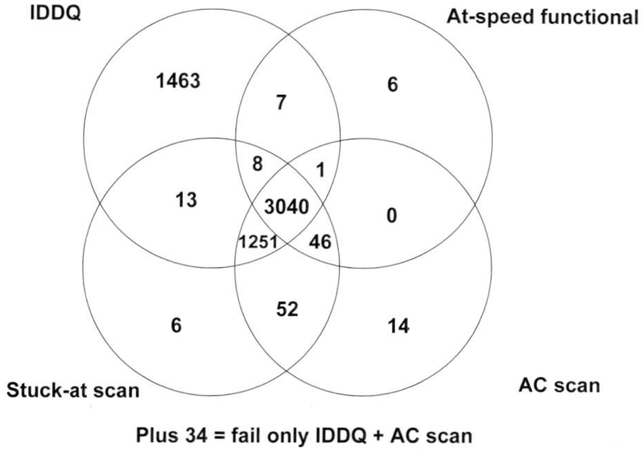

Figure 1-29: Effectiveness of various test types, SEMATECH, 500nA threshold.

Similar results are seen, showing that the general observations regarding I_{DDQ}, at-speed functional and at-speed scan tests could be applied to two chips from different design teams in different processes at different foundries. This confirmation was very valuable.

The burn-in results from this experiment were also very useful. Unique I_{DDQ} fails were viewed as potential reliability risks. An extensive burn-in was conducted on some of these parts, and on a control group, giving an acceleration equivalent to over 3 million hours of operation. The failure rates observed are shown in Table 1-7:

[22] LSSD uses a two phase clock. When both clocks are raised simultaneously, the scan chain becomes a large delay element, so a transition may be timed as it is "flushed" from one end to the other.

Current	Percent burn-in failures
< 5uA	0.5
5-100uA	0.5
0.1-1.0mA	1.1
1-5mA	2.0
> 5mA	3.8

Table 1-7: Burn-in failure percentages versus initial I_{DDQ}.

Again, two results were observed, which lead to two separate conclusions. First, unique I_{DDQ} fails had significantly reduced reliability compared to ordinary parts. Second, the vast majority of unique I_{DDQ} fails were not reliability risks. The data support arguments both for and against I_{DDQ}, but these are mainly settled by economic discussions at the per-chip level: which is more costly, a reliability failure or the yield loss from I_{DDQ} testing? Significant work was begun to reduce the cost of I_{DDQ} testing, in particular that associated with single threshold pass/fail measurements.

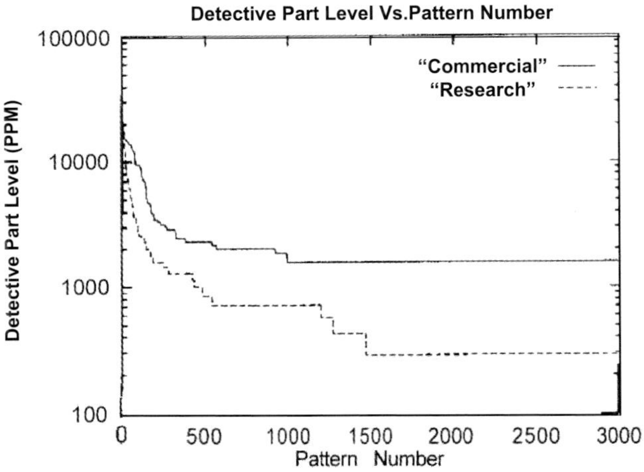

Figure 1-30: Tester fallout for commercial and observation-enhanced ATPG.

Another experiment in Defect-Oriented test that warrants discussion is the work done at Texas A&M University and at TI on the nature of test detection [15], [13]. As mentioned previously, the stuck-at fault model is known to be a marginal predictor of test quality, and test sets that targeted stuck-at faults multiple times (n-detect) were observed to have higher quality than those that targeted them once [13], but these studies looked into the underlying causes. Two key causes were identified. First, the keys to defect detection are not controllability and observability (as used by stuck-at ATPG), but rather activation and observability – some conditions need to be set to cause a defect to fail. Second, for unknown defect behaviors, activation is difficult to deterministically identify, so a defect activation approach should generate large amounts of varying behavior at a defect site, but observability is mainly deterministic and standard stuck-at ATPG techniques can be applied. The team used this approach to develop an ATPG tool and compared it to a commercial version. Their results (see Figure 1-30) showed that higher quality could be obtained at any fault coverage by using the technique described above.

1.6.3 Statistical Post Processing

Some interesting work has been done at LSI Logic and Portland State University [42], [43] on methods for extending single die pass/fail thresholds for various tests into a statistical post-processing operation that gathers data for all die on a wafer and makes decisions based on how individual die relate to their neighbors and to overall trends. An example is shown in Figure 1-31 for a Min V_{DD} screen [43]. The minimum passing voltage is recorded for each die on a wafer, and a *nearest neighbor ratio* (NNR) is calculated. Those die with Min V_{DD} significantly different from their neighbors are isolated as outliers and rejected, even though other die on the wafer with the same reading pass.

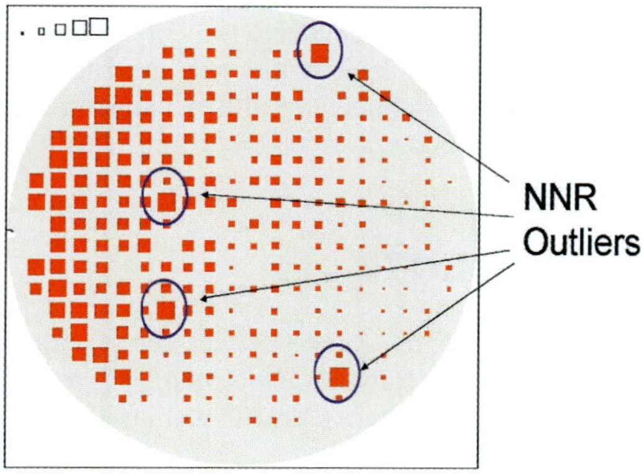

Figure 1-31: Wafer data for Min V_{DD} showing outliers (courtesy R. Madge).

1.7 FUTURE TRENDS AND CONCLUSIONS

All testing strives to continuously improve efficiency, allowing products to be shipped at the required quality levels with lowest cost. Defect-Oriented testing is a natural extension to functional and structural test methods that enables improvements in quality and cost by targeting defects directly, rather than indirectly.

As described in this chapter, the current state of the art in Defect-Oriented test methods are n-detect scan patterns for logic testing, at-speed tests for delay defects, very low voltage testing, stress testing, and some type of current-based test. As geometries shrink, inherent process leakage has been rising as well in order to achieve performance targets [3], [44]. In addition, lithography, chemical, and physical challenges have resulted in increasing variation between manufactured devices, leading to wide variation in process leakage. As a result, conventional single threshold I_{DDQ} testing no longer works in many situations and statistical approaches (e.g. current ratios) have been developed as an alternative.

Statistical methods (outlier identification) are promising for non- I_{DDQ} measurements as well (e.g. min-V_{DD}). These methods identify defects implicitly (by measurement signature), rather than explicitly (failure of a specific single measurement).

Process variation is a catch phrase that encompasses many physical and electrical properties. As technology continues to advance, the classical distinction between defects and variation will begin to disappear. This has happened already in some areas, in particular in SRAM, and will continue in the future. As a result, design-for-manufacturability (DFM) techniques (e.g. [45], [8]) will become increasingly intertwined with DFT methods. New technologies will also be developed to tolerate failures during normal operation (e.g. [46]), in order to account for defects whose behavior changes over time, as well as others which may not have been found during test.

As the distinction between defective silicon and merely different silicon decreases, Defect-Oriented test approaches will need to work increasingly tightly with circuit design, system architecture, and manufacturing processes in order to continue to ensure the continued availability of high quality products at the lowest cost.

ACKNOWLEDGMENTS

The author would like to thank Anne Gattiker for graciously providing materials from our joint tutorial. Chuck Hawkins, Bob Madge, Peter Maxwell, Phil Nigh, and Jerry Soden also contributed. Thanks also to Dimitris Gizopoulos for his patience and help.

REFERENCES

[1] J.P. Shen, W. Maly, and F.J. Ferguson, "Inductive Fault Analysis of MOS Integrated Circuits," *IEEE Design and Test*, Vol. 2, No. 6, pp. 13-36, Dec. 1985.

[2] R. Rodriguez-Montanes, E. Bruls, and J. Figueras, "Bridging Defects Resistance Measurements in a CMOS Process", *Proc. IEEE International Test Conf.*, pp. 892-896, 1992.

[3] R.C. Aitken, "New Defect Behavior at 130nm and Beyond", *European Test Symposium*, May 2004.

[4] M. Tripp, comments at Defect Based Testing Workshop, Monterey, CA, April 2001.

[5] R. Rodrigues-Montanes, P. Volf, J. Pineda de Gyvez, "Resistance Characterization for Weak Open Defects", *IEEE Design and Test*, Vol. 19, No. 5, pp. 18-26, Sept.-Oct. 2002.

[6] W. Maly et al, "Deformations of IC Structure in Test and Yield Learning", *Proc. IEEE International Test Conference*, pp. 856-865, 2003.

[7] F M. Schellenberg, "Sub-Wavelength Lithography Using OPC", *Semiconductor Fabtech Journal*, 9th edition, March 1999.

[8] C. Guardiani et al, "Proactive Design For Manufacturing (DFM) for Nanometer SoC Designs", *Proc. IEEE Custom Int. Circ. Conf.*, pp. 309-316, 2004.

[9] L. Wei, Z Chen, M. Johnson, K. Roy and V. De, "Design and Optimization of Low Voltage High Performance Dual Threshold CMOS Circuits", *Proc. ACM/IEEE Design Automation Conf.*, pp. 489-494, 1998.

[10] R. Aitken et al, "Library Modeling for Effective Leakage Management", *Proc. Synopsys Users Group*, San Jose CA, 2004.

[11] J. Segura, C. F. Hawkins, *CMOS Electronics: How It Works, How It Fails*, IEEE Press (Wiley), 2004.

[12] D. Monticelli, "Solving the Real Challenges of Low-Power SOC Design in 90 Nanometers", *DesignCon 2004*, Santa Clara CA, February 2004.

[13] J. Dworak et al, "Enhanced DO-RE-ME Based Defect Level Prediction Using Defect Site Aggregation – MPG-D", *Proc. IEEE International Test Conference*, pp. 930-939, 2000.

[14] R.D. Eldred, "Test Routines Based on Symbolic Logical Statements," *Journal of the ACM*, Vol. 6, pp. 33-36, 1959.

[15] M.R. Grimaila et al, "REDO - Random Excitation and Deterministic Observation - First Commercial Experiment", *Proc. IEEE VLSI Test Symposium*, pp. 268-274, Apr. 1999.

[16] K.C.Y. Mei, "Bridging and Stuck-at Faults," *IEEE Trans. Computers*, Vol. C-23, pp. 720-727, July 1974.

[17] I. Polian, *On Non-standard Fault Models for Logic Digital Circuits: Simulation, Design for Testability, Industrial Applications*. VDI Fortschritt-Berichte, Reihe 20, Nr. 377. VDI-Verlag, Düsseldorf. 2004.

[18] J.M. Acken, S.D. Millman, "Accurate Modeling and Simulation of Bridging Faults", *Proc. IEEE Custom Integrated Circuits Conference*, pp. 17.4.1-17.4.4, 1991.

[19] P.C. Maxwell and R.C. Aitken, "Biased Voting: A Method for Simulating CMOS Bridging Faults in the Presence of Variable Gate Logic Thresholds", *Proc. IEEE International. Test Conf.*, pp. 63-72, 1993.

[20] R.L. Wadsack, "Fault Modelling and Logic Simulation of CMOS and MOS Integrated Circuits," *Bell Sys. Tech. Jour.*, Vol. 57, pp. 1449-1474, 1978.

[21] T.W. Williams et al, "The Interdependence Between Delay-Optimization of Synthesized Networks and Testing", *Proc. ACM/IEEE Design Automation Conf.*, pp. 87-92, 1991.

[22] S. Chakravarty and P. Thadikaran, *Introduction to I_{DDQ} Testing,* Kluwer Academic Publishers, 1997.

[23] T.W. Williams et al, "Iddq Test: Sensitivity Analysis of Scaling", *Proc. IEEE International Test Conf.*, pp. 786-792, Washington DC, Oct. 1996.

[24] R.C. Aitken "Finding Defects with Fault Models", *Proc. IEEE International Test Conf.* pp. 498-505, 1995.

[25] D. Josephson, M. Storey, and D. Dixon, "Microprocessor IDDQ Testing: A Case Study", *IEEE Design and Test*, Vol. 12, No. 2, pp. 42-52, Summer 1995.

[26] P. Nigh et al, "An Experimental Study Comparing the Relative Effectiveness of Functional, Scan, IDDq and Delay Fault Testing", *Proc. VLSI Test Symp.*, pp. 459-464, 1997.

[27] P. Maxwell et al, "Current Ratios: A Self-Scaling Technique for Production IDDQ Testing", *Proc. IEEE International Test Conf.*, pp. 738-746, 1999.

[28] E. Peterson and W. Jiang, "Practical Application of Energy Consumption Ratio Test", *Proc. IEEE International Test Conf.*, pp. 386-394, 2001.

[29] J. Rearick and M. Sharma, "Method and apparatus for measuring the quality of delay test patterns", U.S. Patent #6708139, 2004.

[30] T. McLaurin, "Debugging and Diagnosing Delay Defects in Deep Submicron Designs", *Proc. Silicon Debug and Diagnosis Workshop*, 2004.

[31] P. Maxwell, I. Hartanto, and L. Bentz, "Comparing Functional and Structural Tests", *Proc. IEEE International Test Conf.*, pp. 400-407, 2000.

[32] J.T.Y. Chang and E.J. McCluskey, "Detecting Delay Flaws by Very Low Voltage Testing", *Proc. IEEE International Test Conf.*, pp. 367-376, 1996.

[33] R. Madge, B.Goh, V. Rajagopalan, "Screening Min V_{DD} Outliers Using Feed-Forward Voltage Testing", *Proc. IEEE International Test Conf.*, pp. 673-682, 2002.

[34] R. C. Aitken, "Test Generation and Fault Modeling for Stress Testing", *Proc. Int. Symp. on Quality in Elect. Design*, pp. 95-99, 2002.

[35] A. Righter, "IDDQ/Burn-in Effectiveness," *SEMATECH Project Report S-88B*, January 1997.

[36] P.C. Maxwell, R.C. Aitken, V. Johansen, and I. Chiang, "The Effect of Different Test Sets on Quality Level Prediction: When is 80% Better Than 90%", *Proc. IEEE International Test Conf.*, pp. 358-364, 1991.

[37] P.C. Maxwell, R.C. Aitken, "The Effectiveness of IDDQ, Functional, and Scan Tests: How Many Fault Coverages Do We Need?", *Proc. IEEE International Test Conf.*, pp. 168-177, 1992.

[38] P.C. Maxwell, R.C. Aitken, K.R. Kollitz, and A.C. Brown, "IDDQ and AC Scan: The War Against Unmodeled Defects", *Proc. IEEE International Test Conf.*, pp. 250-258, 1996.

[39] P. Nigh et al, "An Experimental Study Comparing the Relative Effectiveness of Functional, Scan, IDDq and Delay Fault Testing", *Proc. IEEE VLSI Test Symp.*, pp. 459-464, April 1997.

[40] P. Nigh et al, "Failure Analysis of Timing and Iddq-Only Failures from the SEMATECH Test Methods Experiment", *Proc. IEEE International Test Conf.*, pp. 43-52 Nov. 1998.

[41] C. Thibeault, "Speeding-up IDDQ Measurements", *Proc. IEEE VLSI Test Symposium*, pp. 295-301, 2002.

[42] W. Daasch et al, "Neighbor Selection for Variance Reduction in IDDQ and Other Parametric Data", *Proc. IEEE International Test Conf.*, pp. 92-100, 2001.

[43] R. Madge et al, "Screening MinV_{DD} Outliers Using Feed-Forward Voltage Testing", *Proc. IEEE International Test Conf.*, pp. 673-682, 2002.

[44] T.W. Williams et al, "Iddq Test: Sensitivity Analysis of Scaling", *Proc. IEEE International Test Conf.*, pp. 786-792, 1996.

[45] C. Visweswariah, "Death, Taxes, and Failing Chips", *Proc. Design Automation Conf.*, pp. 343-347, 2003.

[46] D. Ernst et al, "Razor: Circuit-Level Correction of Timing Errors for Low-Power Operation", *IEEE Micro*, 24(6):10-20, November 2004.

Chapter 2

Failure Mechanisms and Testing in Nanometer Technologies

Jaume Segura, Charles Hawkins and Jerry Soden

CMOS technology scaling has been a constant since its initial development in the early 70's as an effort to obtain ICs working at higher operating frequencies that perform more operations per unit area. Each advance in CMOS technology scaling is called a technology generation, or a technology node, and pursues the ability of fabricating smaller transistors. A new technology generation doubles the number of transistors per unit area, increases operating frequency by more than 40%, reduces the energy per transition by more than 60%, while reducing transistor cost. Technology nodes are reached at a constant pace, a rule known as Moore's law, which currently brings one new generation every 18 months. Moore's law is possible thanks to the scalability of the basic unit used to implement digital switches: the MOSFET transistor. Today ICs contain transistors having minimum geometries of 90 nm (1nm = 10^{-9} m), and industry is now rapidly moving into the 65 nm technology node. Chips today contain hundreds of millions of transistors and operate at frequencies on the order of 5 GHz. They incorporate a variety of circuit blocks

D. Gizopoulos (ed.), Advances in Electronic Testing: Challenges and Methodologies, 43-75.

that implement not only digital tasks, but also perform analog tasks and even RF tasks, incorporating massive memory storage. All these different design entities are constructed on the same piece of silicon.

The capabilities that allowed microelectronics industry to build this vertiginous trend require a sustained improvement in three main technology directions: fabrication technology, architecture and design automation, and test and verification. This chapter is focused on the third of these challenges, specifically on testing, although the understanding of their problems and solutions requires knowledge of the other two. We provide a quick historical perspective on what have been the test challenges and solutions in the past, and focus on the threats faced by today industry to test and qualify CMOS ICs fabricated with transistors having dimensions in the nanometer range. Understanding these challenges requires not only a detailed analysis of the physical phenomena that dominate device behavior in this scaled domain, but also of the role of the interconnect system. We first discuss scaling rules for both device and interconnect, and then analyze two of the main effects that impact today test methods: noise and parameter variations.

Once the required physical background is given we focus on the failure modes and defect mechanisms present in nanometer ICs, describing their induced behavior and the symptoms caused at the circuit level. This defect-based analysis is required to understand present test practices developed to screen the defects and extend the techniques to enhance test technology. The essential content of this chapter is for mechanisms that are now important although they were not in previous technologies.

2.1 SCALING CMOS TECHNOLOGY

CMOS IC miniaturization consists in making the active elements, i.e. MOSFET transistors, progressively smaller to allow more computing per unit area. Smaller transistors switch faster and consume less power as the inherent parasitic capacitances at each device node are smaller and can be charged/discharged at higher rates. In addition, smaller transistors have a smaller equivalent resistance between the conducting terminals (drain and source) as the distance between them (defining the effective transistor length, L_{eff}) gets progressively reduced with technology scaling. These two factors (smaller parasitic capacitances and shorter channel length) translate to circuits operating at higher frequencies.

Having more transistors per unit area on a silicon substrate implies a higher complexity on the interconnect system. Each transistor has three terminals that must be wired to other devices to construct logic gates, circuit blocks, functional units, and higher-level functional structures that connected properly constitute the whole circuit. The interconnect system that wires all the elements, from transistors to functional blocks, and brings the supply and ground levels to all the logic gates is constructed using metal (aluminum, and more recently copper[1]) by defining interconnect lines that are laid out over the silicon surface running at different levels. The interconnect of metal lines at different levels is done with vertical structures

---

[1] Copper interconnects are used because of the material's lower resistivity that can provide 30% of sheet resistance reduction for the same pitch [3].

called *vias*, while structures connecting the lower-level metal lines to drain/source device terminals are called *contacts*. Today ICs may have up to nine or ten metal interconnect levels, compared to the two-metal circuits used in the mid 90's, and the number of vias can be one order of magnitude more than the number of transistors. The complexity of the interconnect system, together with the electrical properties of the metal lines, the progressive lower supply voltages and higher current levels (both static and dynamic) of today ICs requires a detailed characterization of this intricate passive structure as it may be the source of circuit functional errors.

The scaling trends of CMOS technology pose several challenges to both device and interconnect operation that is described in detail as they have a direct impact on the circuit behavior and invalidate some of the classic test methods used in present technologies. We also discuss the noise mechanisms exacerbated in nanometer technologies and describe the challenging problem of parameter variations and its impact on circuit design and testing.

2.1.1 Device Scaling

Figure 2-1 illustrates the structure of a MOSFET transistor showing the drain and source terminals diffused over the bulk or substrate and isolated from other devices through a shallow trench isolation (STI). The gate is the control terminal placed over the transistor channel (the region between the drain and source), which is electrically isolated through the thin gate oxide insulator. Ideally a transistor should act as a perfect digital switch having zero resistance when closed (conducting), infinite resistance when opened (off-state), and an instantaneous response when switching between these two states. Real transistors show many deviations from this ideal picture. The gate terminal controls the conducting state of the transistor through the vertical electric field determined by the ratio between the gate-substrate voltage (V_{GB}) and the gate oxide thickness (T_{ox}). If the gate voltage is such that the vertical field creates a channel of mobile carriers, then the horizontal electric field, determined by the ratio between the drain-source voltage (V_{DS}) and the transistor effective channel length (L_{eff}), determine the acceleration of the mobile channel carriers and the device overall current.

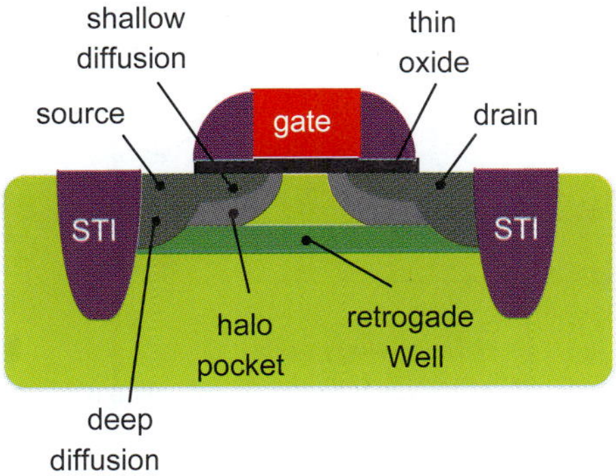

Figure 2-1: Cross-section structure of a MOSFET transistor.

The reduction of device dimensions exacerbates many of the non-idealities of the transistor. Non-ideal effects start to become non-negligible for sub-micron transistors (roughly when channel length goes below 0.5 μm), and they become important when entering the deep sub-micron or nanometer scale (below 0.1 μm). A number of effects like *off-state leakage increase*, *channel length modulation*, *velocity saturation*, *drain induced barrier lowering*, *random doping fluctuations* and *negative bias temperature instability* start to become important. We briefly comment these effects focusing on those that will have a greater impact on test methodologies.

Off-state Leakage Increase

Device scaling implies reducing the device in the horizontal dimension to get a relatively smaller channel resistance (shorter L_{eff}). Scaling the horizontal dimension implies a subsequent reduction of the vertical aspect to maintain a good control of the channel carrier population from the gate terminal. Therefore device scaling implies both channel length, L_{eff}, and gate oxide thickness, T_{ox}, reduction. If the voltage levels remain constant, device scaling increases the electric field across the gate oxide. Reliable oxide operation requires maximum electric fields in the order of 5 – 6 MV/cm[2]. Technology scaling used constant 5V supply voltage (V_{DD}) operation from mid 70's technology generations down to the 0.6 μm node at the beginning of the 90's resulting in a progressive increase of the gate oxide electric field from initial 1 MV/cm up to the reliability limit of 5 MV/cm. Such a scaling approach was referred to as *constant supply voltage scaling*. Once the 5 MV/cm limit was achieved, further device scaling required supply voltage reduction to allow T_{ox} reduction while maintaining the electric field within the reliability limit. This is the scaling approach that we follow today and is known as the *constant electric field scaling* approach. We have gone through supply voltage reduction values of 5 V, to 3.3 V, 2.5 V, 1.8 V, 1.2 V and 1V for the 0.7 μm, 0.5μm, 0.25 μm, 0.18 μm, 0.12 μm, and 0.1μm respectively (Figure 2-2).

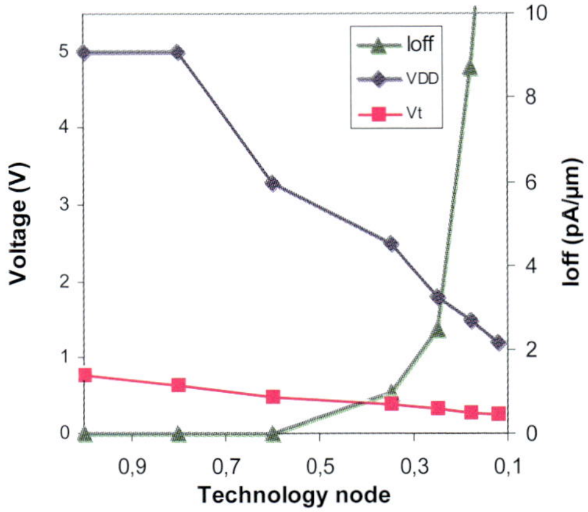

Figure 2-2: Supply and threshold voltage evolution with technology scaling, and impact on I_{off}.

[2] This limit is imposed by reliability engineers to avoid potentially excessive failure rates due to gate oxide breakdown when the electric field increases beyond this value.

Supply voltage scaling has positive and negative side effects on other circuit parameters. On the positive side it helps controlling both static and dynamic power as the first component has an exponential dependence on supply voltage, while the dynamic power has a quadratic dependence. On the negative side supply voltage reduction impacts performance as delay is increased. As a simple, first approach the delay of a CMOS gate, τ_D, can be expressed as [1]:

$$\tau_D = C_L V_{DD} \frac{2L}{W\mu C_{ox}} \frac{1}{(V_{DD} - V_t)^\alpha} \tag{2.1}$$

where C_L is the gate output load, μ is the carrier mobility, C_{ox} is the gate oxide capacitance, V_t is the transistor threshold voltage, L, W is the transistor length and width, respectively, and α is a technology parameter ranging between 1 and 2 for short– and long–channel devices respectively.

Since α is never smaller than 1 (and in general it is larger than that value), a reduction of the supply voltage implies an increase in gate delay. Eq.(2.1) reveals that the only way to compensate for the loss in delay from V_{DD} reduction is to scale the threshold voltage V_t, to increase the gate overdrive (the amount of gate voltage beyond V_t, i.e. $V_{GS} - V_t$)[3]. Threshold voltage reduction certainly improves gate delay and is used by industry when scaling from one technology node to the following (Figure 2-2), even though some penalties are associated. The main penalty associated to threshold voltage reduction is the increase of the device sub-threshold or off-state current. Figure 2-3 shows the drain current (in log scale) vs. the gate voltage for two transistors having different threshold voltages. The off-state current (the drain current at V_{GS} = 0 V) increases over two orders of magnitude, while the maximum saturation current (the drain current at $V_{GS} = V_{DD}$ = 1V) increases around 12% when V_t is scaled. The ideal trend would be the opposite, i.e. significantly increasing the saturation current while reducing the off-state current, since the first component translates to higher frequency operation while the second one only increases static power representing wasted energy. Modern IC designs use two types of transistors with high and low V_t. High V_t devices are used in non critical paths (those paths that do not compromise circuit maximum frequency) for reduced off-state current, while low V_t transistors are used in critical paths for improved performance.

The huge difference of V_t variation impact on the off-state and saturation currents comes from device physics that shows different dependencies of these magnitudes with respect to the threshold voltage. The drain saturation current varies linearly with respect to the threshold voltage (or at most quadratically), while the off-state current has an exponential dependence with the threshold voltage. A detailed analysis of leakage current mechanisms reveals up to eight different mechanisms. A detailed analysis and relative importance for scaled devices can be found in [2].

[3] V_{GS} is the the gate-source voltage.

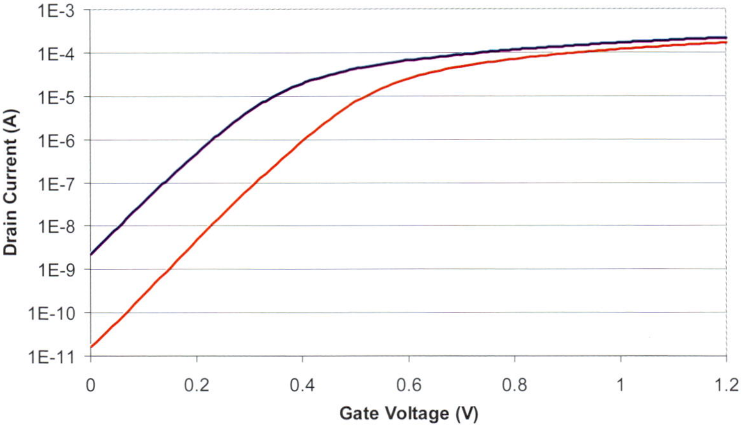

Figure 2-3: Effect of threshold voltage reduction on off-state current [2] (V_t = 0.35V top, V_t = 0.45V bottom).

Later in this chapter we will see that the increase in off-state current at the device level has important effects at the circuit level, and has pushed some test techniques to evolve rapidly to sophisticated methods that are able to cope with this scaling effect.

Drain Induced Barrier Lowering (DIBL)

Short channel effects[4] can be mainly attributed to the *Drain induced barrier lowering* (*DIBL*) effect, which causes a reduction in the threshold voltage as the channel length decreases. A potential barrier exists between the source and the channel region in the weak inversion regime[5] whose height is a result of the balance between drift and diffusion current. The height of this barrier should be controlled only by the gate voltage to maximize transconductance. DIBL occurs when the barrier height for channel carriers at the edge of the source reduces due to the influence of drain electric field, resulting from the application of a high drain voltage and the proximity of this terminal to the source. This increases the number of carriers injected into the channel from the source contributing to an increased drain off-state current. If this happens, then the drain current is not uniquely controlled by the gate voltage, but also by the drain potential. DIBL can be modeled as a parasitic effect causing a threshold voltage reduction depending on the drain voltage and is measured as the amount of V_t lowering per volt of drain increase (expressed in mV/V). The reduction of threshold voltage with channel length scaling is generically referred to as *short-channel threshold roll-off.*

Device manufacturers developed special vertical and lateral non-uniform doping profiles to control short channel effects [3], [4]. These techniques include halo doping or nonuniform channel profile in the lateral direction, implemented by angled

[4] Effects of short-channel devices, i.e. those with α parameter close to 1.

[5] This potential barrier prevents carriers to enter the channel and cause charge flow.

abrupt ion implantation self-aligned to the gate. Shallow junctions and retrograde wells are also performed to assist in short channel control effects. These device design techniques are expected to provide channel length scaling well into the 20 nm regime obtaining off-state currents almost insensitive to channel-length variations [4]. Figure 2-4 shows different threshold voltage roll-off dependencies for devices with different junction depths and doping profiles [4]. These techniques reduce short channel effects by obtaining a device internal rectangle shape (defined by the region formed by the boundary of the gate depletion region, the gate electrode and the source and drain regions) whose aspect ratio has a height smaller than its width, thus getting a long channel-like geometry.

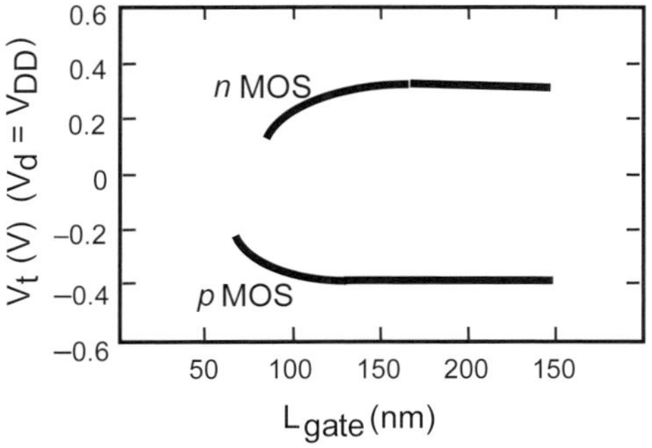

Figure 2-4: Short-channel threshold roll-off for super halo and retrograde (non-halo) doping profiles (from [4]).

In the following section we will see that threshold voltage roll-off translates channel length variations to circuit-level speed and current leakage variations with a significant impact on traditional test methods based on monitoring either circuit delay or quiescent current consumption.

Channel Length Modulation

Ideal MOSFET's should behave as perfect current sources when they are operating in saturation providing a drain current that should be independent of the drain-source voltage for a given fixed gate-source potential. This property is mostly true for long channel devices, while short channel transistors may deviate significantly from this behavior. The reason for this is the relative importance of channel length modulation in short devices.

Channel length modulation refers to the displacement of the carrier-conducting layer at the drain end region when the device is saturated. In this operating regime the horizontal field created from the drain is very intense in the channel region close to this terminal, and its intensity decreases when moving toward the source. As a result, the conducting layer distribution is not uniform along the channel and tends to get thinner significantly when closer to the drain. For large values of the drain-source voltage, the electric field is so intense that the conducting layer is pushed away from the drain and does not "touch" it. Carriers travel from the source to the drain at an increasing velocity that reaches its maximum value when they arrive at the channel end; from this point up to the drain area they move at a constant maximum saturated velocity. The distance between the channel end and the drain terminal, ΔL, (see Figure 2-5) depends on the strength of the electric field and gets higher for increasing drain-source voltages. The net impact of this effect on the device behavior can be viewed as causing a relative increase of the drain-source resistance that depends on the drain-source voltage. If the value of ΔL is not negligible with respect to the channel length L, then this effect is observed on the device I_D-V_{GS} characteristics as the curves not being flat in saturation (see Figure 2-5).

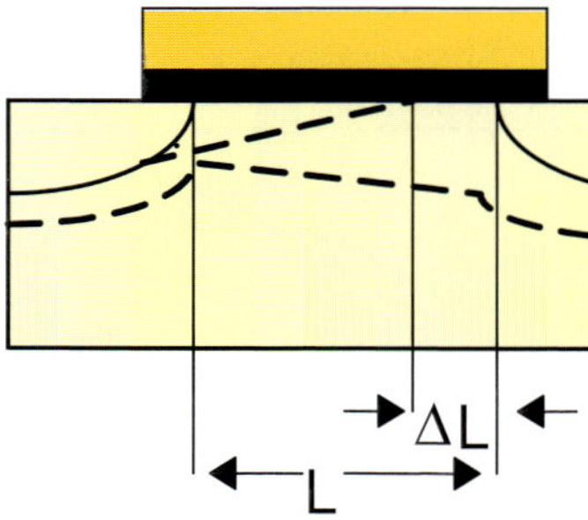

Figure 2-5: Illustration of channel length modulation.

Other Effects

There are other effects that are becoming increasingly important as devices are further scaled down, and have an impact on testing. We briefly comment on gate oxide tunneling, random doping fluctuations and negative bias thermal instabilities (NBTI).

Gate oxide thickness is reduced nearly in proportion to channel length to retain control over the channel in front of the drain terminal influence. MOSFET transistors having channel lengths of 100 nm and below have gate oxide thicknesses below 3 nm that are composed of a few atom layers. These thin oxide layers constitute a barrier that allow quantum-mechanical tunneling between the gate and the channel, giving rise to a non-negligible gate current. This current adds to the drain-source leakage and contributes to the overall static power. High permittivity dielectrics are being considered as possible replacements for silicon dioxide as they would allow thicker gate oxides with equivalent gate oxide capacitances without the problem of direct tunneling current. The potential impact of high permittivity gate dielectrics on device short channel and circuit performance must be fully understood. Challenges exist in replacing traditional silicon oxide by other materials since parasitic outer and internal fringe capacitances are modified in addition to the gate-to-channel capacitance. Lower parasitic outer fringe capacitance is beneficial for the circuit performance, while the increase in internal fringe capacitance and the decrease in the gate-to-channel capacitance will degrade the short channel performance contributing to higher DIBL, drain leakage, and lower noise margin [7].

As channel surface gets scaled because of transistor length reduction the distribution of dopants in the channel may start to have a direct impact on the device characteristics. This is especially important for memory cells, where transistors are designed to have minimum length and width dimensions. Recent studies show that *random doping fluctuations* play a role on transistor threshold voltage variation, adding another source of variability to this parameter. More variation tolerant techniques must be adopted to account for such variations [5].

A recent oxide reliability issue appeared that impacts short channel *p*MOS transistors with their *p*-doped polysilicon gates. It is called *Negative Bias Temperature Instability* (*NBTI*), and it is a wearout mechanism with positive charge buildup at the channel interface of *p*MOS transistors. It causes threshold voltage absolute magnitude increase and reduction in I_Dsat. The damage has been referred to as caused by "cold holes", and is identified as getting worse with scaling of oxide thickness. The instability in its title refers to the time variation in V_{tp} (the *p*MOS transistor threshold voltage) and I_Dsat [6]. The affected *p*MOS transistor has higher V_{tp} than a transistor in normal inversion.

2.1.2 Interconnect Scaling

Interconnect scaling is closely related to device scaling and can be mainly described through three parameters: interconnect pitch (the minimum wire width plus the minimum wire spacing divided by two), number of interconnect levels, and aspect ratio[6] of interconnect lines (that depends on the level at which the interconnect is located). Other parameters determine the electrical properties of the interconnect system as the resistance, capacitance and inductance. For a given line geometry, wire resistance depends only on the interconnect material (aluminum or copper), although metal liners and barriers may play a role on line resistance when defects appear. Capacitance depends not only on the interconnect metal itself, but also on the surrounding dielectric properties, metal inter- and intra-level separations, and topology. Inductance depends mainly on metal surrounding topology and manifests when the rate of change in current delivery is very high.

Technology scaling has brought up to 9-10 interconnect levels, with metal pitch defining the technology node (i.e. a 100 nm technology as a 100 nm pitch), and aspect ratios that have changed from wide signal lines in the 0.7 μm technologies to the narrow interconnect of modern technologies. Table 2-1 lists the metal pitches, thicknesses and aspect ratios for a 6-metal 130 nm technology [3].

Layer	Pitch (nm)	Thickness (nm)	Aspect Ratio
Metal 1	293	280	1.7
Metal 2,3	425	360	1.7
Metal 4	718	570	1.6
Metal 5	1064	900	1.7
Metal 6	1143	1200	2.1

Table 2-1: Layer pitch, thickness (nm) and aspect ratio for a 130 nm process [3].

[6] The quotient between the height and the width of the contact or via.

Figure 2-6 shows cross-sections of two ICs having six and eight levels of metal interconnect, respectively, illustrating the increasing complexity of the wiring system. The cross-section on the left illustrates the differences in pitch and aspect ratio of the interconnect metals depending on the metal level. Low-level metal layers are used to interconnect local dense transistor areas and require narrow lines to allow higher density; these lines are not relatively long as they target short distances. Higher-level metals are used to connect relatively far away blocks and cover higher distances, thus requiring wider lines to achieve overall low resistance.

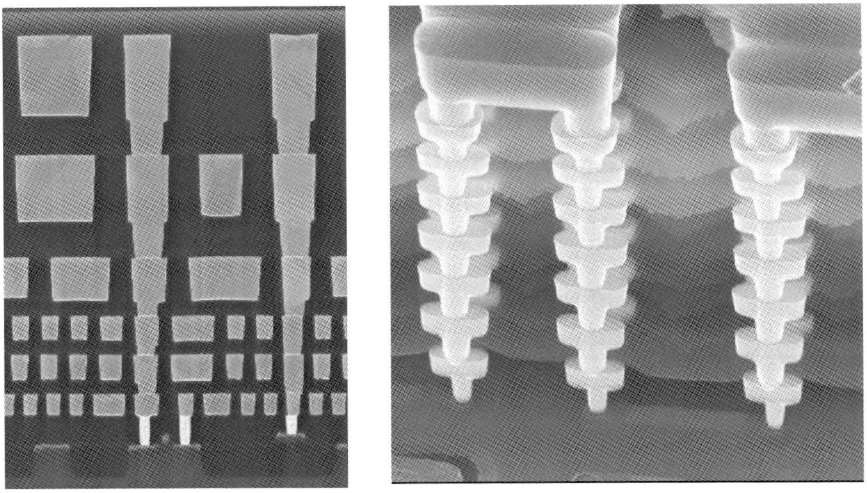

Figure 2-6: Cross-section of metal structures for two different process technologies [1].

The challenge of the interconnect system is to provide reliable signal integrity as this subsystem is an important source of noise within the IC if proper design guidelines are not followed. The next section shows that an important number of fault mechanisms are originated at the interconnect system. Therefore the higher the interconnect system complexity, the more difficult its overall integrity verification task.

2.1.3 Parameter Variations

Parameter variations refer to the fluctuation of any circuit physical magnitude, either geometric or electric, from its nominal targeted value. Parameter variations are an inherent characteristic of any manufacturing process that involves translating geometric features from a design entity to a real physical structure. The issue of parameter variations is related to the ability in controlling these fluctuations within safe bounds ensuring that the logic and electrical characteristics of all circuits fabricated from the same design are not distinguishable in practice either logically or electrically.

The reality of today nanometer ICs is far from following this trend. Current microelectronic circuits suffer from large parameter variations that impact within-die, die-to-die, wafer-to-wafer, and lot-to-lot circuit parameters. There are two main categories of parameter variations: manufacturing and environmental. The first relates to repeatability of the structures constructed during the manufacturing process, while the second relates to the environment conditions at which the circuit operates like supply voltage, temperature, noise, etc. We discuss both types in detail.

Manufacturing Variations

Variation in individual transistor and interconnect parameters come from optical effects during lithography patterning processes, while the metal interconnect system is also subjected to variation coming from chemical-mechanical polishing (CMP) now used as the primary technique for planarizing interlayer dielectrics (IDL).

Optical effects may result in wafer images that can be very different from those drawn on the layout (Figure 2-7). In particular, optical proximity effects such as pitch dependent *critical dimension* (CD) variation and line shortening can degrade transistor parameters or even lead to catastrophic defects (shorts or opens) when occurring in the polysilicon layer. The loss of pattern fidelity may happen during mask making, wafer imaging, and/or etch steps. CD variation in polysilicon and interconnect metal lines has been extensively studied, and different techniques such as *optical proximity correction* (*OPC*) or *phase shift mask* (*PSM*) have been proposed. These methods require modification of the physical design layout on the photo-mask to compensate for the proximity effects [8]. Although some of the variations can be corrected using OPC, significant discrepancies between measurements and models even after corrections still exist [9].

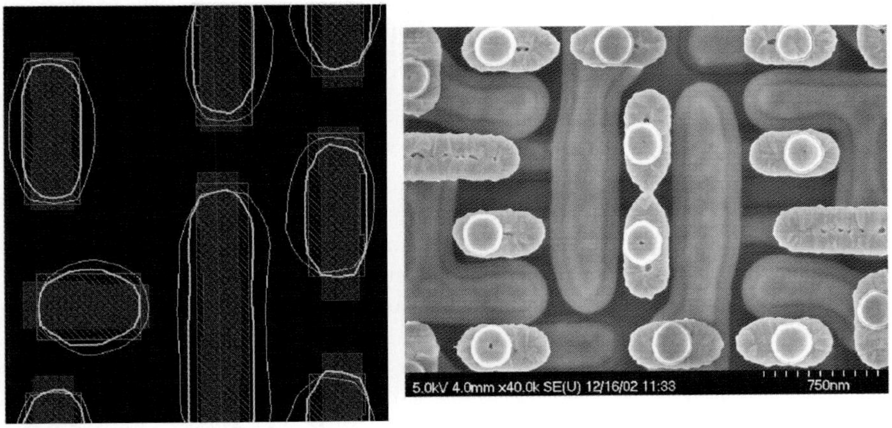

Figure 2-7: Optical Proximity effects result in wafer images different from the drawn layout, and may even induce defects (Courtesy Bob Madge, LSI).

CMP current applications include important process steps like shallow trench isolation (STI) and multi-level inlaid copper interconnection. However, it has been observed that the post-CMP topography shows an important variation that is strongly dependent on the layout pattern. This causes certain regions on a chip to have differences in dielectric layers thickness depending on the underlying topography. A method to reduce this layout pattern dependent dielectric thickness variation is to fill large metal-free areas with dummy metal, with a consequent complexity on modeling [1].

The main geometric and electric parameters suffering from variation for devices and interconnect are discussed briefly. For a detailed analysis we refer to [1].

The main parameters that determine the transistor drive properties are:

- Channel length variation
- Channel width variation
- *n*MOS to *p*MOS length ratio variation
- Effective gate oxide thickness variation
- Doping variation – threshold voltage and diffusion resistance

We focus on those that have a first order impact on transistor characteristics: channel length variation and effective gate oxide thickness variation.

As was discussed earlier in this chapter short channel effects result in threshold voltage roll-off translating channel length fluctuations to threshold voltage variations. Threshold voltage variation has a primary impact on transistor saturation current variation, and therefore on the whole chip operating frequency. Transistor length variations impact circuit operation speed not only through V_t variations, but also directly as was shown in the gate delay Eq. (2.1). Transistor channel length variation also impacts off-state current variation through Vt variation as they are exponentially related.

Gate oxide thickness has a first-order impact on device performance, directly affecting transconductance, threshold voltage, and device drive current. Today ultrathin gate oxides contain only a few layers of SiO_2 molecules, and the absence or presence of a portion of a monolayer can cause T_{ox} to discretely vary by 15-20% in local oxide regions [1], with a consequent impact on local electric field strength.

Regarding interconnect, process variations can induce worst-case sheet resistance fluctuation ranging from 10% in metal lines to about 38% in polysilicon as reported for a 0.8 μm CMOS process [10]. Line width metal variations of about 10% and area capacitance variations from 17% to more than 25% were measured for inter-metal lines. These parameter fluctuations were measured in a 0.8 μm process, but deep submicron processes have larger process-induced variation, especially below 180nm.

Environmental Variations

Device and interconnect characteristics may also vary over different regions of the same circuit, or from circuit to circuit due to environmental conditions. The main environmental variables that impact circuit operation are supply voltage fluctuations and temperature variation.

Supply voltage fluctuation has a direct impact on circuit speed. Different noise mechanisms either external or from within the IC can induce supply voltage variation. The noise mechanisms impacting the circuit supply voltage are described in the following section and may be distributed all over the IC having a different impact on different circuit blocks. It has been reported that submicron ICs may show about 2 MHz of frequency speed shift per mV of supply voltage variation [11]. This is a significant source of delay variation within the IC if different circuit regions are at different supply voltage values due to noise fluctuations.

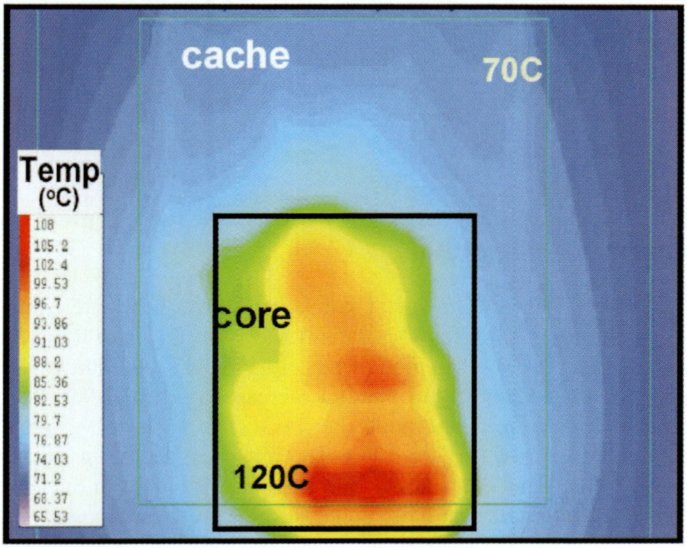

Figure 2-8: Hot spots on a microprocessor [12].

Temperature gradients are originated when different circuit blocks have different activity. The higher activity blocks dissipated a larger average power, which translates to a higher local temperature. Within die temperature gradients of more than 50°C have been experimentally reported for high performance applications (Figure 2-8) [12]. Temperature increase degrades circuit speed as a result of two opposed processes: mobility degradation and threshold voltage reduction. Temperature increase has a negative impact on carrier's mobility due to the increase of silicon atoms thermal agitation that provoke more collisions of such atoms with moving electrons (or holes). On the other hand, the absolute value of the transistor threshold voltage is reduced as temperature increases thus increasing the transistor saturation current. The net result of these two competing effects is an overall transistor speed reduction as mobility reduction dominates with an increase of leakage power due to threshold voltage reduction. Therefore within die thermal gradients will slow down circuit blocks being at higher temperatures with respect to those being at lower values [13].

2.1.4 Noise

Noise in today CMOS ICs comes mainly from circuit internal activity that is coupled and/or amplified through the interconnect system due to non-idealities. Signal node voltage spikes and supply/ground bounces come from the parasitic resistance, capacitance and inductance components of the wiring interconnect.

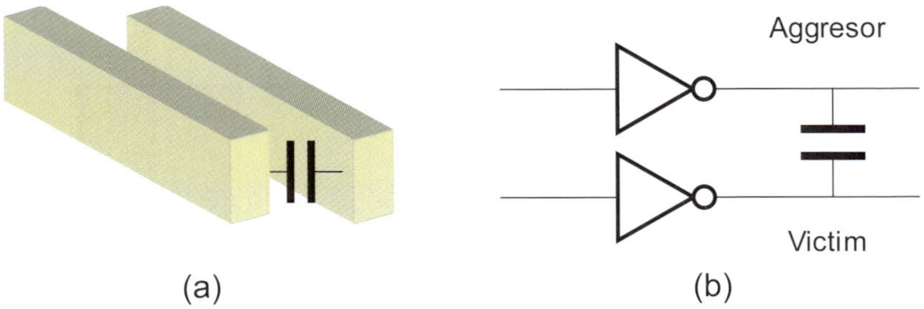

Figure 2-9: Illustration of capacitive crosstalk (a) coupled lines, and (b) circuit electrical equivalent.

The high aspect ratio of the metal wires used for signal interconnect favors *capacitive coupling* (or capacitive crosstalk). Two metal lines separated by an insulator form a capacitor whose value increases as metal separation is reduced, and the length of both lines running in parallel increases (Figure 2-9a). The parasitic capacitor injects charge from one node that makes a transition (aggressor) to the coupled neighbors (victims) (Figure 2-9b). The impact of the noise mechanism will depend on if the victim node is static or it is also transitioning. In the first case, the effect of capacitive crosstalk is a noise spike (or glitch) whose impact on the circuit behavior will depend on if such a glitch travels over a path and reaches a memory element flipping its value. When the victim line is also transitioning, capacitive crosstalk impacts the delay of the transition, either speeding it up or down [1].

Another coupling mechanism exists between signal nodes due to *inductive coupling*. This mechanism impacts only relatively large and low resistance wires carrying very high frequency signals that radiate an electromagnetic field which is the source of the coupling phenomena. These characteristics restrict this noise mechanism to global wiring lines like clock and bus signal. The spatial extension of this noise mechanism depends on the return current paths found by the high frequency signal, and tends to distribute over a large area if specifically designed return paths are not provided. The computation of return paths is very complex as inductive effects are difficult to model and tools are not available. This is further aggravated by the long-range effect and the fact that inductive coupling is design and technology dependent [14].

The resistive and inductive parasitic components of the interconnect metals have an impact on the supply/ground voltage levels arriving to the circuit gates. The resistive component is responsible for what is known as *IR drop*. This effect degrades the supply voltage range to a gate or group of gates that have a sustained activity for a given time and demand an elevated average current. If the supply system is not properly sized for such an average current demand, resulting in a high effective interconnect resistance, then the group of gates are effectively powered at a supply voltage below the nominal value due to the voltage loss caused by this ohmic component.

The inductive component of the supply interconnect translates sudden current demands of circuit gates making transitions to instantaneous voltage fluctuations at the supply/ground nodes of these gates. This noise mechanism is known as *Ldi/dt noise* and is increasingly important in scaled technologies because the di/dt component is increasing as it is related to the gate output voltage transition time that is aggressively reduced. The most practical way to minimize Ldi/dt noise is to reduce the inductance of the supply interconnect. This cannot be done in practice by adopting specific wire design techniques, but using decoupling capacitors. These capacitors are placed close to the gate and provide the instantaneous current needed for the gate transition. The effective line inductance is reduced because the supply current to the gate does not come initially from the far supply voltage source.

Although both IR and di/dt noise impact the supply voltage "seen" by a CMOS gate, they can be distinguished through their frequency components. IR noise is associated to a collective sustained current demand and has a lower frequency band. On the other hand, di/dt noise is related to the short output gate voltage transitions and has much higher frequency components. Another difference is that IR noise always provokes negative voltage spikes (the effective supply voltage gets reduced), while di/dt noise induces both positive and negative noise spikes. Both mechanisms have in common that they impact the delay of the gate since the effective supply voltage varies.

2.2 FAILURE MODES IN NANOMETER TECHNOLOGIES

Many of the present nanometer technologies failure modes have their roots in previous technologies. However, some failure modes that were occasional occurrences, such as parametric failures, are now major challenges to detect and locate. Bridge and open defects were the two main defect mechanisms in old technologies and basically have not changed their electronic behavior in nanometer ICs, with the exception of gate oxide short bridge defects and low voltage (V_{DD}) operation that have new effects that will be described in detail in the following sections. Open circuit defects see a serious increase in the partial or weak opens, particularly with regard to vias and contacts. The third form of failures is the parametric ones that have long been recognized, but are a serious detection problem. We will review knowledge about bridge and open defect behavior, and then describe the important parametric failure modes and mechanisms.

2.2.1 Bridge Defects

Bridge defects (Figure 2-10) are unintended connections between two or more interconnect lines. They can occur in combinational or sequential circuits, between gate, drain, and source nodes of a transistor, between V_{DD} and V_{SS}, or between signal nodes of different logic gates. The rail-to-rail (power supplies) bridges generally don't affect the Boolean performance of a chip, but can decrease lifetime of low power, portable products such as watches, cell phones, or pocket computers. Defect Pareto charts[7] usually show bridges as the number one most numerous defect in most fabs. They remain a dominant defect in nanometer technologies. The electronic

[7] Defect Pareto charts present defects by frequency of occurence, (usually in descending order), identifying the most important defects.

effects of bridges on digital circuits remains as it was for older technologies with the exception of the electronic behavior of gate oxide shorts, and an increased concern with metal slivers. Both of these variants of bridge defects will be discussed after a review of classic bridge defect behavior.

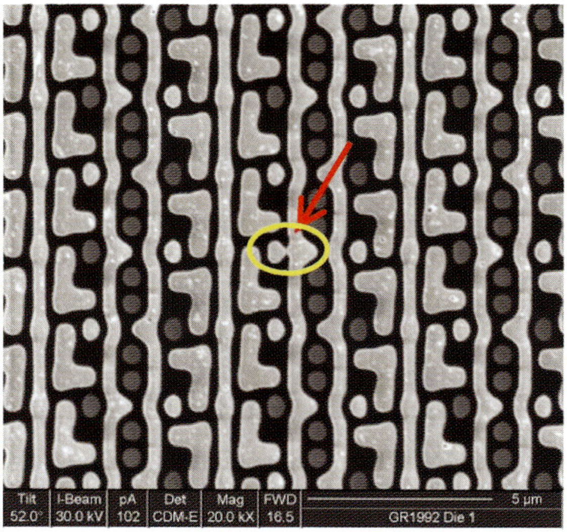

Figure 2-10: Bridging defect (courtesy of Bob Madge, LSI).

A bridge defect must have opposite polarity voltages across it in order to activate its presence. When identical logic states exist across the bridge, then it is as if no defect was present. Bridge defects across the nodes of a transistor, or bridges across signal lines, or bridges from a signal node to one of the power rails have one thing in common. There is a defect resistance value above which Boolean failure will not occur. The value is called the *critical resistance* R_{crit} [15]. It is easy to envision a large resistor, e.g., 1 GΩ, having no effect on Boolean behavior or circuit speed. It is also easy to envision that a small resistance, e.g., 10 Ω, as having a strong effect on the signal integrity. Somewhere in between these ranges there is a resistive value that divides the pass region from the fail region. The critical resistance is a function of the pull up and pull down strengths of the *p*- and *n*-transistors. The transistor current drive ratios and the absolute value of K_p and K_n[8] give R_{crit} a range from about 100 Ω to 2 kΩ. R_{crit} may go higher or lower than these values depending on the degree of transistor mismatch and magnitude of K_n or K_p.

An example analysis will show this effect: assume a bridge defect between the inverter gate output node and $V_{DD} = 1.5$ V (Figure 2-11). The logic threshold voltage (V_{TL}) is about 0.5 $V_{DD} = 0.75$ V. We compute the critical resistance assuming that $K_p = 50\ \mu A/V^2$, $K_n = 125\ \mu A/V^2$, $(W/L)_n = 2$, $(W/L)_p = 4$ and $V_{tn}=0.4$ V (the threshold voltage of the *n*-channel transistor). The defect will only be activated when $V_{IN} = 1.5$ V. The *n*-channel transistor is fully on, and the *p*-channel transistor is fully off. The *n*MOSFET tries to pull V_{OUT} to 0 V, but the power supply, V_{DD}, tries to pull the node

[8] $K_n = \mu_n C_{ox}$, $K_p = \mu_p C_{ox}$

high through the defect bridge. The result will either be a correct, but weak logic-0, or an erroneous weak logic-1. The problem is to calculate the critical resistance R_{crit}.

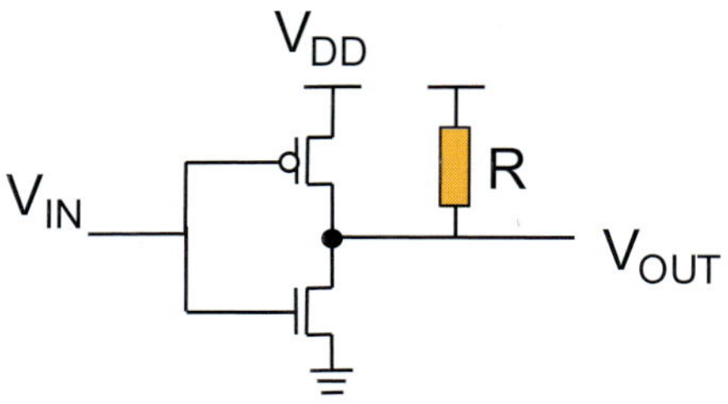

Figure 2-11: An inverter with an ohmic bridge defect between output node and V_{DD}.

When V_{OUT} is at the pass-fail boundary defining the critical resistance, then $V_{OUT} = V_{TL} = 0.75$ V. We know gate, drain, and source voltages, and conclude that the transistor is in non-saturation from $V_{GS} > V_{DS} + V_{tn}$, or $1.5 > 0.75 + 0.4$. Since the transistor is in the linear bias region we can calculate the drain current. I_D also goes through the bridge defect, so we can use Ohm's Law to calculate R_{crit}[9].

$$I_D = 125\ \mu A/V^2\ (2)\ (1.5 - 0.4\ V)^2 = 302.5\ \mu A$$

Then,

$$R_{crit} = (1.5 - 0.75)\ V/\ 302.5\ \mu A = 2.5\ k\Omega$$

Any defect resistance below 2.5 kΩ will cause a Boolean failure. If the current drive, K_n (W/L) was increased, then R_{crit} would decrease as seen in the previous equation. We can also see from these two equations that if V_{DD} was decreased for the same V_t values, then I_D would decrease, and R_{crit} would increase. This means that a larger fraction of bridge defect values would fall below the critical resistance and would have better chance of detection at the lower V_{DD}.

A similar analysis can be done for line-to-line bridges (i.e. a bridge between two gate outputs) where now two contending transistors (one per logic gate) are involved. In this case, the resistance node attached to the stronger transistor determines voltage threshold for the failure and the operating condition for the two contending transistors needs to be computed. For a detailed exhaustive analysis we refer to [1].

Gate oxide ruptures and metal slivers are bridge defect forms that are discussed under the parametric failures section below because they have a more massive behavior.

---

9 A simple expression for the saturation current in long-channel transistors is $I_Dsat = \mu_o C_{ox}(W/L)(V_{GS}-V_t)^2$. Short channel transistor equations are more complicated but the dependence on the channel length is similar [1].

2.2.2 Open Circuit Defects

There are six open circuit defect behavioral forms [16]. The major variables are the size of the open (large or a narrow crack), or the location -gate, drain, source, single gate lead, or an open affecting pairs of complementary transistors. These six forms still exist in nanometer technologies and are the following:

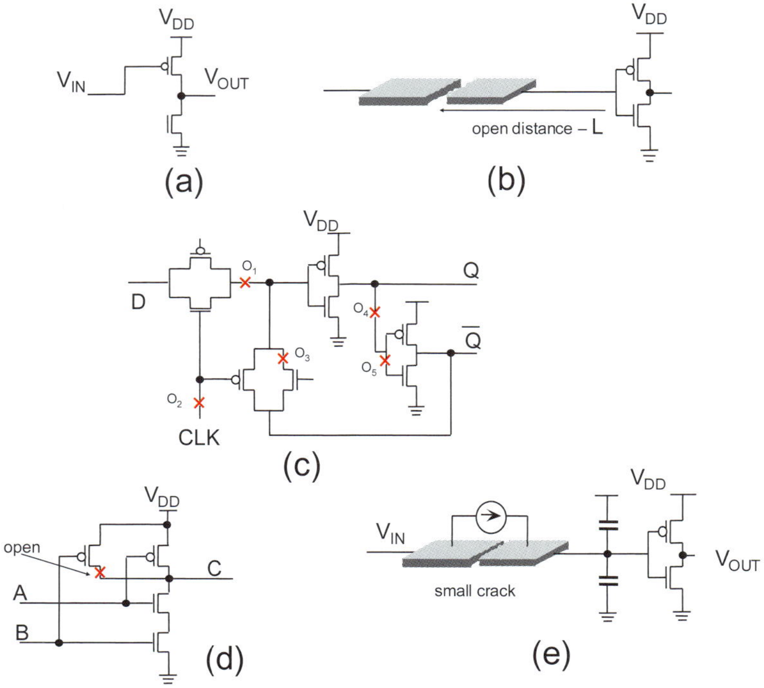

Figure 2-12: Open defect forms.

- Open transistor gate (Figure 2-12a) is the *first* open defect type: If a contact is missing between the polysilicon gate and the metal above the gate, then there is no apparent drive to the gate terminal. However, a capacitive voltage divider exists between the drain to gate and gate to source. The gate voltage is affected by the drain to source voltage. If the gate voltage acquires a value above V_t then that transistor turns on supporting drain current. That transistor acts somewhat like a sluggish resistor, and correct logic operation is supported. Noise margin, speed of operation, and quiescent power supply current are compromised, but the circuit will function.
- Open to a logic gate (Figure 2-12b) defines the *second* and *third* open defect types affecting two complementary transistors: This open has a complex response to its topological environment. The open node is isolated, but acquires a DC floating node voltage. This fixed voltage then causes the logic gate output node to fix a constant value. This

defines a true stuck-at fault behavior (stuck-at-1 with V_{DD} and stuck-at-0 with ground). Two different open defect forms are defined whether the floating node goes strongly to one of the rail voltages, or it floats to an intermediate value turning on both affected complementary transistors. The result is that I_{DDQ} is elevated for the intermediate state.

- A *fourth* open response is categorized with sequential circuits (Figure 2-12c) such as flip-flops. Several location possibilities exist and many responses are similar to those described above in the previous cases. Opens in one of the parallel paths of a CMOS transmission gate will degrade signal voltages and also slow the response.
- A *fifth* open defect type is called the CMOS stuck-open fault (Figure 2-12d): If an open defect occurs in a drain or source lead, then that transistor can undergo carrier inversion in the channel, but charge is not transferred outside the transistor. This leads to floating nodes for certain vectors. A floating node in response to a test vector will read the previous logic state. That result may be correct for one two-pattern test sequence and incorrect for another sequence [18], [23]. This is a difficult defect situation to detect or locate.
- The *sixth* open defect form occurs when the open is a small crack (Figure 2-12e) in the interconnect line. Narrow cracks can support *electron tunneling*[10], and ICs have been known to function correctly in the hundreds of MHz region. The cracks may be the interconnect flatlines but are more commonly associated with the metal in vias and contacts. Since metal shrinks when cooled, the crack will widen and electron tunneling is reduced. These tunneling opens overlap another category of nanometer technology failure modes called parametric failures. They are discussed next.

2.2.3 Parametric Failures

Parametric failures do not behave as bridge and open defects do. These failures are speed related, and are functions of temperature, V_{DD}, and clock frequency. Windows of pass-fail may appear in Shmoo plots. Parametric failures can occur in ICs with or without the presence of defects, but in both cases it is a parameter alteration that causes a speed failure. The *intrinsic* parametric failures are caused by an unlucky interaction of electrical parameters, not by any defect. Interactive effects of two or more parameters can cause failure although each individual parameter may be within the specifications [19]. A statistically long L_{eff} may have a larger V_t than normal, and that transistor drives an interconnect with high sheet resistance. It is the adverse sum of the three parameters that causes a failure to appear at specified clock frequency, temperature, and V_{DD}.

The next sections will review the inherent variations found in IC electrical parameters that affect intrinsic parametric failures.

---

[10] Tunneling is a quantum effect by which a particle that has no enough energy to jump over a given potential barrier can statistically go through it.

Intrinsic Parametric Failures

V_{DD} Variation and Performance

The IC performance can be measured by plotting the maximum operating frequency, F_{MAX}, versus V_{DD} like the plot shown in Figure 2-13 for the Pentium 4 processor [3]. Such plots show that a 1 mV change in V_{DD} can cause frequency adjustments from hundreds of kHz to MHz [1]. The shift in V_{DD} can be static, such as when a printed circuit board voltage regulator is off target or dynamically induced by variations in V_{DD}, temperature, or power supply noise. Dynamic shifts in V_{DD} occur constantly due to Ldi/dt or IR drops in the power lines. The current rise times in power supply lines are in the order of tens of amps per ns, so that the inductive drops require careful design. The V_{DD} and GND lines undergo instantaneous changes virtually with each clock pulse and current surge. These surges (positive and negative voltages) have an instantaneous affect on speed performance.

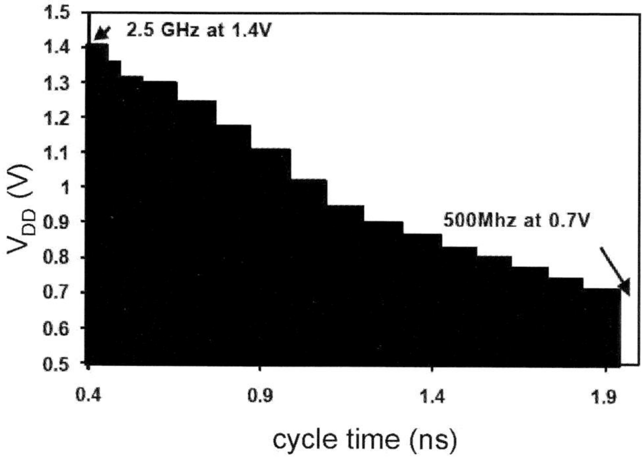

Figure 2-13: F_{max} Schmoo plot for the Pentium 4 processor [3].

Temperature Variation and Performance

The die or junction temperature of a high performance IC is now well above 100°C. Hot and cold spots exist on the die reflecting its distribution of computation at any instant. Hot spots on the order of 120°C are reported [12]. Temperature has a strong affect on transistor delay times. The carrier mobility gets worse quickly as temperature rises, and threshold voltage decreases its absolute value. The dependence of die F_{MAX} versus the circuit temperature ranges from tens to hundreds of kHz per °C. The temperature across a die is a constantly changing parameter and is not uniform resulting in a thermal map that changes depending on the particular circuit operating conditions. The formation of a non-uniform thermal map in the circuit results in different regions operating at different speeds which can induce delay-related faults. This effect can be especially important for the clock distribution circuitry since it is distributed over the whole die [13].

L_{eff} Variation and Performance

The effective channel length of a transistor, L_{eff}, is the parameter whose statistical variation most affects the F_{MAX} variation. A primary reason is that L_{eff} lies in the denominator of the $I_D sat$ equation[11]. Small changes in L_{eff} cause larger changes in $I_D sat$. It is difficult to exactly control the critical length dimension of a MOSFET. The increase in $I_D sat$ means that load capacitances can be charged and discharged faster thus allowing the IC to run at a higher clock frequency. Another factor as L_{eff} shortens is that drain induced barrier lowering causes a reduction in V_t further speeding the transistor.

Threshold Voltage Variation and Performance

Threshold voltage, V_t, has its own statistical distribution. Its affect on speed is through the gate overdrive function ($V_{DD} - V_t$). As V_t becomes smaller, the overdrive increases thus increasing $I_D sat$. Threshold variations arise from non-homogeneous implants and dimensional variation in L_{eff}, and in W_{eff}.

Cross-talk Variation and Performance

Crosstalk was introduced in Subsection 2.1.4 and illustrated in Figure 2-9, and it can slow down or speed up a signal. A fast rise time in an aggressor line can influence the timing of the pulse in the victim line. These events are difficult to predict. There is great pressure to understand and apply design rules that can avoid crosstalk as well as Ldi/dt noise and IR drop in the power lines.

This description of the die under operational conditions illustrates the dynamic nature of an IC. Even the concept of F_{MAX} is vague when we examine it closely. F_{MAX} might better be defined for a small 20 ps window in terms of the instantaneous power line bounces, the die temperature distribution, and the $I_D sat$ variation across all active transistors in the critical path. This of course is not possible. F_{MAX} is a gross but useful measurement of speed capability of an IC.

Extrinsic Parametric Failures

The second form of parametric failure occurs when certain subtle defects are present. This is called an *extrinsic* parametric failure. It is also sensitive to temperature, V_{DD}, and clock frequency parameters. Some examples of extrinsic parametric failures are given below.

Resistive Vias and Contacts

A common example of defect related parametric failures are the resistive vias and contacts. These occur when the metal structure is imperfect and has an abnormally high resistance with voids, poor adhesion to the dielectric wall, or imperfect interfaces between the basic metal and the barrier metal used in both Cu and Al technologies (see Figure 2-14 for examples). Temperature has a strong influence on the structure due to the thermal coefficient of expansion (TCE). An increase in temperature of most resistive vias will shrink the voids, lowering the resistance, and giving a faster RC time constant response. So if speed performance is compared at

---

[11] Again, the saturation current in long-channel transistors is given by the simple expression $I_D sat = \mu_o C_{ox}(W/L)(V_{GS}-V_t)^2$.

two temperatures, then we find that the IC runs faster at the hotter temperature. This is a strong signature for resistive vias and contacts. The failure is designated not by a logic error, but by the temperature speed performance signature.

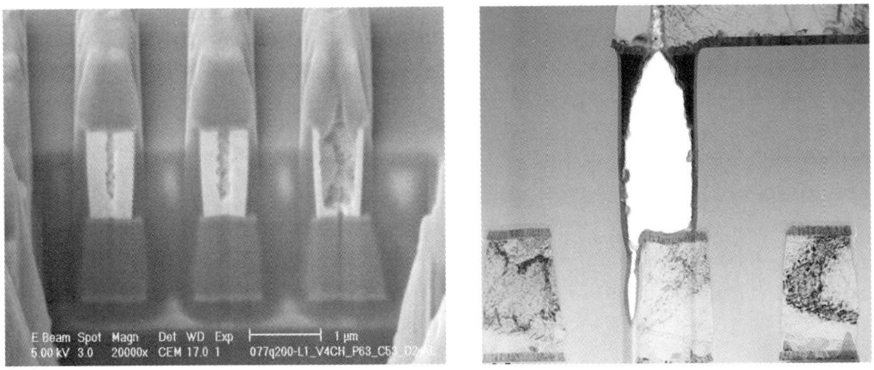

Figure 2-14: Imperfect vias (courtesy of Bob Madge, LSI).

Metal Slivers

Metal slivers are small particles that lie between two signal interconnect lines as shown in Figure 2-15. The sliver may not even touch the signal lines to be a concern. All metals used in IC construction have a high resistance oxide film at their surface. When temperature is elevated on a die, then the sliver can expand, and even touch the signal interconnect. When opposite polarity voltages are present across the sliver, then the thin oxide can be punctured and the signal lines and sliver can atomically bond. When temperature is decreased, then a permanent bridge exists. Slivers have been with the industry virtually from the beginning, but the recent use of Chemical Mechanical Polishing (CMP) and the very small spacing between interconnects has increased their occurrence. It is not possible with current test technology to detect a metal sliver before it bonds to the interconnect lines next to it. Slivers are burn-in activated and that is one approach to reducing the problem because during burn-in they will turn to bridges.

Figure 2-15: Metal sliver (courtesy of Bob Madge, LSI).

Gate Oxide Ruptures

Gate oxide ruptures in older technologies with the transistor thin oxide greater than about 60 Å were a serious concern. The energy to cause severe damage to the oxide was supplied by the $0.5\,CV^2$ capacitive storage on the gate. The thermal event caused the gate material to physically merge with the silicon crystal underneath the oxide leading to parasitic diodes and low ohmic resistance [21]. As technologies shrunk, so did the transistor dimensions and V_{DD}, so the $0.5\,CV^2$ energy became over a thousand times smaller. The result is that the ultrathin oxides (< 30 Å) have a softer form of breakdown. An oxide rupturing event creates a path of molecular level defects (traps) that support high ohmic conduction [1]. These soft breakdowns appear to cause only a small increase in gate current and electrical noise. A ring oscillator with seven gate ruptures functioned with a slight degradation in frequency [22]. More data are required, but it appears that a large fraction of the gate shorts that were a reliability concern in older technologies may not be in nanometer technologies. A hard breakdown can be caused with ultrathin oxides if a high voltage event happens, such as an Electrostatic Discharge (ESD) pulse. In this case, parasitic diodes or low ohmic resistors would be formed and be a concern.

Metal Mousebites

Defect nicks can be taken from the normal width of a metal interconnection. The remaining strip of metal left from the mousebite (Figure 2-16) usually has a small resistance so that RC delay is negligible. However the remaining strip of metal carries an increased current density that poses an electromigration risk. The detection of the mousebite defect is not possible with modern test techniques, so that the failure found is an open induced by electromigration.

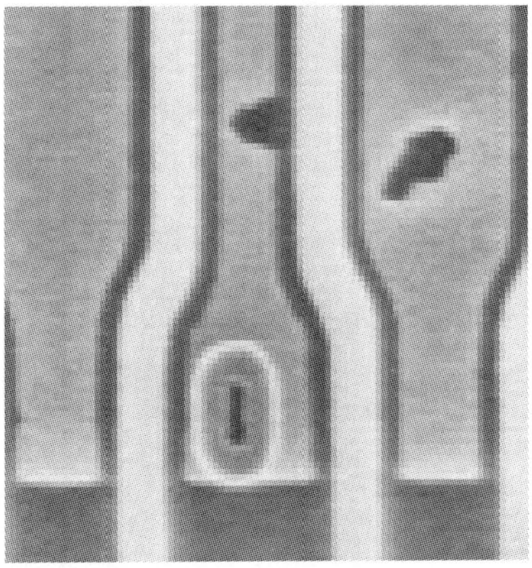

Figure 2-16: Metal mousebite (courtesy of Bob Madge, LSI).

2.3 TEST METHODS FOR NANOMETER ICS

Chapter 1 of this book elaborated on test technology concepts and historical development while presenting a detailed analysis of the Defect-oriented Testing concept. Defect-oriented Testing is the most suitable strategy to test CMOS ICs as it analyzes first the impact of defects on the circuit behavior, and then determines the best methodology to detect these defects. From another point of view, test methods can be divided in two main categories: logic- and parametric-based test techniques. Logic-based testing checks for the logic correctness of the circuit operation verifying digital values at circuit nodes, while parametric-based test methods check for the particular value of a given magnitude (like delay or quiescent current). We analyze the impact of technology scaling on these techniques, and present the solutions devised to cope with these challenges.

2.3.1 Impact of Technology Scaling on Testing

Deep submicron structures don't affect test and diagnosis just because they are small. They primarily impact test because the manufacturing parameters are not tightly controlled as they were in the past. This problem arises from the fact that CMOS IC critical dimensions scale faster than the ability of controlling them. Therefore, wide variances exist in circuit parameters as explained earlier and these variances give a wide distribution to the performance of each IC. As a first order, test limits must be set at the worst-case values for the intrinsic part performances. Logic-based test methods are not significantly impacted by parameter variations in deep-submicron circuits. The main impact of technology scaling on these test methods, like stuck-at based test, is related to the use of large amounts of tester memory to store test vectors and the expected responses. The slow scan delivery of test vectors in the 50 – 400 MHz range is a deficiency. Comparative test method studies show that stuck-at fault testing has the weakest defect detection efficiency of the major test methods [1].

The most popular parametric-based test methods are delay-based testing and current-based testing or I_{DDQ}. Delay-based test methods rely on checking the time taken for a signal to travel between two nodes. This delay time is compared against a reference value to determine if there is any delay-related fault in some element (gate or node) of that path. I_{DDQ} measures the current drawn from the power supply during the quiescent states of the circuit, and compares this value to a reference or maximum limit. Both test philosophies are compromised by submicron IC parameter variations due to the spread in either delay or leakage resulting from die-to-die fluctuation. Parametric-based test methods were very efficient for non-submircon technologies because there was a clear gap in the distribution of the test parameter (either delay or I_{DDQ}) between fault-free and faulty circuits. The distribution showed a clear outlier population corresponding to ICs with defects, thus allowing a clear setting of the pass-fail threshold value without a compromise on yield loss.

In addition to the challenge of parameter variation, I_{DDQ} testing faced the additional difficulty of increased background current. The off-state leakage increase due to threshold voltage scaling at the device level gave an increase of the overall IC current leakage. Although the extrapolation from the device level to the circuit level is not direct because gate-level transistor topologies, like transistor stack, may reduce significantly the gate leakage depending on the inputs, intrinsically good parts started to show hundreds of miliamperes of current in the quiescent state. The statement that good circuits had almost no quiescent current became invalid. Different techniques were initially devised to cope with circuit-level leakage current increase trying to reduce it during test mode by increasing the threshold voltage through reverse body-bias[12], testing at lower temperatures, or reducing supply voltage. These techniques, although reducing the large current background levels, did not solve the parameter variation problem, and some of them were found not to scale properly with technology [2].

The key advances that allowed test industry to benefit from the advantages of parametric test methods like current-based techniques rely in developing parameter-variation tolerant strategies. We provide an overview of the progress made in this field, and focus on the test methods adopted by industry and those proposed more recently.

2.3.2 Dealing with Background Current Increase

We describe two test techniques that were pioneer in proposing methodologies that could be applied to leaky circuits since they do not check for the actual value of the quiescent current, but for relative changes or differences on a signature or on a sequence of measured values.

Current Signatures

The method of *current signatures* [23] measures the quiescent current values for the whole test vector set (a set that contains a number of test vectors that completely exercise the circuit) and then generates a current signature by rank ordering all the I_{DDQ} measurements from the smallest to the largest value. The method looks for jumps in the current signature. A smooth plot of the current value indicates a fault free circuit, while significant jumps or discontinuities in the current signatures indicate a faulty behavior (Figure 2-17).

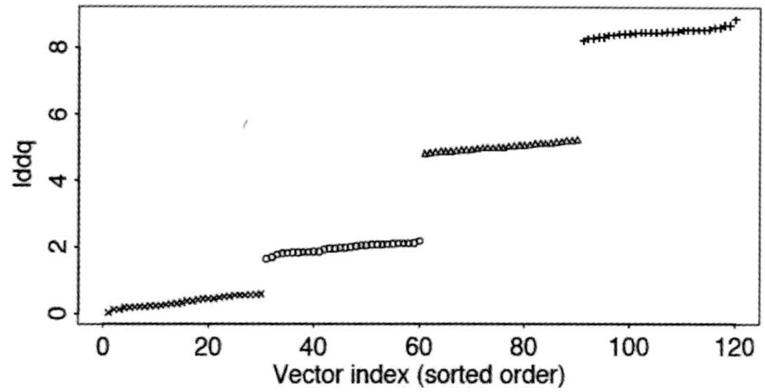

Figure 2-17: Illustration of the discontinuities in the current signature approach [23].

This technique introduced a new concept in the adoption of quiescent current testing since a given circuit is not rejected at first fail indication during testing, but

---

[12] Typically the substrate, or body, of the transistor is connected to ground and it is not used as a signal terminal. Since the threshold voltage absolute value increases with the absolute value of the source–to–substrate voltage, these techniques used this terminal to increase the threshold voltage during the test mode.

only after all tester data are collected. The main limitation of this post-test data processing technique is related to its sensitivity to parameter variations since it requires setting a pass/fail threshold not for the absolute I_{DDQ} value, but for the magnitude of the jump at the signature. This threshold should be valid over a range of process variations.

Delta I_{DDQ}

A similar conceptual method called *Delta* I_{DDQ} in which the test observable is not the absolute I_{DDQ} value, but the differences in I_{DDQ} among successive test vectors was proposed in [24]. This difference is treated probabilistically to determine if a given circuit is defective or not, and can reduce the test limit value by an order of magnitude of about 20X as shown in Table 2-2, and reported from applying delta I_{DDQ} to commercial memory and processor circuits with a mean leakage current of 2 mA [25]. These circuits were not appropriate for single threshold I_{DDQ} since the single threshold I_{DDQ} three-sigma limit was 22 mA posing a significant I_{DDQ} yield loss. The study showed that for a reduced population of samples sent to burn-in, more than 50% of the units that failed Delta I_{DDQ} but passed functional testing and single threshold I_{DDQ} became functional failures after the experiment. Delta I_{DDQ} has a parameter variation limitation similar to current signatures.

	Traditional I_{DDQ} Test	Delta I_{DDQ} Test	Ratio (T/D)
Min current	1.5 mA	0 uA	–
Max current	10.4 mA	540 uA	19
Mean current	2.3 mA	43 uA	54
Std deviation	1.4 mA	94 uA	15
Three sigma	6.6 mA	330 uA	20

Table 2-2: Limit setting comparison for single threshold (Traditional) I_{DDQ} and Delta I_{DDQ}

2.3.3 Noise-tolerant Techniques

The two techniques discussed in the previous section must set an absolute threshold either for a jump in the signature or the delta value, therefore suffering from variation in the quiescent current from die-to-die or lot-to-lot. We look now to two test techniques that can be also applied to leaky parts, but set a limit that can be adapted to each die.

Current Ratios

A test technique called *Current Ratios* was proposed by Maxwell et. al., and uses a concept similar to current signatures with the added features of tolerating parameter variations [26]. This is done by setting a specific quiescent current limit for each die that remains valid for any vector of the test set. This limit is computed individually for each part once the first minimal quiescent current vector measure is taken. Therefore, the first I_{DDQ} value (obtained from the expected vector giving the smallest intrinsic I_{DDQ} value) establishes a range of quiescent current values that are acceptable for that part.

This technique is based on the observation that the slope of the rank ordered quiescent current signatures for dies having significant different absolute I_{DDQ} values are similar (Figure 2-18a).

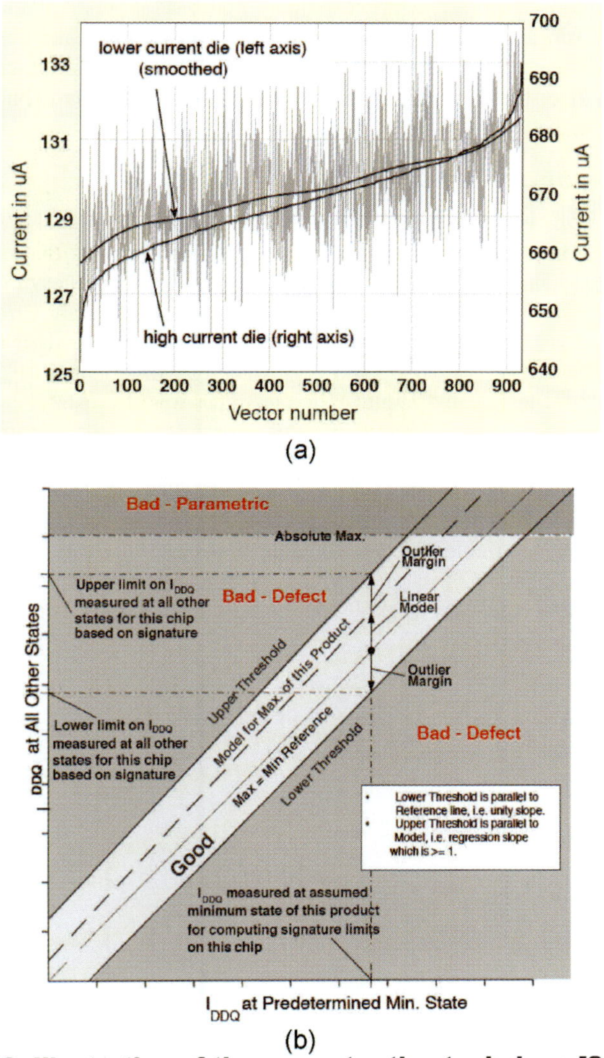

Figure 2-18: Illustration of the current ratios technique [26].

The test limit for a given die is determined from the ratio of the maximum to minimum I_{DDQ} value and the slope of the rank ordered currents. This was observed to remain reasonably constant for all the devices no matter the mean of the I_{DDQ} measurements for each die. This ratio is determined from a small population of devices having as wide a spread of current as possible, and is done through an iterative process in which the maximum current is plotted versus the minimum current for all dies. The minimum current is found by characterizing the response of many die, and identifying the vector that typically gives a minimal reading for each die. The maximum current of a die under test (DUT) is computed from the measured minimum I_{DDQ} vector and the slope (Figure 2-18b).

This technique overcomes the two main limitations of other quiescent current techniques since it tolerates both high background currents and parameter variations. Another benefit of this technique is its suitability for production testing environments since the test limits are adjusted for each part once the first measurement is taken.

Nearest Neighbor Statistics

The nearest neighborhood, post-test statistical analysis method is a powerful demonstration of adjusting I_{DDQ} limits to the quality environment that each die sees in its immediate vicinity [27]. Nearest neighbor test techniques measure parameters in the neighboring die on a wafer. It assumes that defects and fabrication variation cluster on the wafer, so that die in close proximity on the wafer will have similar average chip values of L_{eff}, V_t, I_{DDQ}, and F_{MAX}. The measurements of F_{MAX} and I_{DDQ} are referenced to the nearest eight neighboring chips to establish a level of performance consistent with nearest chips. An expression is to distinguish "bad die in good neighborhoods, or good die in bad neighborhoods" (More sophisticated techniques also correlate parameters in the rows and columns). The nearest neighbor test method was demonstrated using I_{DDQ}, V_{DDmin}, and F_{MAX}, but it has general applicability to other variables. It is a valuable tool in parametric failure detection, especially as it evaluates IC data at varying V_{DD} and temperature. Nearest neighbor testing is a production test method at LSI Logic Corp. Figure 2-19 visualizes the concept of neighbor distribution.

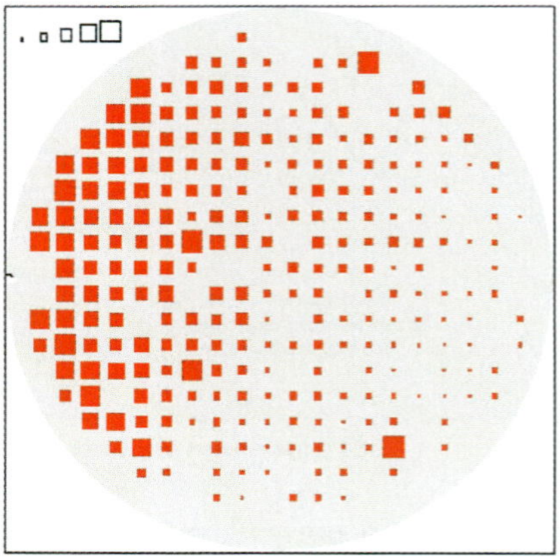

Figure 2-19: Nearest neighbor techniques establish a test limit based on die wafer distribution.

Multi-parameter Testing

The techniques explained in the previous sections monitor a single parameter and develop a method to adjust a threshold value based on a max-to-min ratio or the information from other dies in the same wafer. Another technique was proposed in [28] that uses a different approach based on exploiting the fundamental correlations that exist between circuit parameters. Leaky devices are not necessarily faulty (in terms of logic faults), they are transistors that have shorter gate lengths and therefore will give higher drain saturation current and leading to gates that switching faster with respect to other gates having larger devices. At the circuit level this implies that parts showing higher leakage should run faster, and less leaky circuits should run slower. Such a correlation led to the concept of two-parameter testing. The decision of a circuit being good or bad is not taken depending on the actual value of the quiescent current (I_{DDQ}), but on its relation to the maximum speed of the circuit (F_{MAX}). This correlation is based on fundamental device physics and must be followed by intrinsically good parts. A circuit drawing a high leakage current with respect to the population, but showing a F_{MAX} value lower than other leaky parts is most probably a defective part.

The concept of two-parameter testing can be extended to muliparameter testing by correlating more than two magnitudes for which we know their fundamental relationship. A third parameter that can be added to the two-parameter F_{MAX} vs. I_{DDQ} plot is temperature. We know that circuits run faster at lower temperatures as mobility gets increased, while leakage is reduced because the threshold voltage increases at lower temperatures. A three-dimensional plot of these magnitudes leads to a distribution for intrinsic parts from which outliers can be pulled out and recognized (Figure 2-20).

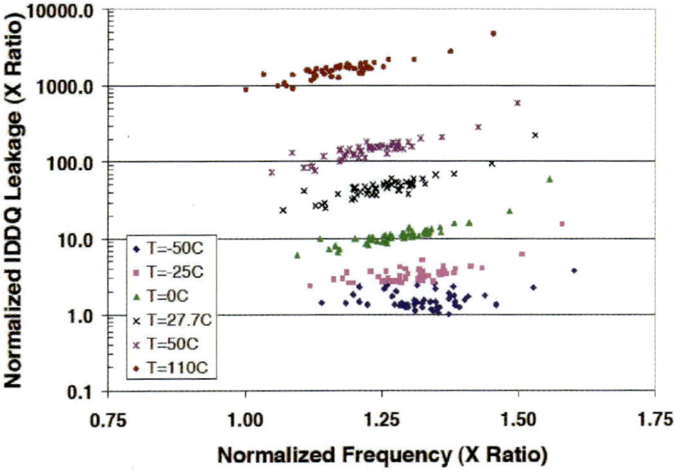

Figure 2-20: Example of multi-parameter testing [28].

It is important to notice that these techniques represent a significant change in the way that circuits are tested with respect to the traditional methods, since the decision of a particular circuit being good or bad is not taken on the tester, but once all data has been gathered. This requires the adoption of an efficient tracking method for each part during the production test flow.

2.3.4 Impact of Variation on Delay

The abundant activity experienced in the last few years to search for test methods that deal with leakage-increase and variance-tolerant current based methods is also being extended to delay-based test methods. The challenge is very similar to the one faced by current testing with respect to variation. An ideal delay-based test technique would adjust the maximum delay of each critical path set (the set of paths that limit the maximum frequency of the circuit) to each part dynamically, thus accounting for parameter variation. Critical path determination has been traditionally considered a design problem as part of circuit optimization. Parameter variation is starting to displace the problem of critical path determination to the test field. We describe two techniques related to variation in delay testing; one is related to correlation, while the other incorporates statistical methods to critical path determination.

Low-V_{DD} Testing

Low V_{DD} testing can be categorized as another parameter correlation technique as it evaluates the correlation between the power supply voltage and timing [29]. Supply voltage has a strong influence on gate delay, and therefore on the overall circuit speed. The low V_{DD} testing concept lowers the supply voltage at a given clock frequency, and measures the minimum value of the supply voltage at which the circuit will still function. This technique is referred to as $MinV_{DD}$. The reduction of the IC supply voltage decreases the drive strength of the transistor gates making them more sensitive to defects impacting timing such as resistive shorts or weak opens. The drawbacks of this technique for production testing relate to the increase in test time, and the cost to do a search test for the minimum voltage at which the die still passes the test. To overcome the search time penalty of the MinV_{DD} test, a 3–step process was proposed based on MinV_{DD} to detect outliers and a posterior statistical process for outlier screening [30]. This technique uses a reduced vector set instead of the full vector set to search the MinV_{DD} value, and then applies the full set of test vectors to the die using the minimum voltage found in the previous step.

Delay Testing using Statistical Analysis

A simple way to detect timing related defects models delay defects as large gate delay faults and then checks for the timing of any path that passes through that gate. This approach is known as transition fault testing and a high fault coverage using this method does not guarantee high delay-defect coverage since small delay defects cannot be detected on short paths, and path delay fault testing for a number of selected critical paths is necessary [31]. The selection of a small set of critical paths is necessary because of cost and complexity, and is challenging since this process depends on tool's model accuracy and the way that parameter variations are taken into account. Worst case analysis is becoming less efficient because of parameter variations, and a poor selection may lead to excessive yield loss or to poor test quality.

It has been shown that the incorporation of statistical analysis can help in accounting for parameter variations and incorporate possible differences in the critical path set from chip to chip [32].

2.4 CONCLUSION

Technology scaling into the nanometer regime has a direct impact on test methods not because there are significantly new defect mechanisms but because of the significant increase in variance and noise mechanisms. This has affect the most effective test methods developed for non-submicron technologies based on checking non purely logic parameters. Parametric failures are a difficult failure mechanism, although they have always been present in ICs, no matter the technology, but the very small dimensions of the die and the rapid logic transition times have brought this failure mode to a concerned level. The relative increase of parametric failures has driven a significant effort to develop sophisticated methods to enhance test technology.

The most effective approaches to detection of parametric failures are to analyze statistical parameters at the various temperature, supply voltage, and clock frequencies. The adoption of statistical-based tools will be required in future scaled technologies to capture variation, and test methods will rely on parameter correlation and decisions will be made off-testers once data is collected.

More research is needed to understand and characterize the various parameters involved in circuit characterization, determining their correlation, and more efficient techniques to quickly pull outliers from the intrinsic population are required.

REFERENCES

[1] J. Segura and C. Hawkins, *CMOS Electronics: How it Works, How it Fails*, IEEE Press – John Wiley & Sons, NJ, 2004. ISBN0-471-47669-2.

[2] A. Keshavarzi, C. Hawkins and K. Roy, "Intrinsic leakage in low-power deep submicron ICs", *IEEE Int. Test Conference*, pp. 147-158, 1997.

[3] S. Thompson, et. al., "130 nm logic technology featuring 60 nm transistors, low-K dielectrics, and Cu interconnects", *Intel Technology Journal*, vol. 6, no. 2, pp. 5-13, 2002.

[4] Y. Taur, "CMOS design near the limit of scaling" *IBM Journal of Research & Development*, vol. 46, no. 2/3, March/May 2002.

[5] S. T. Ma, A. Keshavarzi, V. De, and J. R. Brews "A Statistical Model for Extracting Geometric Sources of Transistor Performance Variation", *IEEE Tran Electron Devices*, vol. 51, no. 1, pp. 36-41, January 2004.

[6] W. Abadeer and W. Ellis, "Behavior of NBTI under ac dynamic circuit conditions," *International Reliability Physics Symposium (IRPS)*, pp. 17-22, April 2003.

[7] N. R. Mohapatra, M. P. Desai, S. G. Narendra, and V. R. Rao, "The Effect of High-K Gate Dielectrics on Deep Submicrometer CMOS Device and Circuit Performance" *IEEE Tran Electron Devices*, vol. 49, no. 51, pp. 826-831, May 2002.

[8] W. Grobman, et. al., "Reticle enhancement technology: implications and challenges for physical design," *Design Automation Conference*, June 2001.

[9] V. Mehrotra and D. Boning, "Technology scaling impact of variation on clock skew and interconnect delay," *IEEE Int. Interconect Technology Conference*, pp. 122-124, 2001.

[10] S. Natarajan, M. Breuer and S. Gupta, "Process variations and their impact on circuit operation," *International Workshop on Defect and Fault Tolerance in VLSI Systems*, pp. 73-81, November 1998.

[11] J. Segura, A. Keshavarzi, J. Soden, and C. Hawkins, "Parametric failures in CMOS ICs – A defect-based analysis," *IEEE International Test Conference (ITC)*, pp. 90-99, October 2002.

[12] S. Borkar, T. Karnik, S. Narendra, J. Tschanz, A. Keshavarzi, and V. De "Parameter Variations and Impact on Circuits and Microarchitecture" Design Automation Conference (DAC), pp. 338-342, 2003.

[13] S. Bota, M. Rosales, J.L. Roselló, A. Keshavarzi and J. Segura "Within Die Thermal Gradient Impact on Clock-Skew: A New Type of Delay-Fault Mechanism" *IEEE Int. Test Conference*, 2004.

[14] A. Deutsch, et al., "On-chip wiring design challenges for gigahertz operation," *Proceeding of the IEEE*, Vol. 89, No. 4, April 2001.

[15] J. Segura, V. Champac, R. Rodríguez, J. Figueras, and A. Rubio, "Quiescent current analysis and experimentation of defective CMOS circuits," *Journal of Electronic Testing: Theory and Applications*, Vol 3, pp. 337-348, 1992.

[16] C. Hawkins, J. Soden, A. Righter, and J. Ferguson, "Defect classes – An overdue paradigm for testing CMOS ICs," *IEEE International Test Conference (ITC)*, pp. 413-424, October 1994.

[17] R. Wadsack, "Fault modeling and logic simulation of CMOS and MOS integrated circuits," Bell Systems Technical Journal, pp. 1449-1488, May-June 1978.

[18] J. Soden, R. Treece, M. Taylor, and C. Hawkins, "CMOS IC stuck-open fault electrical effects and design considerations," pp. 423-430, *International Test Conference (ITC)*, pp. 302-310, August 1989.

[19] T. Turner, "A step-by-step method for elimination of burn-in as a necessary screen," *96IRM Final Report,* pp. 82-86, 1997.

[20] J.S. Suehle, "Ultra-Thin gate Oxide Breakdown: A Failure that we can live with?" Electronic Device Failure Analysis, Vol. 6, No. 1, pp. 6-11, February 2004.

[21] J. Segura, C. De Benito, A. Rubio, and C. Hawkins, "A detailed analysis of GOS defects in MOS transistors: testing implications at circuit level," *IEEE International Test Conference (ITC)*, pp. 544-550, October 1995.

[22] R. Degraeve, B. Kaczer, A. De Keersgieter, and G. Groeseneken, "Relation Between Breakdown Mode and breakdown Location in Short Channel NMOSFETs," *International Reliability Physics Symposium (IRPS)*, pp. 360-366, May 2001.

[23] A. Gattiker and W. Maly, "Current signatures: application," *IEEE International Test Conference (ITC)*, pp. 156-165, 1997.

[24] C. Thibeault, "An histogram based procedure for current testing of active defects," *International Test Conference (ITC)*, pp. 714-723, October 1999.

[25] A. Miller, "I_{DDQ} testing in deep submicron integrated circuits," *International Test Conference (ITC)*, pp. 724-729, October 1999.

[26] P. Maxwell, P. O'Neill, R. Aitken, R. Dudley, N. Jaarsma, M. Quach, and D. Wiseman, "Current ratios: A self-scaling technique for production I_{DDQ} testing," *International Test Conference (ITC)*, pp. 738-746, October 1999.

[27] R. Daasch, K. Cota, J. McNames, and R. Madge, "Neighbor selection for variance reduction in I_{DDQ} and other parametric data," *International Test Conference (ITC)*, October 2001.

[28] A. Keshavarzi, K. Roy, M. Sachdev, C. Hawkins, K. Soumyanath, and V. De, "Multiple-parameter CMOS IC testing with increased sensitivity for I_{DDQ}," *IEEE International Test Conference (ITC)*, pp. 1051-1059, October 2000.

[29] H. Hao and E. McCluskey, "Very-low voltage testing for weak CMOS logic ICs," *IEEE International Test Conference*, pp. 275-284, 1993.

[30] R. Madge, B. Goh, V. Rajagopalan, C. Macchietto, R. Daasch, C. Shuermyer, C. Taylor, and D. Turner, "Screening MinVDD outliers using feed-foward voltage testing analysis," *IEEE International Test Conference (ITC)*, pp. 673-682, 2002.

[31] T. M. Mak, A. Krstic, K. T. Cheng, and Li-C. Wang, "New challenges in delay test of nanometer multigigaherts designs" *IEEE Design & Test of Computers*, pp. 241-247, May-June 2004.

[32] J-J. Liou, A. Krstic, Y-M. Jiang, and K-T. Cheng, "Modeling, Testing, and Analysis for Delay Defects and Noise Effects in Deep Submicron Devices" IEEE Trans. On Computer Aided Design of Integrated Circuits and Systems, Vol. 22, No. 6, pp. 756-769, June 2003.

Chapter 3

Silicon Debug

Doug Josephson and Bob Gottlieb

3.1 INTRODUCTION

Silicon debug, the stage in the chip development cycle that starts with the arrival of initial silicon and often continues well after a product has gone into volume production, is perhaps the most exciting and challenging stage of the integrated circuit development process. After the chip design is complete and the design database is frozen, the masks are generated and sent to the fabrication facility, and the chip is manufactured. Once manufacturing of the first chips is complete, the chips are sent back to the design team, and post-silicon debug can commence. This is the highly anticipated time where design engineers can finally test out the designs they've worked on for the last several months (or even years).

The purpose of debug is to identify and fix any problems (also known as *bugs*) that exist in the design to ensure that the product operates correctly for customers over the product specification range. The term "bug" has been used for well over a century to describe mechanical or electrical problems. Thomas Edison wrote in a letter in 1878: "*It has been just so in all of my inventions. The first step is an intuition, and comes with a burst, then difficulties arise—this thing gives out and* [it is] *then that "Bugs"—as such little faults and difficulties are called—show*

D. Gizopoulos (ed.), Advances in Electronic Testing: Challenges and Methodologies, 77-108.

themselves and months of intense watching, study and labor are requisite before commercial success or failure is certainly reached." An interesting example of "de-bugging" was in 1945 when a computer failure was traced down to a moth that was caught in a relay between contacts (Figure 3-1).

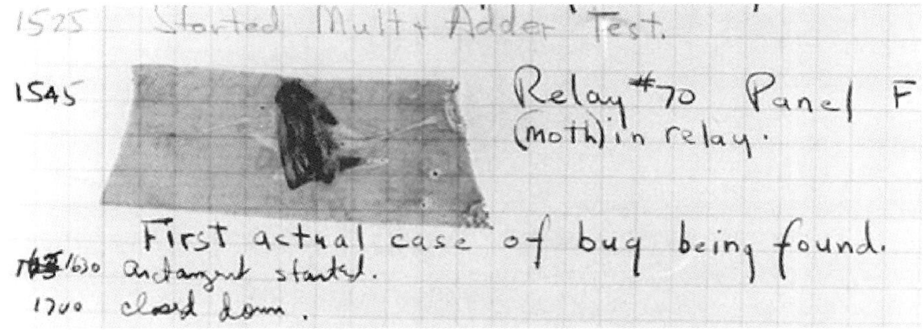

Figure 3-1: Computer "de-bugging" circa 1945.

Functional bugs, also known as logic bugs, occur when the logic was not defined or implemented correctly and thus the device does not usually work under any combination of frequency, voltage, temperature or processing. These can also be the result of equivalence issues – where a schematic/artwork implementation does not match the higher level RTL (register transfer language) description.

Electrical bugs occur when a circuit does not behave according to specifications; there is both a passing window of operation and a failing window of operation (i.e. the device operates correctly for at least one operating point). These are the result of marginalities in the circuit design. Another type of bug is the yield or manufacturability bug – the design may be sensitive to manufacturing variations which require that changes be made to circuits or the layout of the design to improve the yield (the ratio of good vs. defective parts) of the device during manufacturing testing.

All of these bug types are analyzed and debugged differently as a result of the differences between them. Throughout the rest of the chapter, as we review different tools and features to aid the process of analyzing and fixing the bugs, we will discuss which features assist debug of functional, electrical and yield bugs.

There is usually very high pressure to "get the product out the door" as any product delays that occurred earlier in the program are hopefully absorbed in a shorter debug cycle. Also, there is no safety net to silicon debug; mistakes made during the silicon debug phase can be extremely expensive. Extra chip revisions (steppings), product delays, or even worse, product recalls are possible if silicon debug is not done correctly. Table 3-1 shows the relative costs of finding mistakes at different stages of the design cycle. As can be seen from this table, successful debug can easily make the difference between success and failure of a product.

Design Stage	Cost of Fix	Effort Involved
Initial Design	$10	5 minute fix
Design Review	$100	1 hour re-work
Layout	$1000	10 hours – schematic, layout, simulation
Tape release	$10,000-$100,000	50 hours rework, validation, masks aborted
Early silicon	$100,000-$1,000,000	200 hours debug/fix, equipment costs, new masks
Sampling	$1,000,000	Delay product launch
Volume production	$10M-$500M	Product recall – look for new job!

Table 3-1: Cost of mistakes at each stage of the design cycle[1]

In this chapter, we will give an overview of silicon debug. We will first discuss the process of debug and the tools and design features used when debugging. We will then go through a case study. Finally, we'll conclude the chapter with our thoughts on what new challenges might appear in the area of silicon debug. As both of the authors of this chapter work on microprocessor design and debug, our discussions and case studies are focused on microprocessor issues. However, the chapter content is generally applicable to all types of debug, including that done in other processes (e.g. gallium arsenide or silicon on insulator).

3.2 SILICON DEBUG HISTORY

The process of silicon debug, along with all other aspects of integrated circuit design, has changed drastically over the past twenty years. In the mid 1980s, transistor budgets were several orders of magnitude smaller than what they are today. For example, the Intel386™ microprocessor had just 275,000 transistors, compared to 1.72 billion on the latest Itanium® microprocessor. These relatively small budgets were both a big help and a big hindrance to the silicon debug process.

Because early chips were relatively simple compared to today's designs, it was likely that several design engineers had a fairly complete understanding of the entire chip, from the logic design all the way down through the layout. This knowledge was required in order to debug effectively, because there was no silicon space available to add additional debug features that are now taken for granted on most chips. Also, *computer-aided design* (*CAD*) navigation databases did not yet exist, so physical probing of the device required manual analysis of pages and pages of layout

[1] Source: Paul Ryan, Intel Corporation, personal communication.

plots; the complete layout of the 80486 was printed out on 65 rolls of paper for use in debug.

In some ways, silicon debug is now simpler and more automated than ever before because of added on-chip debug features. However, more complex failure modes are encountered due to the increasing power consumption and operating frequency of designs and the decreasing feature sizes of current semiconductor processes. These failure modes will be explained below, along with how to detect and debug them. As transistor budgets and device performance continue to increase, and as process features continue to decrease in size, new debug features, tools, and methodologies will be required to keep pace.

3.3 SILICON DEBUG PROCESS

3.3.1 Post-silicon Validation

The introduction discussed how important it is to complete the debug process as quickly as possible while being extremely cautious to avoid the possibility of missed bugs. To be able to accomplish this, the debug team must execute a detailed plan that defines all of the criteria required for the debug process to be complete. This plan is often referred to as a *validation plan*. There are two main portions of the validation plan. The first is the *functional validation plan*. The functional validation plan is the list of activities used to confirm that all of the features of the product work as they were intended to. Taking the case of an arithmetic logic unit (ALU), a functional validation plan would list all of the operations that the chip can perform (add, subtract, shift, etc.) along with all possible combinations of data. Even in a case like this, the complete validation space is almost infinite; the simple act of adding two 64 bit numbers has 2^{128} different data combinations possible – an impossibly large validation space to exhaustively cover. Therefore, a lot of planning needs to go into defining the validation patterns to ensure complete coverage of all functionally unique possibilities. Functional validation is usually done in systems that are representative, or better yet, equivalent, to the actual systems where the product will be used.

The second plan is the *electrical validation plan*. The intent of this plan is to ensure that the part meets the data sheet specifications, and that the device's operation is robust across frequency, voltage, temperature, and across the variation of parts coming out of the fabrication facility. Figure 3-2 shows a simple two dimensional plot of the voltage versus frequency windows of a device. The device specification window is the range of voltage and frequency across which the device is guaranteed to work [1]. The characterization window is a superset of the specification window, and the debug team will not consider the electrical validation complete until the device operation is clean across the entire characterization window. The difference between the two windows is the *operating margin* of the device. Margin accounts for any inaccuracy in the testing and characterization of the design.

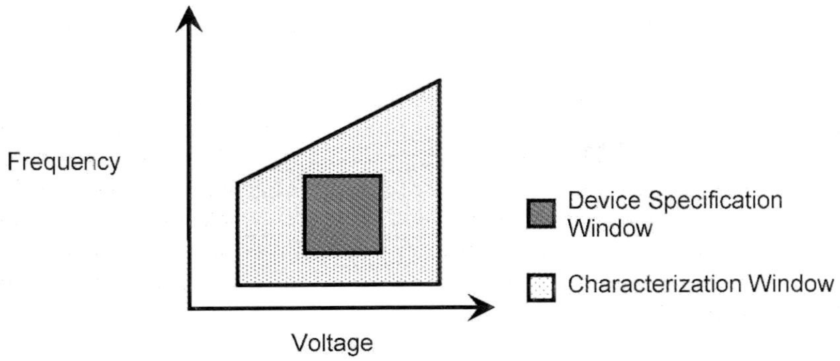

Figure 3-2: Specification window and characterization window for electrical validation.

For electrical validation, the validation patterns are similar to functional validation patterns; however, test cases that are interesting for electrical validation are likely very different from those that are interesting for functional validation. For the ALU example, a CMOS dynamic circuit implementation may perform differently electrically if there are two add instructions executed in consecutive clock cycles than it would if there was a long period of inactivity between the add instructions due to precharge and evaluation of the circuit. However, from a functional validation point of view, these cases would be identical. Covering all the corner cases of pattern combinations and initialization makes electrical validation far more difficult.

Another difference between the functional and electrical validation is the type of platform where the validation is done. There are two main types of platforms for doing validation. The first is the system environment which is similar to the target system for the product. For microprocessor validation, this system environment is a modified PC or server-based platform. These systems are ideal for running large volumes of code quickly, but it is difficult to precisely control the environment (voltage and temperature). The second platform is the tester platform (also referred to as ATE or automated test equipment). This is a specialized machine that on every clock cycle applies and compares values on the device input/output pins against simulated results. Testers are very useful for validation and debug as it is simple to control the environment. However, it takes a long time to generate tests for this platform as pin values need to be simulated in order to determine what values to send and compare at each clock cycle. The tester platform is also used in manufacturing testing to screen parts for manufacturing defects.

As discussed earlier, functional validation is almost exclusively done in system platforms where the goal is to run a lot of code in an environment as close to the actual product environment as possible. Some electrical validation is done in this environment because of the need to cover such a large code space. However, the tester platform is used for electrical validation as well because of the ability to accurately control voltage and temperature.

3.4 DEBUG FLOW

The four steps of the silicon debug flow are as follows:

1. Control the failure
2. Isolate the failing circuit
3. Root cause the failure
4. Try to expand the problem

3.4.1 Step 1: Control the Failure

The first step is to control the failure through experimentation, i.e. applying inputs and observing outputs of the circuit being debugged. For functional failures, this involves trying to find a minimal sequence of code that can cause the failure to appear (and to disappear). Using the ALU example, understanding that the adder fails whenever there is a carry out of bit 31 of the adder puts control around the failure. To make the failure appear, the debug engineer can write a code sequence to have the adder generate this carry. To make the failure disappear, the code can be modified to not generate this carry. This is known as a *light switch* test – a simple test that can turn on or off a failure with minimal modification. For electrical failures, controlling the failure first involves understanding the electrical conditions required to make the failure appear. A useful tool to discover these conditions is the *shmoo plot* [2], [3].

Shmoo Plots

Because electrical bugs have a passing and failing region, a two dimensional plot of how the device performs across voltage and frequency can be taken to control the failure, as well as have a first pass look at the nature of the failure. Other types of plots exist where two different voltages are varied and plotted against each other. Temperature is often also a common variable to control when exploring an electrical failure. More complicated plots may utilize multiple variables such as temperature and several voltages together. Such plots are referred to as shmoo plots. Figure 3-3 shows some examples of frequency vs. voltage shmoo plots.

The "normal" shmoo shows a traditional voltage versus frequency curve for a CMOS circuit. As voltage increases (left to right on the shmoo plot), the device works at a higher frequency, due to the increase in performance of the field effect transistors (FETs) as the voltage is raised.

The *wall* shmoo plot fails at a certain voltage, irrespective of frequency. This is often indicative of problems related to noise coupling between interconnect lines (inductive or capacitive), charge sharing, or races. Noise is aggravated by higher voltage since dv/dt (for capacitive coupling) and di/dt (for inductive coupling) increases. Higher voltages also mean faster circuits, which can lead to violating hold time to latch elements (a race condition, also known as a "mintime" or "early mode" failure), as well as more charge sharing between circuit nodes. Noise failures may be worse at low (cold) or high (hot) temperatures based upon the nature of the circuit that is failing. High temperatures reduce circuit speed (and thus di/dt and dv/dt which leads to less noise) but also reduce V_{TH} (the transistor threshold voltage) which makes circuits more prone to incorrect evaluation since devices can turn on more easily. Conversely, low temperatures increase coupling noise by increasing dv/dt and di/dt as well as reducing resistance, but also increase V_{TH}, making circuits less likely to fail due to a noise event.

```
400    *    *    *    *    *    *
450    *    *    *    *    *
500    *    *    *    *
550    *    *    *
600    *    *
650    *

      1.0  1.1  1.2  1.3  1.4  1.5
```

Normal
Well behaved shmoo
Typical Speedpath

```
400    *         *         *
450         *         *         *
500    *         *         *
550         *         *         *
600    *         *         *
650         *         *         *

      1.0  1.1  1.2  1.3  1.4  1.5
```

Brick Wall
Bistable
Initialization

```
400                   *    *    *
450                   *    *    *
500                   *    *    *
550                   *    *    *
600                   *    *    *
650                   *    *    *

      1.0  1.1  1.2  1.3  1.4  1.5
```

Wall
Fails at a certain voltage
Coupling, charge share, races

```
400         *    *    *    *    *
450              *    *    *    *
500                   *    *    *
550                        *    *
600                             *
650                             *

      1.0  1.1  1.2  1.3  1.4  1.5
```

Reverse speedpath
Higher V reduces frequency
Speedpath, leakage

```
400
450
500
550
600    *    *    *    *    *    *
650    *    *    *    *    *    *

      1.0  1.1  1.2  1.3  1.4  1.5
```

Floor
Works at high but not low frequency
Leakage

```
400
450
500              *    *    *    *
550                        *    *
600
650

      1.0  1.1  1.2  1.3  1.4  1.5
```

Finger
Fails at a specific point
Coupling

Clock period in ps on the left; frequency increases going up
Voltage on the bottom, increases from left to right
* - indicates a failure

Temperature effects
Cold: faster FETs, decreased R, higher V_{TH} - *mintime races, noise*
Hot: slower FETs, increased R, lower V_{TH} - *noise, leakage, speed, charge sharing*

Figure 3-3: Sample shmoo plots.

The *floor* is a failure that occurs only at lower frequencies of operation. Failures that manifest themselves as frequency floors are typically due to leakage problems in a circuit, for example with dynamic nodes. At lower frequencies, when leakage is present and no active circuits are working to combat it, enough time between clocks may exist to cause circuit state to "leak" away. Higher temperature always makes leakage failures worse, as heat increases subthreshold leakage in FETs.

The *brick wall* is indicative of an initialization problem for circuits. This occurs when a circuit resets itself at every other point that the shmoo is run at, which changes state in the device and makes it go from a passing to a failing state. Such problems are usually cured by adding a hardware reset signal to circuits that need to be initialized.

Some more complex variations of shmoo plots are shown in the last two examples. The *reverse speedpath* is an example of how a circuit path with a significant RC delay may act when it is shmooed. Even though voltage is being increased, if a significant fraction of the total propagation delay of the path is due to RC delay, the RC delay will dominate the frequency of operation. The increase in the driving transistor performance cannot overcome the penalty of the significant RC delay in the path. Increasing voltage actually makes the problem worse due to adverse changes in the trip points of the circuits relative to the edge rates of the RC dominated part of the path. Increased voltage also results in increased power and thus temperature. For an RC dominated path, this will cause the parasitic resistance to be higher, which will slow the path down as well. In addition, increased heat increases subthreshold leakage and thus overall power as well, which further increases temperature and may also cause additional voltage droop to circuits. Finally, the aptly named *finger* may be a signature of inductive and/or capacitive coupling. There may only be a small window of frequency where a coupling event can affect a downstream circuit – if the coupling event is too early or late, the downstream circuit may never see the glitch.

As temperature is increased, transistor performance is reduced and resistance goes up, which leads to lower frequencies of operation. Colder temperatures decrease leakage and resistance and improve transistor performance. Knowing the effects of temperature on circuits can expose different kinds of circuit failures as shown at the bottom of Figure 3-3. Taking multiple shmoo plots at different temperatures is very helpful in analyzing circuit failures.

3.4.2 Step 2: Isolate the Failing Circuit

Once a sequence of code and conditions has been found that makes a failure repeatable, the next step is to try and isolate what circuit on the chip is failing. This step requires gathering as much information as possible about what is happening internally on the chip at the time of the failure. Design teams often add internal debug features to a design to increase the visibility of the internal state of the chip [4]-[13]. The goal of all of these features is to gather as much information as possible from the device, while being as minimally intrusive as possible to the actual device operation. If the debug feature is too intrusive, it may no longer be possible to reproduce the original failure. It is also important to minimize the cost of additional debug hardware (in area, power and speed, as well as design time required) since these can add cost to the manufacturing of the chip. However, such features are usually well worth the additional cost since they aid in resolving bugs more quickly, thus leading to quicker time to market. The most important of these debug features will be covered in the following subsections.

Internal Logic Analyzer

The heart of the debug feature set on many complex chips is an internal logic analyzer that enables the debug engineer to define *triggers* to send to the rest of the chip. These triggers are used to enable the features that will be described below. On a tester failure, defining an event is easy, because the test is completely deterministic. The trigger can usually be defined as some number of clock cycles from the beginning of the test. To enable this, the debug block simply requires a programmable counter. For complex failures, the act of defining the event of interest is much more difficult. Often, a very specific sequence of events is required for a failure to occur. Flexible debug blocks provide the ability to define a trigger event by these very complex sequences of events. For example, in a system, a debug engineer may wish to match on a certain instruction opcode type being issued, wait 3 cycles, see if a certain data pattern shows up on a bus, and then generate a trigger. This sort of complex triggering capability is necessary to narrow down failures in systems, but adds complexity in the design of the debug blocks.

On Die Clock Shrink (ODCS)

An *on die clock shrink* (ODCS) mechanism provides the ability to adjust the internal frequency or duty cycle/phase width the chip clock is running at for one or more cycles. For example, an entire test can be run at 2 GHz, with one cycle in the middle of the test run at 2.2 GHz. Figure 3-4 shows two examples of ODCS. In the first example, in response to a debug block trigger, two cycles of the core clock are shrunk (run faster) than the rest of the cycles. As long as the trigger fires, the clock is manipulated. Only the rising edge of the clock is actually manipulated. In the second example, two clock pulses are modified, similar to the first example. However, in this case, both the rising and falling edges of the clock are modified.

With the ability to modify both rising and falling edges independently, and an ability to control the amount that the edges are moved, it is possible to perform complex clock manipulations that can help in identifying cycles that are critical to failures in the device when it is running. For electrical failures that are repeatable, deterministic, and have frequency dependence, the test can be run repeatedly, each time with the trigger moved by a single cycle, in an effort to narrow the failure down to a single point in time. For example, if there is a speed path that fails at 2.1 GHz, the core clock can be set to 2 GHz. Then, the test can be run repeatedly, each time shrinking one cycle to 2.1 GHz. The test will only fail when the exact cycle of the speed path is shrunk.

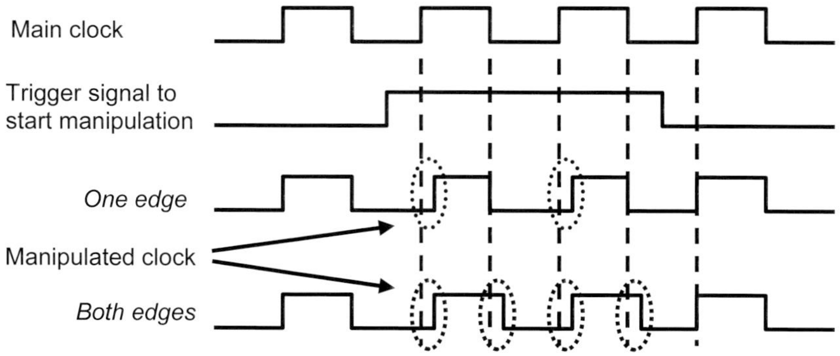

Figure 3-4: ODCS examples.

In addition to isolating the exact point in time when a failure is occurring, there are some additional benefits to ODCS. One, if a test has multiple failures at a given frequency (say 2.1 GHz), running the entire test at 2 GHz and sequentially shrinking a single clock cycle to 2.1GHz will find all of the failures, not just the first one. Two, because the rest of a test is run at a more relaxed frequency, the failure often becomes more stable.[2] A stable failure makes the debug process simpler as the failure is easier to control. Three, by shrinking different edges, the debug engineer can determine whether the failing circuit is a ½ cycle long path, a full cycle path, or a multiple cycle path.

Scan

In addition to understanding when in time (temporal locality) a failure occurs, the debug engineer needs to understand where the device is failing internally (spatial locality). One of the features that provide internal state visibility is scan. Most complex chips contain scan chains in order to more easily test the device during manufacturing and to aid diagnosis of failures in the field. However, scan can be used for debug as well. Scan chains are merely existing chip flip-flops connected in a serial chain, which allows serial access to all the state of the device via a simple interface (often based upon the IEEE 1149.1 standard, also referred to as JTAG[3]). Figure 3-5 shows a traditional flip-flop with and without scan support.

In the Figure on the right, data can be shifted into the storage element when Shift is high, and is shifted out to SO when Shift is low. The chain of latches is connected together by SO from one latch connecting to SI of the next latch. Care must be taken to ensure that CLK does not let data into the storage element while the scan is occurring.

[2] A stable failure is one which can be recreated consistently. Sometimes, on the threshold of failure for electrical failures, slight variations in temperature, voltage, clocking, or initialization may cause a bug to be unstable – it may pass or fail from run to run.

[3] Joint Test Action Group.

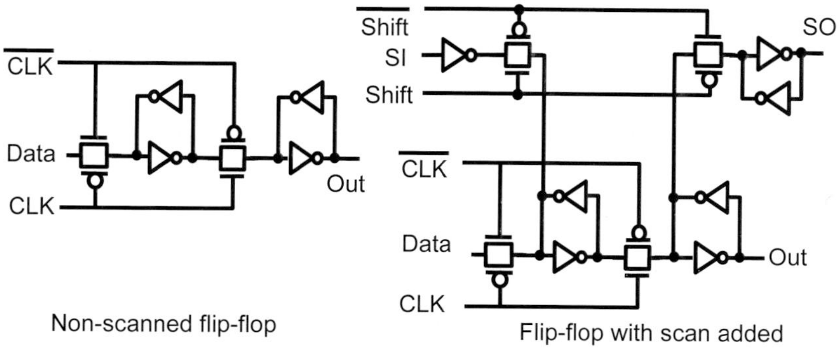

Figure 3-5: Flip-flop without and with scan support.

Using scan as a debug feature works as follows; chip operation proceeds normally, with the storage elements capturing their data value at each clock cycle. At a certain cycle of interest, a trigger (from the on-chip debug logic analyzer) stops the clock of the chip. At this point in time, all of the storage elements contain the value that they captured on the previous cycle. The scan chains can now be used to shift out all of the values of the storage elements. In this way, a very good picture of the state of the device at the time of interest can be obtained. If this feature is used in conjunction with ODCS, it can be an extremely valuable debug tool for isolating electrical failures. ODCS is first used to identify the "critical clock" – the cycle when the speed path is occurring. The test is then re-run several times and scan data is captured for several cycles around the failure in both the passing and failing condition. The passing and failing scan data can then be compared, which should indicate the first scan latch that sees incorrect data. If all of the flip flops on the chip are connected in these scan chains, this should completely isolate the failing circuit.

One drawback with this type of scan methodology is that it is destructive; after the values are shifted out of the chain, it may be difficult to "restart" the clock and have operation continue depending on the complexity of the design and its external interfaces. Because the expected externally visible (i.e. on the pins of the device) failure signature occurs later in time than the internal failure, stopping the clock around the time of the internal failure makes it impossible to know for sure whether the test would have failed if it were allowed to continue. If the failure is deterministic and stable, this is not a problem – the engineer can be confident that the scan data obtained is of interest. However, if the failure is not consistent (perhaps failing 10% of the time)[4], it is difficult to know whether the failure would have occurred if the test had been allowed to continue.

To get around this limitation, many designs add read-only scan latches (ROSLs – "raw-zuls"), also sometimes called *shadow scan*. These scanlatches are added

[4] Sometimes failures are not consistent because of initialization issues, environmental variations, or dependence on semi-random events. For example a microprocessor may have different behavior from run to run due to memory refresh events causing timing of events to vary.

strictly for debug capability and are not used for normal functional operation. They are able to capture state in response to a trigger, similar to stopping the clock and scanning out internal scan state from functional latches. However, they are not destructive, so the test can continue to run. This can be particularly effective when debugging a chip in the actual environment it will be used in (for example in a computer system vs. on the manufacturing tester). Figure 3-6 shows an example of how ROSLs can be connected to functional latches or logic in order to gain non-destructive observability of some functional signals in a design.

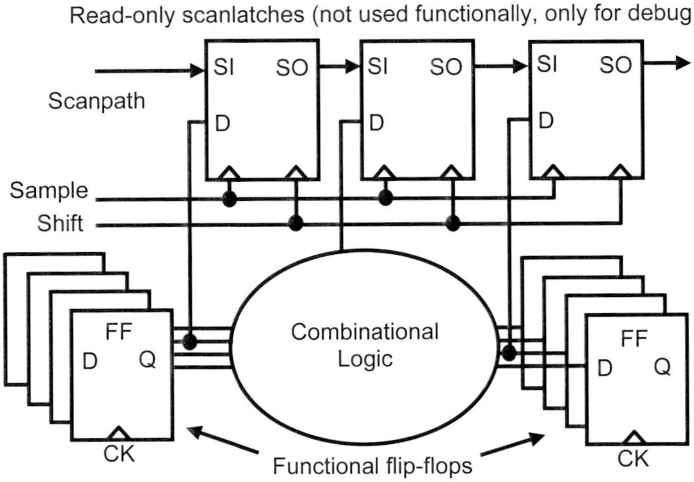

Figure 3-6: Example of flip-flops with attached ROSLs.

When the debug unit sends the trigger (the Sample signal), the value of D (a signal of interest from functional flip-flops or logic) is stored into the read-only scanlatch. Because there is an additional storage element for this value, the chip clock does not need to be stopped, and internal chip state is not altered by the capture. The test can be continued without reinitializing the device, which is useful when debugging in systems. If a failure occurs, the data captured is valid and can be used to debug the problem by toggling the Shift signal and scanning the data out of the chain through the SI (scan in) and SO (scan out) ports of the ROSLs, without corrupting state of the functional flip-flops. If a failure did not occur, the test can be re-run to try and get new data at a different voltage or frequency point. Because of this ability to check if the test failed as intended, ROSL debug is especially valuable for non-deterministic system failures. The downside of the ROSL latch is the area overhead and the additional load on the data signal.

Local Clock Skew

One final debug feature to discuss with respect to isolating the failing circuit is local clock skew [9], [14]. Like ODCS, this debug feature is valuable for electrical failures that have frequency dependence. However, unlike ODCS, which isolates when in time a failure is occurring (temporal locality), local clock skew helps localize which circuits contain the electrical bug (spatial locality). Complex integrated circuits often internally distribute their clock in a tree. Figure 3-7 shows a sample clock distribution network. The concept of local clock skew is to add variable delay elements to the unique branches of the clock distribution network to allow adjustment of the clock skew between regions. This can be an effective aid for debug, and can also be used to improve frequency performance of a chip, since zero skew between regions is not always the most optimal setting for the highest frequency of operation.

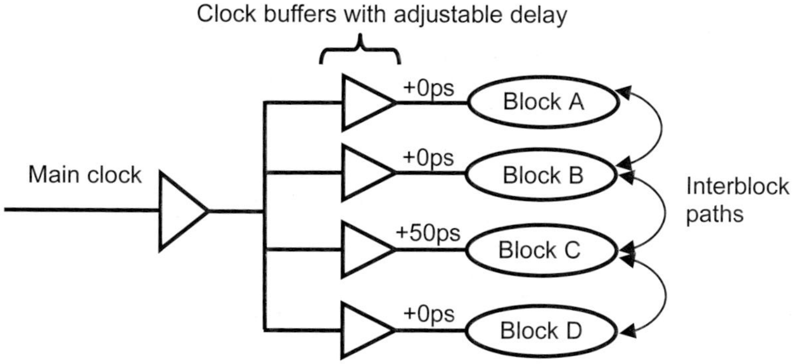

Figure 3-7: Sample clock distribution network.

In this example, 50 ps of delay is added to all of the logic that is in Block C of the design by adjusting Block C's clock buffer. As a result of this extra delay, any path whose driver is in Block B and whose receiver is in Block C will have 50 ps of extra timing margin when signals travel from Block B to Block C. However, any path going from a driver in Block C to a receiver in Block D will have 50 ps *less* margin, since Block D's clock is 50 ps *earlier* relative to Block C's clock.

By operating the device right at the boundary of the passing and failing point and running the test repeatedly, it is easy to find out which region is the driver and which region is the receiver for a failure. One side benefit of this feature is the ability to get around critical issues by "permanently" altering or even optimizing the clock skew to the different regions of the device, which can increase overall performance. The effectiveness of this is usually limited by the granularity of the regions that can be controlled – if the regions are too large, settings that are optimal for signals traveling from one region to another may not be good for signals traveling the other direction. Also, if both the driver and the receiver are in the same region (block), this debug feature has no effect.

3.4.3 Step 3: Root Cause the Failure

After the failing circuit is identified using the debug features described above, the next step in the debug process is to identify why the circuit is failing. For some failures (most functional failures for example), isolating the failing circuit (logic) is enough to fully understand the cause of the failure. For other issues, mainly electrical, understanding which circuit is failing may not yield any information about the ultimate cause of the failure.

For example, with the ALU example, just knowing that the adder is failing to add correctly when there is a carry out of bit 31 at certain voltages and frequencies doesn't go very far in the identification of the root cause of the failure. An engineer with good knowledge of the design may be able to write and run code experiments to determine if the failure is sensitive to noise on lines adjacent to the failing circuit. Similarly, temperature experiments might provide some information as to the cause of the failing circuit.

There are also several physical tools available to facilitate this process. These include probing tools such as the Laser Voltage Probing (LVP) [15] and Picosecond Imaging Circuit Analysis (PICA) [16], [17] (also known as Time Resolved Emission (TRE)) that can take waveforms of the internal nodes of the circuit with tens of picoseconds of resolution, similar to how an oscilloscope is used to debug circuits at a macroscopic level. Other probing tools such as the Emission Microscope (EMMI) [18] and Laser Assisted Device Alteration (LADA) [19] can help further isolate the failing circuit to a more localized area than was possible with the tools outlined above. These tools all make use of the fact that silicon is transparent to infrared light, and/or that switching transistors on the chip emit infrared light. These techniques are utilized from the "backside" of a chip, which is common with flipchip packaging.

LVP and LADA are active techniques. An infrared laser beam is used to either observe voltage levels on particular transistor diffusions (in the case of LVP) or induce electrical changes in the behavior of particular transistors (in the case of LADA) by photo-injection of carriers. Observing voltage waveforms with LVP allows the debug engineer to "see" signal waveforms similar to an oscilloscope as well as timing relationships between signals. LADA can be used to speed up individual transistors, which can be useful in debugging speedpaths. Another use for LADA is to "search" for failure by rastering the laser across the die – when a transistor involved in the failure is hit by the laser beam, it changes behavior and the test may go from passing to failing or vice versa.

TRE and EMMI are passive and rely on the fact that transistors emit infrared photons when in the saturation region. Photons emitted by circuits in operation are collected and a plot of emission intensity is created that can be overlaid on the chip layout to identify areas of interest. Often this can be very instructive in indicating a behavior of the circuit that is otherwise very difficult to determine. Another advantage of the passive technique is that the simple act of "probing" the circuit with a laser beam (LVP or LADA) may disrupt circuit operation. The passive techniques do not have this limitation since they simply observe circuit operation.

An older technique is the use of electron beam (e-beam) probing. E-beam probing is only effective for probing metal lines. As a result, it is often used for wirebonded designs, where only the "frontside" of the chip is available (i.e. only the upper layers of metal are accessible). An electron beam can be focused on a particular area of the circuit in question. Secondary emission of electrons occurs based on the voltage of the metal line being probed; these electrons can be collected and analyzed to examine the voltage changes as the circuit switches. The result is an oscilloscope-like waveform from the node being probed. E-beam probing has several drawbacks – it must be performed in a vacuum, and heat removal from the circuit is often a problem as a result.

Another even older technique is that of "micro-probing", where a micro-manipulator is used to place a very fine tipped metal probe onto the metal lines of circuit nodes, which can then be directly observed with an oscilloscope. This can damage the circuit if care is not used, and the additional parasitic loading of the probe itself may be enough to change the behavior of the circuit under observation. Active amplifying probes exist that can limit parasitic loading by the probe to tens of femtofarads. However, this technique remains one of the most invasive methods of probing and requires significant preparation to ensure good results.

Another vital tool for the debug engineer is the Focused Ion Beam (FIB) machine that enables the engineer to perform "microsurgery edits" on an existing device to test out theories as to the cause of a failure. For example, cutting away adjacent lines to lower cross-capacitance, or connecting a new buffer to a wire to increase the drive strength, can change the behavior of the device and may offer insight into the failure. Such edits may also be used to actually "test out" a possible fix in advance of actually making a new mask to fix the design. This offers tremendous benefits in reducing the time it takes to get designs to market. Testing out a fix early can both save additional delays if the fix isn't "right", and also can unmask other bugs that may be "hiding" behind a bug[5] that can be resolved with a FIB edit.

3.4.4 Step 4: Try to Expand the Problem

Once a failure cause is understood, it is very important to try to *expand* the problem. This activity has two goals: ensure that the worst case variation of the problem is known, and ensure that no other place in the design is susceptible to the same problem.

It is extremely important to ensure the worst case situation has been applied to a failing circuit – this guarantees that the problem is completely understood and can be fixed and tested to the worst conditions ever expected. For example, the values going to the ALU may be coming from many different sources (different registers, memory locations, or bypasses from recent results), and any of these sources might exhibit the worst case bug. Other variations can come from different data patterns, different data on adjacent lines (that could make coupling worse), or even differently processed silicon.

Finding the worst case pattern or condition for a specific bug often involves a lot of test writing and code experiments, as well as circuit simulation, to ensure that the failure is well understood. Failing to find the worst case version of a particular bug may result in an incomplete circuit fix that may not remove the failure – or, even worse, inadequate testing resulting in defective parts getting shipped to a customer.

It is also important to ensure that there are no other possible bugs in other areas of the design caused by the same problem. This part of the process usually involves trying to find out what happened in the pre-silicon process that allowed this bug to be missed. For example, a portion of the circuit may have been mismodeled in

---

[5] For example, another speedpath which cannot be tested because the first speedpath does not allow the second speedpath to be exercised. In this case, a FIB edit to remove the first speedpath may be critical in exposing the second speedpath.

simulation. If so, the entire design may need to be re-simulated once the modeling error is corrected to ensure that there are no other bugs related to the modeling problem. This step is not complete until there is correlation between the pre-silicon environment and the post-silicon failure.

3.5 CIRCUIT FAILURES

Given the many different types of circuits used in CMOS designs [20], [21][6], it should come as no surprise that there are many ways in which these circuits can fail as voltage, frequency, temperature and processing are varied. Circuit failures can range from the merely benign (e.g. a speedpath in an infrequently used circuit) to the catastrophic (a design that does not even operate!) In this section we describe some of the most common failures that are seen when debugging CMOS integrated circuits. Failures can often be a combination of these basic failures, which can greatly complicate finding the root cause of the failure.

Many different topologies of CMOS circuits exist and have been used with success in integrated circuit designs. In this section we will review some types of common CMOS circuit topologies and also how they can fail.

3.5.1 Speedpaths

Simply put, *speedpaths* are the slowest circuit paths on the chip, and govern how fast the entire design can operate. Speedpaths are also known as critical paths, maxtime paths, setup or "late mode" failures. A speedpath is the result of the propagation delay of a particular circuit path being longer than the clock period used for the circuit. For example, if the clock period is 1 ns, but the propagation delay of a particular circuit path is 1.05 ns, data propagating through the circuit path would fail to arrive at the destination prior to the next clock edge and would thus be "late".

Often the speedpaths that are found during silicon debug are not the same speedpaths that are predicted during the design cycle. This is because of limitations or mismodeling that exists in the pre-silicon environment. These speedpaths directly determine the frequency at which a part can be sold (binned).

A significant amount of the debug effort of a design can be spent debugging speedpaths and fixing them in order to increase the frequency of operation of the device and thus increase overall performance. Speedpaths are often among the easiest failures to detect in a device through the use of ODCS and scan techniques. Speedpaths are often also easy to fix as well. Usually a combination of increasing transistor sizes, changes to interconnect wires, delaying or advancing a clock edge, or implementing the logic in a different way can make up for any deficit in timing.

3.5.2 Mintime Races

Mintime race conditions, also known as mintime paths, hold-time or "early mode" failures, can occur when a signal arrives earlier than it should in a circuit, as opposed

---

[6] For example, static, dynamic/domino, pseudo-NMOS, pseudo-dynamic, self-resetting, pulsed, annihilation gates, sense amps, mousetrap, zipper, ratioed, pass gate, low voltage swing...

to speedpaths, where signals arrive too late. Mintime races can cause a circuit to change its state "too early" relative to when it logically should. In addition, even for circuits that may not retain state, races can cause unanticipated effects such as additional power consumption for periods of time.

In Figure 3-8 there are two transparent latches clocked by clocks C1 and C2 that have a delay of Δt_1 between them. There is also some logic between the latches, indicated by delay Δt_2. For the circuit to operate properly, the clock C2 should fall before new data arrives on node N1 as the result of C1 rising. If the delay Δt_2 is greater than the delay Δt_1, then there is no race, because C2 will fall before the data on N1 can change due to C1 rising, and the proper data will be latched on node N2 (as indicated by the dashed lines of N1 and N2).

If, however, the delay Δt_2 is *less* than the delay Δt_1, new data will arrive at N1 as the result of C1 rising before C2 falls. Thus when C2 falls, the incorrect value will be latched in the second latch (as indicated by the solid lines of N1 and N2).

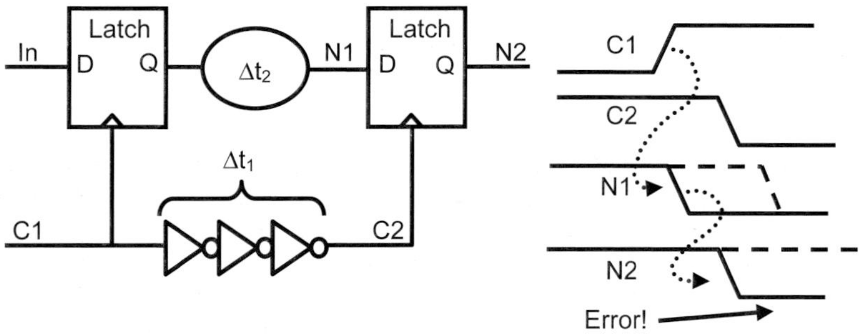

Figure 3-8: Example race condition.

This failure is due to the relationship between the timing of the clock waveforms and the amount of time for the signal to get through the latch and subsequent logic. Because neither of these conditions changes with frequency, these failures often show up on a shmoo plot as a (frequency insensitive) voltage wall (at high or low voltage). Mintime races are very dependent upon the particular clocking and latching methodologies used in a design. Design teams tend to be conservative in developing such methodologies to ensure that mintime races do not occur. Unlike speedpaths, where a reduction in frequency can make the circuit work again, mintime races can fail at any frequency of operation. This can greatly impact debug progress, particularly if a race involves a circuit that is vital to operation of the design. Such a problem can halt debug of the design until it is fixed. Therefore, design teams fear mintime races more than speedpaths, and go to great lengths to avoid them. Sometimes it is possible to change voltage (thus changing the respective delays of the clocks and data) and "win" the race. Also, if C1 and C2 from the example are on different branches of the clock tree, local clock skew, as discussed above, can be used to "fix" the race condition.

3.5.3 Charge Sharing

Charge sharing is the unintentional transfer of charge between nodes in a circuit. The charge is stored in the parasitic capacitances of individual transistors in a circuit. Charge sharing manifests itself through a high capacitance node transferring charge to or from a low capacitance node. It commonly affects circuits that retain state – e.g. latches and dynamic circuits. However, it can also affect even static circuits by slowing node transitions down. CMOS pass gate circuits can also be affected significantly by charge sharing. This type of failure often has either unusual or no frequency dependent behavior, as the charge sharing event may or may not change over time.

If the ratio of capacitance between the nodes is high enough, the lower capacitance node can "flip" – i.e. change voltage significantly, which will lead to inversion of its logic value. In the event that the ratio of capacitances is more equivalent, charge sharing may result in simply slowing down a node transition instead of changing the state of the node.

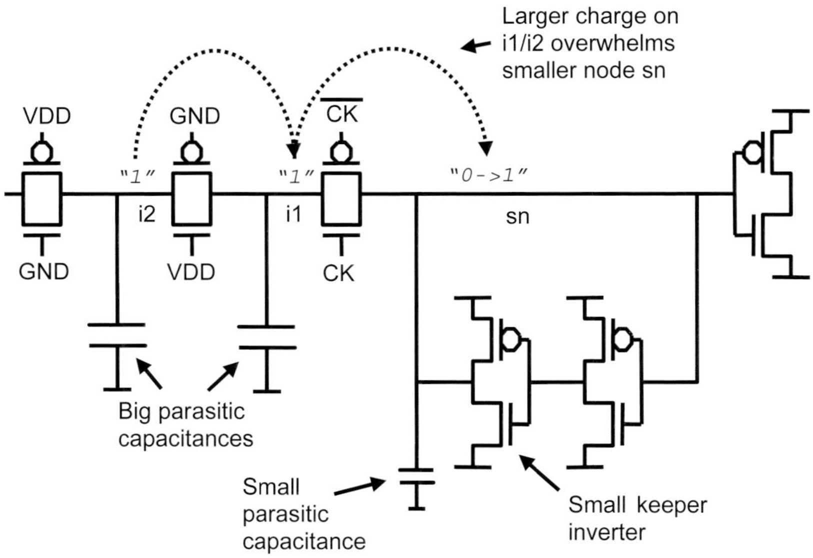

Figure 3-9: Charge sharing example.

An example of charge sharing is shown in Figure 3-9. In this figure, a latch is shown with several serial pass gate inputs. Assume that the two nodes i1 and i2 have been previously charged by past operations to the supply voltage, but that the clock CK has not gone high when this happened, and that the storage node of the latch sn is currently holding a value of zero from the last time that CK went high. Also assume that the leftmost pass gate is shut off, so that the normal latch input data should not propagate through nodes i1 and i2 into storage node sn when the clock signal CK fires. Note that the nodes i1 and i2 have large parasitic capacitance on them due to the relatively large transistors on the pass gates, and that sn has a relatively small parasitic capacitance.

When the clock signal CK fires, since i1 and i2 are at a high voltage, and only a small parasitic capacitance exists on storage node sn, charge is transferred from the sn node, which can cause the node voltage to change. If the charge transfer is high enough, the small keeper inverter in the latch may actually be unable to hold the node value, and the sn node could inadvertently be "flipped" to the incorrect value. Note that the latch should not be updating since the initial pass gate is off when CK goes high.

Possible solutions to this problem would be to increase the size of the keeper inverter, reduce the pass gate parasitic capacitance through FET size reduction or by changing the topology of the input structure of the latch, or increasing the "good" parasitic capacitance on the latch sn node to make it harder to "flip" it (although adding "good" capacitance will also slow the latch down).

Another common circuit implementation is dynamic logic, also known as a *domino* circuit. Figure 3-10 shows a domino NOR followed by a domino NAND. In domino circuits, all of the gates are pre-charged when the clock CK goes low, causing their respective outputs after the inverters to fall. When the evaluation phase starts with clock CK rising, outputs OUTA/OUTB will either stay low or go high (falling in sequence like dominoes) depending on the values of A/B/C. Note that small keeper PFETs exist to stabilize the value of the internal storage nodes of the dynamic gates (e.g. SN) against possible noise events, leakage and charge sharing.

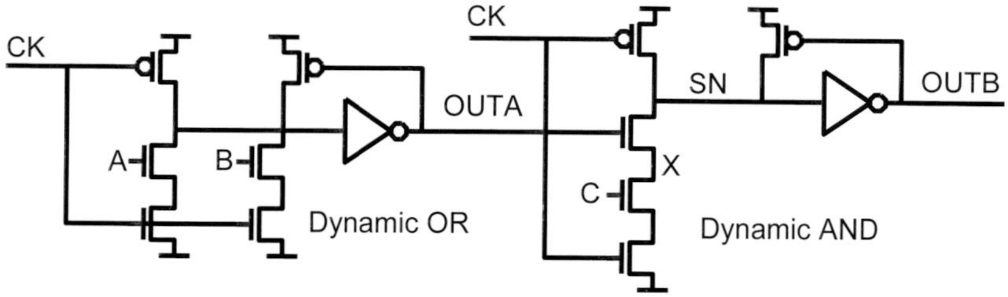

Figure 3-10: Example CMOS dynamic gates ("domino logic").

This type of circuit is susceptible to charge sharing. Assume node X was discharged (i.e. evaluated to a zero) in a previous high cycle of the clock CK. Clock CK then goes low, causing node SN to precharge and the value of OUTB to go low. Clock CK then goes high to evaluate the gate. Assume the value of C is zero during the evaluation cycle (indicating that the dynamic gate should not discharge) but OUTA goes high due to evaluation of the previous gate. If the parasitic capacitance on node X is large enough, it may be enough to overcome the precharged value on SN and cause the output OUTB to change inadvertently by discharging SN due to charge sharing. This failure would be a function of the size of the keeper PFET in the dynamic AND gate, the ratio of the PFET to NFET in the forward inverter of the dynamic AND, the size of the transistors in the evaluation NFET stack, and any additional parasitic capacitances on the SN and X nodes. By adjusting these, potential for failure in the gate can be reduced.

3.5.4 Interconnect Noise

Interconnect noise is the result of parasitic capacitive and/or inductive coupling between wires in the design, and/or problems in power delivery to a circuit as the result of parasitic inductance and resistance in the package or chip power network. A schematic example of noise coupling between several interconnect lines is shown in Figure 3-11.

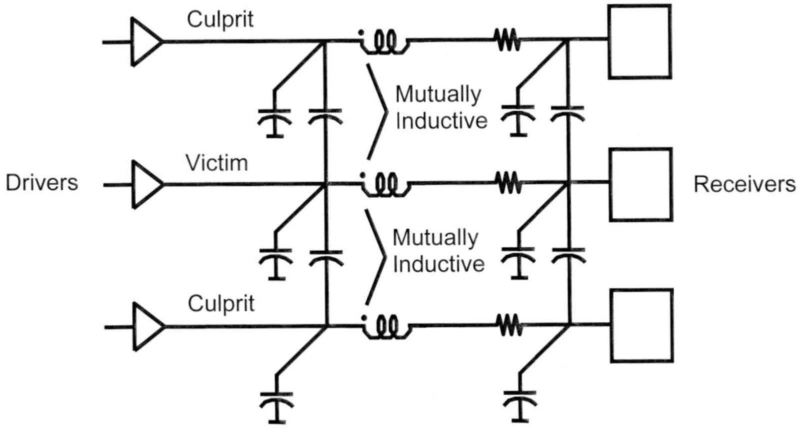

Figure 3-11: Noise coupling between interconnect lines.

In this figure, parasitic capacitance and inductance between the *victim* line and the *culprit* lines can cause the victim line to be disturbed when the culprit lines switch. The effects of such noise are dependent upon the resistance of the culprit and victim lines, the total amount of capacitance on the victim line, the parasitic capacitance and inductance between the lines, the strength of the respective drivers, and the sensitivity of the receiving circuit to the noise induced. Test pattern experiments are very helpful in diagnosing this type of failure as changing data values on neighboring lines that should not have any direct effect on the failing circuit can cause the performance of the circuit to change drastically due to coupling.

An example of how parasitic coupling noise can cause a circuit failure is shown in Figure 3-12. A latch input is driven by a small inverter a long distance away, through significant resistance. Adjacent to the victim line, two culprit lines (Cul) with strong drivers are capacitively coupled to the victim line (Data). Assume that the latch is holding a value of "1" on the internal storage node N1 after Clk transitions high and shuts off the latch pass gate. When the culprit lines go low after the latch has closed, capacitive coupling occurs to the Data line, and the small inverter on Data is unable to hold the value stable through the resistance present to the input of the latch pass gate. As a result, the Data node is coupled below ground. If Data drops lower than a V_{TH} below ground, the NFET in the pass gate will turn on, and it is possible that the N1 node can be pulled down, thus corrupting the correctly stored value in the latch.

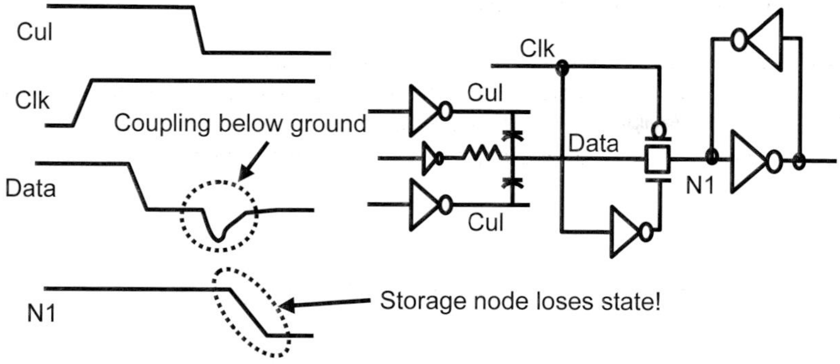

Figure 3-12: Noise failure due to capacitive coupling.

A power supply noise problem is shown in Figure 3-13. A driver sends a signal to a dynamic gate that is located 1000 μm away. As the result of the separation between the driver and dynamic receiver, there can be differences in the local voltage supplied to the driver and receiver, due to switching of circuits adjacent to each (e.g. perhaps there are a large number of other drivers in the vicinity of the driver or receiver). Due to parasitic impedance in the power supply network, large di/dt events due to local circuits switching may cause the local ground or supply rail to rise or droop, respectively. This results a voltage higher than ground being supplied to the input NFET of the receiving dynamic gate. Assume the V_{TH} of the receiving NFET is 350 mV, and that the local CK is high and the dynamic gate is waiting to evaluate. If the local ground of the dynamic receiver is zero volts, but the local ground of the driver is 350 mV due to power supply noise, a spurious discharge of the dynamic receiver could occur, even though a "logical" zero is being driven by the driver. A possible complication for this situation is that noise from switching of adjacent neighbors of the driven line could add to the voltage, making the failure even worse.

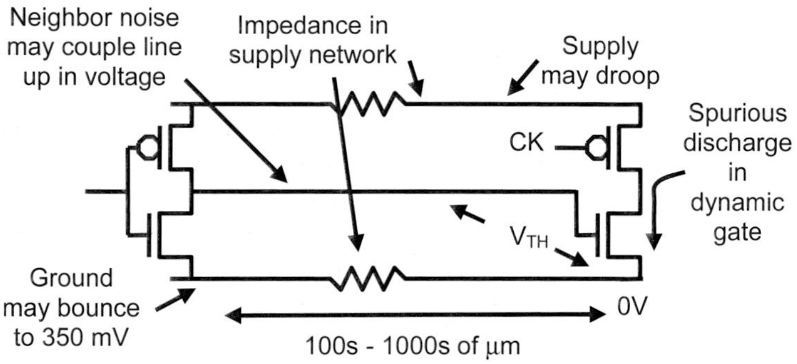

Figure 3-13: Power supply noise example.

Circuit failures due to noise are often among the most difficult problems to debug. This is because they can be very sensitive to environmental conditions (voltage, frequency, temperature[7]) as well as small variations in silicon processing. It is also difficult to create characterization patterns that cause the worst case noise conditions to occur at a particular point in time. Such patterns are often hand-written after a noise problem is identified as the result of random patterns used to characterize a design.

The effects of noise must be taken into account during the design of high-performance integrated circuits. Parasitic net extraction can be used to determine the capacitive coupling between adjacent lines for both timing purposes and for noise analysis. Results from net extraction can be utilized by noise analysis tools to determine circuit sensitivity to noise, so that modifications may be made to circuits to make them more noise immune[8] prior to design fabrication. Armed with the information from such design tools, the design team can use additional buffering and adjustment of transistor sizes to mitigate the effects of such noise effects.

3.5.5 Leakage

Parasitic leakage of CMOS FETs has become a major issue for design teams using advanced nanometer processes. There are three major types of leakage that affect CMOS circuits: reverse biased diode leakage, subthreshold leakage, and gate leakage, as shown in Figure 3-14.

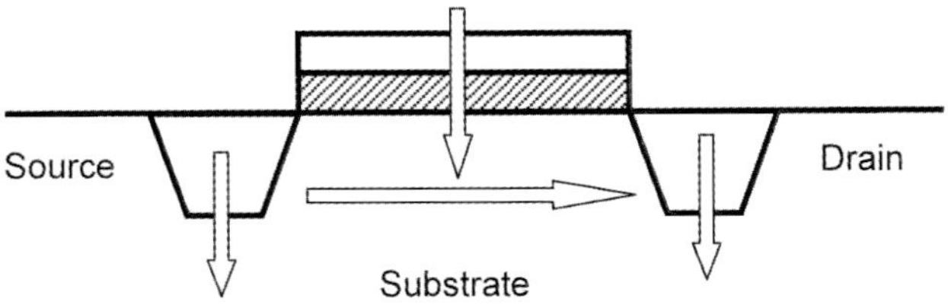

Figure 3-14: FET leakage.

Reverse biased diode leakage has existed since CMOS was invented and is the result of the parasitic junctions formed by the transistor drain and source junctions to the bulk substrate or wells. Over the past decade, subthreshold leakage (between the source and drain) has been growing in magnitude exponentially as transistor gate length has been reduced. Reduced gate length increases drive current and thus circuit speed, but at the expense of increased subthreshold leakage. Designs of only a

[7] For example, if temperature increases, noise is worse because threshold voltages are reduced, making circuits more susceptible to noise failures. Increasing voltage increases noise as well. Frequency of operation plays a part as well by aligning coupling events with other events (e.g. setting latches at a point when noise is occurring).

[8] Common methods of making circuits more noise resistant are to adjust the trip points of circuits to make them less likely to evaluate during noise events, as well as adding buffers to circuit inputs. Adding buffers aids in rejecting noise, as long as it is not severe enough to trip the buffer as well. Adding buffers adds to circuit delay of course, illustrating the tradeoff of speed vs. robustness.

decade ago may have had nanoamps of leakage, where designs of today can have tens of amps of leakage – an increase of over a billion times. This is a consequence of both the growth in number of transistors from thousands to billions, as well as the exponential increase in subthreshold leakage of each transistor with FET length reductions. Finally, gate leakage (between the gate and the substrate) has become an issue in the last few years as processes use ever thinner gate dielectrics that are only tens of nanometers thick – on the order of 5-10 molecular layers! Such thin dielectrics are susceptible to electron tunneling from the gate through the dielectric.

Leakage can affect both state retention and speed of circuits, and commonly affects structures that store state like dynamic logic and latches. Good circuit design techniques account for leakage by adding "keeper" FETs in dynamic logic and latches that provide active feedback against leakage. However, addition of such keeper transistors can adversely affect performance as well, since they will actively "fight" any state transition – whether due to leakage or a desired logical transition. Leakage effects can also be ameliorated through the use of non-minimum length gates (for subthreshold leakage) and multiple gate oxide thickness use (to control gate leakage). "Leaky" transistors can then be used only in circuits that require the highest performance. "Stacked" devices – where multiple devices are "stacked" on top of each other (for example in the NFET evaluation tree of a dynamic gate) are also useful in controlling leakage effects.

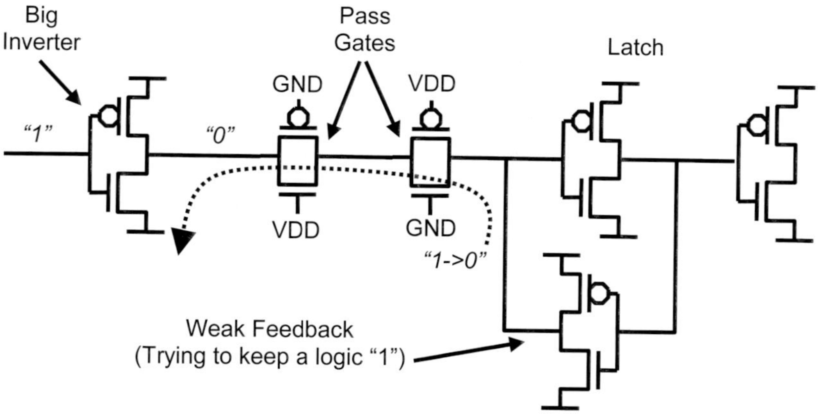

Figure 3-15: Leakage failure example.

An example of a leakage failure is shown in Figure 3-15. In this example, a latch has just been set to a logic one value. Two pass gates exist in front of the latch; the pass gate closest to the latch has just been shut off to hold a logic one value in the latch. A small "keeper" inverter, several times weaker than the forward inverter in the latch, exists in the latch to maintain the value by feedback. Now assume that the inverter driving the latch changes the input value to the latch and is thus driving a "0". Since the pass gate closest to the inverter is fully on, a logic zero is present on the pass gate input of the latch. Even though the latch pass gate is logically "off", subthreshold leakage through the latch pass gate may be enough to overcome the keeper inverter in the latch, thus corrupting the value of the latch by spuriously "flipping" the held value from a one to a zero. The leakage is a function of how big the pass gate is and whether minimum length gates are used in the pass gate, as well as the size of the keeper inverter in the latch.

Bugs involving leakage are becoming more prevalent in designs due to the overall increase in the various types of parasitic leakage. Leakage failures are typically relatively easy to diagnose as to the cause of the failure. If a failure gets worse by decreasing frequency (giving nodes more time to leak away) and by increasing temperature (which increases parasitic leakage), it is a definitive sign of a leakage failure. Although it is easy to diagnose the general cause of a leakage failure, actually discovering the mechanism for leakage in a circuit can be challenging. Use of scan techniques may isolate the failing circuit. Once it is found, SPICE simulations may be helpful in determining if the circuit is failing due to leakage. Emission microscopy can be useful in determining where leakage is occurring, as leakage can lead to light emission that can help in localizing the failing circuit.

3.5.6 Manufacturability

As process technology has advanced, process variation has become a significant issue. Not only do designs need to be manufactured without defects, but they also must be manufactured to tight tolerances to ensure that high performance circuits will work properly. Manufacturability is the measure of how robust a design is to manufacturing variations and how easily the design meets its specifications.

Designs must take into account the significant variation that can occur between two otherwise identical transistors. This is particularly important in analog circuits used in digital devices (e.g. I/O drivers, clock circuits like PLLs and DLLs, sense amplifiers for memories). Device characteristics such as threshold voltage may vary anywhere from 5-30% across a single large integrated circuit depending on the process technology used. Such variation can be particularly problematic in large circuits like memory arrays that can span tens of thousands of microns. Otherwise identical transistors may be "printed" differently depending on their particular location or orientation on the die. Such variation may be random or systematic (an effect that occurs as a result of a repeated topology for example), and can be global (random across a large area) or very local (dependent upon topology of adjacent circuitry). An example of such local variation is shown below in Figure 3-16, a microphotograph of a 180 nm design showing several transistors. The upper metal layers have been removed to expose the transistors for analysis. In this example, two FETs that are otherwise identically drawn in length and width are significantly different from each other, even though they are only a few microns apart.

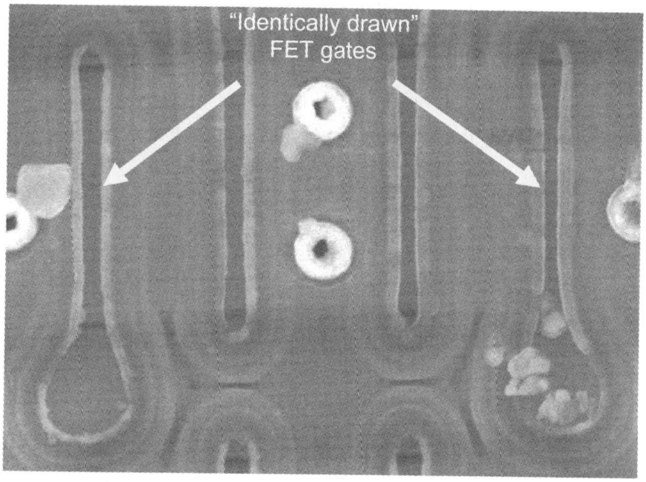

Figure 3-16: Example of variation in lithography.

Such variation can be reduced through a combination of design team effort and process "tricks". Optical proximity correction (OPC) [22] is commonly used in nanometer processes to ensure that devices are actually printed as laid out by the design team. Since features being printed today are actually smaller than the wavelength of light used to print them, they are often "rounded" or are otherwise inaccurately printed unless special measures are utilized. OPC adds special shapes to vertices and edges of design data before masks are printed. These added shapes on the mask ensure that the constructive and destructive interference of the light used to "print" transistors and metal results in the exact shapes desired when the chip is manufactured. Inadequate correction can result in very small transistors having increased leakage or being weaker than intended.

In addition to such process tricks, the design team may add "fake" structures (otherwise unused transistors or metal) to circuits to ensure that the edges of large circuit structures are not affected by variation as they might be if real functional circuits were at the edge of the circuit instead of "fake" structures. These also may be added to ensure that the area density of metal or other materials in a given area is roughly equivalent across a die. This results in more uniform etching and polishing during manufacturing and thus increases control over variation.

3.6 A CASE STUDY IN SILICON DEBUG

To tie all of the previous information together, we present a real life bug example and how various debug features, processes and methods described previously are used to "root cause" the failure mechanisms behind a complex bug. This example also illustrates several different circuit failure modes described above and how very complex interactions are sometimes necessary to cause a design to actually fail. Several other interesting industry debug case studies can be found in [1], [7], [23].

This failure was discovered while performing validation of a microprocessor design in a system. While running an application on the system and checking for expected results, a discrepancy was found on some parts in the values returned by the ALU in the design when shmooing the voltage and frequency of the chip in the system, particularly at high temperature and high voltage. The shmoo showed a reverse speedpath, where the part slowed down as voltage was increased.

Since this failure occurred in a high level application running on a computer system, it was necessary to isolate the particular area of the failure. Once a failing section of code was identified, it was possible to examine the assembly language instructions and data being used in the code and begin experiments to try to find sensitivities to the failure. A small sequence of assembly instructions was discovered that were causing the problem. Once the sequence was isolated, it could be simulated in the RTL model and tester patterns were extracted from the simulation to allow the device to be tested on a tester instead of a system.

Once the failing sequence had been ported to the tester, ODCS was used to find the critical cycle of failure. The failure was noted to occur in the second instruction of a two instruction sequence where the instructions were the same opcode (thus using the same circuit twice in a row). Passing and failing scan data were taken by programming the debug trigger unit to fire when the specific opcode and data sequence used in the second instruction was in the pipeline. The data were then compared to determine what circuit was failing on the given cycle. Shmoos run on various parts with different processing indicated that the failure was much more likely to occur on faster parts.

After examining the scan data to identify possible failing circuitry, and by considering what phase and cycle of the clock the failure was sensitive to, a suspect circuit was identified by examining the ALU circuitry upstream of the failing scanlatches. The suspect circuit is shown in Figure 3-17. This is a dynamic multiplexer that picks between two sets of data d0 and d1 based on the select lines s0 and s1 (s0 selects d0, and s1 selects d1).

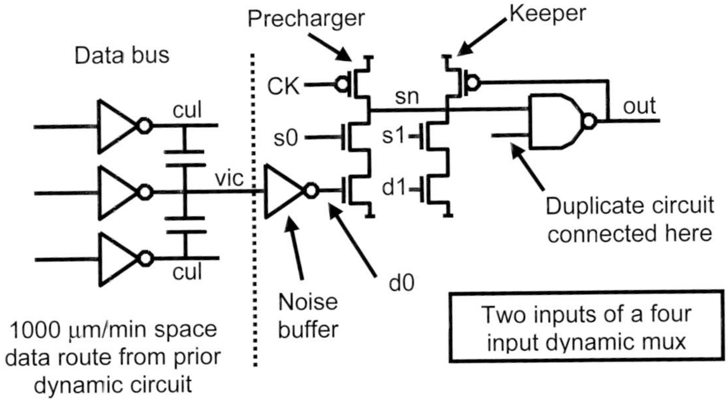

Figure 3-17: Suspect dynamic circuit.

Initial suspicion was that the failure was related to noise due to the fact that high temperature and voltage made the failure worse (high temperature lowers V_{TH}, making dynamic circuits more susceptible to spuriously discharging, and high voltage increases dv/dt, increasing coupling noise). Investigation into pre-silicon noise analysis tool results showed that while the circuit passed electrical rule checks for noise problems, it had only a small amount of margin. Also, it was noted from experimenting with data patterns used in code experiments on the system and tester that the failure appeared to be related to noise induced on neighboring lines of the failing data bit in the ALU; these culprit lines were the results returning from the previous instruction execution. However, simulations done in SPICE showed that noise alone was not enough to cause the observed failure.

The dynamic gate did not have clocked NFETs in the evaluation tree below the s0/d0 and s1/d1 NFETs. This was a conscious tradeoff to gain speed in the circuit. However, due to the lack of a clock to interrupt any existing evaluation during precharge, when CK went low to precharge the gate, if s0/d0 and/or s1/d1 were still high, a drive fight could exist between supplies. This could locally cause the power grid to droop in the vicinity of the circuit (especially since this was in a datapath and had many such circuits in a small area), and could partly explain why two similar instructions in a row activating the same circuit might be needed to cause the failure.

Indeed, when a simulation was performed (as shown in Figure 3-18) it was discovered that in the initial instruction, a drive fight would indeed occur for part of the previous cycle, which would cause local droop on the power grid (the power grid was not modeled in the simulation). The simulation also showed that results returning from the previous instruction could couple to the d0 data line significantly, causing it to spuriously glitch upward during the second evaluation cycle where the failure occurred (even though it was supposed to remain a zero).

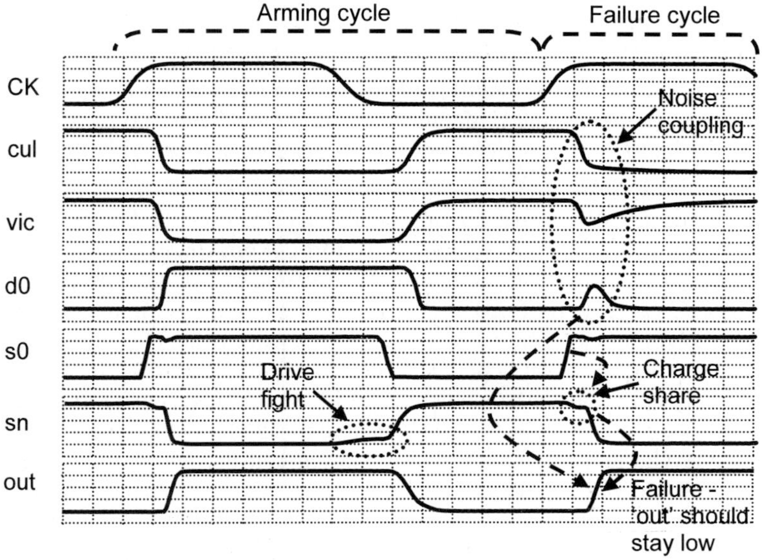

Figure 3-18: SPICE simulation of failing circuit.

Finally, the simulation also showed that during the first instruction execution, the gate correctly evaluated, discharging the sn node through the s0/d0 NFET evaluation tree. During the second instruction execution however, the gate was not supposed to logically evaluate through the s0/d0 path. The initial discharging of the sn node during the first instruction also caused the interstitial capacitance between the s0 and d0 FETs to be fully discharged. This meant that during the second instruction, in order to properly function, the gate had to precharge the sn node, and also was "armed" to allow charge sharing to happen between the sn and s0 nodes if s0 was high during the second evaluation.

As a result of the drive fight on the previous cycle, if the local power grid drooped, the precharge of the sn node on the subsequent precharge cycle could be "weak" since it could not be fully charged. In addition, when the s0 node went high early in the second evaluation phase, charge sharing could further cause the sn node to droop even further. Finally, when the results returned from the previous instruction, the coupling from the result culprit lines to the d0 input was enough for the d0 NFET to turn on enough to discharge the sn node (since it was already in a "weakened" state from inadequate precharge and charge sharing). Note that in evaluation the PFET precharger is off and there is only a small PFET keeper to maintain the high state of the sn node.

These theories for several different possible failure mechanisms all fit the data. In addition, laser voltage probing of the failing circuit showed both the noise event and the charge sharing affecting the sn node, as well as evidence of weak precharging of the sn and vic nodes due to the drooping power grid. Through modifications in test patterns it was possible to isolate the various different contributors to the failure (noise coupling, power supply droop from a drive fight leading to inadequate precharge, and charge sharing) and determine that none of them in isolation was significant to cause the failure. Only when all of them interacted together would the failure manifest itself.

Modeling of the noise and charge sharing in the SPICE simulation showed that the circuit was much closer to failure, but still did not fail without going to extreme conditions that did not match the conditions of failure in silicon. It was believed that inability to model the power supply network accurately made the SPICE simulation look "better" than silicon reality.

Based on the simulation and probing results and theories for failure, a fix was proposed to make the circuit more robust. The PFET pre-charger could simply be tied to ground, converting the circuit from a dynamic circuit to a pseudo-NMOS circuit. This had the benefit of making the gate much more noise immune – without a clock, the PFET precharger would always be on and available to fight against spurious discharging of the sn node due to noise and charge sharing. It also had the benefit of being only a metal fix, which could be much more quickly executed and tested. A focused ion beam edit was performed to try the fix on the most susceptible bit in the datapath, and when completed it showed that the failure was completely eliminated. The edit was then made for all the bits in question in the design database and a new mask was fabricated to fix the problem permanently. One drawback of this was increased power – whenever the circuit evaluated, it would cause a drive fight to occur between the supplies, but this tradeoff was acceptable given the greatly increased noise margin from the fix.

From the initial sighting of the bug to root cause and a fix took about a month. Most of this time was spent isolating the failing code sequence in a system. Once the failing code sequence was examined, it did not take long to hypothesize a failure, propose a FIB to check the theory and implement the fix (perhaps a week). While getting to root cause took longer than a typical bug (which may take less than a day if it is a simple speedpath), this was still excellent progress given the complexity of the bug, which was one of the more difficult ones encountered in this particular design. Several engineers were involved: a pair working on the system to isolate the code sequence, several people writing tests and running shmoos on a tester, and a few design engineers who examined the circuit for possible failures and proposed a FIB to check the hypothesis for failure as well as the final fix.

This example illustrates the difficulty in debug of ensuring that a complex design is free from electrical robustness issues. Tools existed to monitor all three of the contributors to the failure (noise, power droop and charge sharing), but no method existed to consider all of the contributions together. Additionally, the failure was extremely pattern sensitive – in order to create the right combination of local power grid droop in the first cycle followed by the right coupling noise on the second cycle, precise timing of results returning was required in addition to the right values occurring to cause maximum noise on both the power grid and on the victim data line. Such examples indicate a growing need for better analysis tools that can realistically assess the impact of multiple contributors to failure, as well as pattern generation tools that can create worst case patterns to exercise the design.

3.7 FUTURE CHALLENGES FOR SILICON DEBUG

As designs grow in complexity, so does the process of debug. While great strides have been made in the process of silicon characterization, on-chip debug hardware, and failure analysis tools, many challenges lie ahead.

One issue is power dissipation. As CMOS FETs become smaller, they become leakier, with even gate leakage becoming a dominant factor. Leakage power can contribute 30% or more of total overall power in a 90 nm process. In addition, higher clock rates continue to push power upwards as well. Thermal density and the ability to remove heat from the device, particularly in debug situations where devices are deprocessed to facilitate failure analysis tool use (like LVP, PICA and LADA), are becoming difficult issues to solve.

Designs must change to be more power-aware in the future if they are to continue to advance in performance. Since $P = CV^2F$ (where P = power, C = capacitance, V = voltage and F = frequency) for dynamic power in CMOS designs, dynamic power can only be modulated through reduction in capacitance, voltage, or frequency. Leakage power can be modulated through process techniques but only at the cost of reduced transistor performance. Even FET gate leakage is becoming a problem.

Voltage reduction clearly offers the greatest benefit in reducing power, but voltage scaling is reaching its limits, and frequency continues its ever upward trend. Designs of the future will need to dynamically control their voltage and frequency environments to control power, which will create difficult problems for debugging. Devices that change frequency and voltage can lead to asynchronous, unrepeatable behavior. Such behavior could become a nightmare to debug.

Another problem is external interface speeds. Automated test equipment (ATE) is barely able to keep up with today's high pin count, high frequency devices. ATE will have to undergo a transition in the future to support more on-chip DFT features for testing, as it will simply become economically impractical to support high-speed test with accurate parametric control. Such issues may lead to more reliance on custom test solutions and DFT embedded within devices that allow them to be tested and debugged in the systems that they were designed for [8].

Reliability of circuit designs over process variation in leading edge processes is a growing concern. Variation in transistor performance across even a single die can cause robustness issues if not carefully planned for. Thermal variation in clock networks across a die must be taken into account to lower clock skew to acceptable levels. The ability to actively "tune" clock networks for such variation will be needed to support future debug. The ability to internally determine signal timing through debug hardware (e.g. on-chip timing analyzers or even oscilloscope-like functionality) will become increasingly important for debugging.

The addition of on-chip debug hardware will become increasingly important as designs become more complex, particularly in the areas of high-speed interfaces and "system on a chip" (SOC) devices. One area for debug hardware innovation is in the active injection of various errors in hardware to explore how the design reacts to failures that may be infrequent in real application and thus very hard to validate during debug. This may expose design mistakes that might not otherwise be visible without a specific set of conditions. Other opportunities lie in adding inexpensive circuitry to monitor things like on-chip power droop, device variation across the die, and on-chip temperature variation.

Additional work is needed in tools to support the identification of "sensitive" areas in designs and to target testing at these areas. This is similar to the process of ATPG (automated test pattern generation), but is more difficult since it must consider aspects of parameters such as signal timing, voltage variation, noise margin in circuits, etc., to determine how to expose weaknesses in the design. Another possible direction along this line is the addition of special "watcher" circuits that check activity of signals on the device and indicate if they have been toggled during validation, or if errors occur on the signals from what was expected. Such specialized extra circuits for debug may also be useful for monitoring things like process variation across a die.

3.8 CONCLUSION

The evolution of integrated circuits is nothing short of astonishing. Over a few decades devices have gone from handfuls of transistors to over a billion, and frequency of operation has increased incredibly over the same period. This great rise in complexity challenges the engineers who design and debug these devices. Many innovations have been made in on-chip debug hardware, probing methods, and debug infrastructure to address this growing complexity. Additional innovation will be required in the area of silicon debug over the coming years to keep pace with the amazing progress being made in the design of ever more complex devices. In addition, failure modes are becoming more complex as multiple failure modes interact to cause design failures. The role of the debug engineer has never been more secure!

ACKNOWLEDGEMENTS

The authors wish to thank Jason Stinson, a Principal Engineer at Intel Corporation, for providing several of the figures, as well as for a valuable review of the content of the chapter, and editor Dimitris Gizopoulos for his careful review and editing of the chapter. The authors are also indebted to numerous colleagues at Hewlett-Packard, Intel, IBM and Philips for their valuable indirect contributions to the content of this chapter through discussions and interaction with the authors during the debug of numerous designs and participation in industry conferences.

REFERENCES

[1] D. Josephson, "The Manic Depression of Microprocessor Debug", *Proc. IEEE International Test Conf.*, pp. 657-663, 2002.

[2] K. Baker, J. V. Beers, "Shmoo Plotting: The Black Art of IC Testing", *IEEE Design and Test of Computers*, Vol. 14, No. 3, pp. 90-97, July/September 1997.

[3] J. Bockhaus, R. Bhatia, C. M. Ramsey, J. Butler, D. Ljung, "Electrical Verification of the HP PA8000 Processor", *Hewlett Packard Journal*, pp. 32-39, August 1997.

[4] H. Balchandran, et al., "Facilitating Rapid First Silicon Debug", *Proc. IEEE International Test Conf.*, pp. 628-637, 2002.

[5] B. Vermeulen, et al., "Core-based Scan Architecture for Silicon Debug", *Proc. IEEE International Test Conf.*, pp. 638-647, 2002.

[6] X. Gu, et el., "Reusing DFT Logic for Functional and Silicon Debugging Test", *Proc. IEEE International Test Conf.*, pp. 648-656, 2002.

[7] C. Pyron, et al., "Silicon Symptoms to Solutions: Applying Design-for-debug Techniques", *Proc. IEEE International Test Conf.*, pp. 664-672, 2002.

[8] T. Litt, "Support for Debugging in the Alpha 21364 Microprocessor", *Proc. IEEE International Test Conf.*, pp. 584-589, 2002.

[9] D. Josephson, et al., "Debug Methodology for the McKinley Processor", *Proc. IEEE International Test Conf.*, pp. 451-460, 2001.

[10] T. Wood, "The Test and Debug Features of the AMD-K7™ Microprocessor", *Proc. IEEE International Test Conf.*, pp. 130-136, 1999.

[11] A. Kinra, et al., "Diagnostic Techniques for the UltraSPARC™ Microprocessors", *Proc. IEEE International Test Conf.*, pp. 480-486, 1998.

[12] A. Carbine, D. Feltham, "Pentium® Pro Processor Design for Test and Debug", *Proc. IEEE International Test Conf.*, pp. 298-299, 1997.

[13] M. Bass, et al., "Design Methodologies for the PA7100LC Microprocessor", *Hewlett Packard Journal*, pp. 23-35, April 1995.

[14] J. Stinson, S. Rusu, "A 1.5GHz third generation Itanium 2 processor", *Proc. of Design Automation Conf.*, pp. 708, June 2003.

[15] M. Paniccia, et al., "Novel Optical Probing Techniques for Flip Chip Packaged Microprocessors", *Proc. IEEE International Test Conf.*, pp. 740-747, 1998.

[16] J. Tsang, et al., "Picosecond imaging circuit analysis", *IBM Journal of Research and Development*, Vol. 44, No. 4, 2000, pp. 583-603.

[17] D. Knebel, et. al., "Diagnosis and Characterization of Timing-Related Defects by Time-Dependent Light Emission", *Proc. IEEE International Test Conf*, pp. 733-739, 1998.

[18] M. Bruce, V. Bruce, "ABCs of Emission Microscopy", *Electronic Device Failure Analysis*, pp. 13-20, Volume 5, Issue 3, August 2003.

[19] T. Eiles, et al., "Critical Timing Analysis in Microprocessors Using Near-IR Laser Assisted Device Alteration (LADA)", *Proc. IEEE International Test Conf.*, pp. 264-273, 2003.

[20] A. Chandrakasan, et al., *Design of High-Performance Microprocessor Circuits*, Wiley-IEEE Computer Society Press, 2000, ISBN: 078036001X.

[21] K. Bernstein, et al., *High Speed CMOS Design Styles*, Kluwer Academic Publishers, 1998, ISBN: 079238220X .

[22] F. Schellenberg, "A Little Light Magic [Optical Lithography]", *IEEE Spectrum*, pp. 34-39, Volume 40, Issue 9, September 2003.

[23] W. Huott, et al., "The Attack of the 'Holey Shmoos': A Case Study of Advanced DFD and Picosecond Imaging Analysis (PICA)", *Proc. IEEE International Test Conf.*, pp. 883-891, 1999.

Chapter 4

Delay Testing

Adam Cron

4.1 INTRODUCTION

4.1.1 Why Delay Testing

For as long as chips, boards, and systems have been designed, they have been put together to run at a targeted rate of speed. In the past, delay testing has insured that they did work at that rated speed. But today, delay testing is also used to detect other defect types. Delay testing methods lead to the detection of manufacturing defects that are difficult to detect with other means because they affect the performance of a circuit and not its basic logical behavior.

This chapter explores how delay tests are generated and used in the chip manufacturing process. It also provides background on why these tests are important for the detection and removal of defective components in the manufacturing flow. The reader should consider this chapter an introduction to delay test methods used today and supported in the design and test synthesis and test generation tools offered by Electronic Design Automation (EDA) companies.

D. Gizopoulos (ed.), Advances in Electronic Testing: Challenges and Methodologies, 109-139.

4.1.2 Why Now

With the advent of Moore's Law [1], it is a forgone conclusion that technology will shrink, allowing more circuit functionality to occupy a smaller and smaller footprint on the semiconductor die. As we enter into the age of deep and very deep sub-micron design, different defect types and populations are plaguing the manufacturing process. For example, to improve the speed characteristics of the metal layers of these chips, copper has replaced aluminum as an interconnect medium. The ITRS (International Technology Roadmap for Semiconductors) [2] points out that 10 levels of metal may be the norm for the 90nm technology node. *Copper voids* are a typical manufacturing defect causing a slow-down in signal transmission. Only a delay-based test can detect such defects.

Another defect type is actually design related (copper voids are manufacturing related defects). With more wiring in chips comes more potential for *crosstalk*[1]. In fact, coupled with the edge rates of today's design styles, timing-based defects could well be on the rise. Crosstalk causes both speed-ups and slow-downs of signal propagations. Only a delay-based test [3] can detect these issues, if not routed out in the design process.

Leakage currents are also on the rise as geometries shrink. I_{DDQ} testing, once an indicator of possible wafer defects, now has many design and test tricks associated with its use in order to keep it viable in the deep sub-micron age. The difference in leakage currents between a working component and one containing a short, for example, is very small and difficult to measure. Without serious consideration in the design process for I_{DDQ} implementation, delay testing has, of late, become more important to help detect and filter defective parts in the factory.

Despite the long list of defects that delay tests *cannot* catch [4], [5], there is plenty of proof that delay testing *can* detect many defect types [6], [7] including pure delay defects [8], gate oxide shorts [9], unfilled and resistive vias [10], resistive opens and shorts [11], bridges [12], opens [13], high leakage circuits [14], and random particle defects [15]. More specific data is available, but this should give the reader a good start and taste for the potential of *at-speed testing*[2]. More information can be found in the conclusion to this chapter.

4.2 DELAY TEST BASICS

A delay in a circuit is defined as a failure for signals to propagate from a source node to a destination node within the specified time. If a typical source node is a sequential element like a flip-flop, and a typical destination node is also a flip-flop, then it is easy to see that the "specified time" is the time between clock pulses. The first clock pulse, typically called the *launch pulse* or *release pulse*, begins the propagation of the signals through connecting logic to the destination. The second

[1] Crosstalk is the undesired electrical coupling between adjacent circuits in a system.

[2] Testing a device at its rated operational frequency.

clock pulse, called the *capture pulse*, defines the time at which these signals should have completed their function, and arrived at the destination sequential element.

Figure 4-1 depicts a typical group of logic for illustration. We will use this Figure to define some additional terminology.

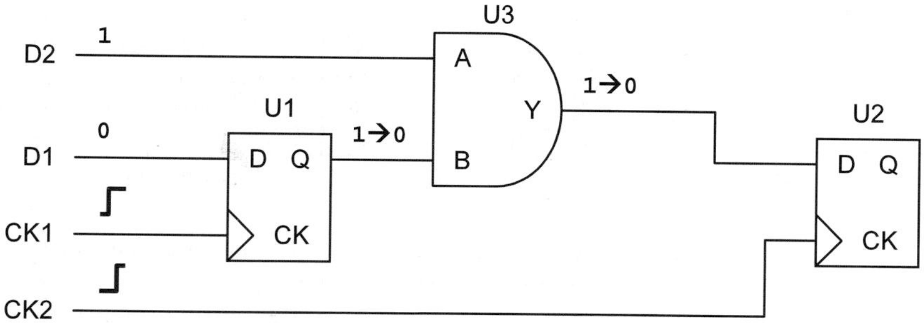

Figure 4-1: Delay test basic illustration.

In Figure 4-1 CK1 is called the launch clock, as mentioned earlier. U1/Q is called the launch node. It is, in this case, initialized to one (1). This one along with any other initial constraints is sometimes referred to as V1 or T1. It is the initial state of the signals involved in the test prior to the launch clock. D1 is sensitized to cause a transition to occur at U1/Q upon arrival of the launch clock, CK1. D1, then, is set to apply V2 or T2. After the launch clock is applied, an *at-speed time*, or *delay*, is made before the application of the capture clock, CK2. The transition arrives at the capture node, U2/D. If the zero (0), in this case, is captured into U2 by CK2, then the test passes. If a one (1), in this case, is captured into U2, then the test fails. Note that for the transition to propagate through the combinational logic, U3, a one (1) must be held on U3/A by setting D2 to one (1). U3/A is called an off-path or side input. These are signals that may serve to sensitize the path of interest, but are not actually *on* the path of interest.

A passing test indicates that there is no defect causing a delay larger than the cycle time of the path. A failure of the test indicates a defect resides along the path. These defects are modeled as faults. A *slow-to-rise* (str) fault is a failure for a low-to-high transition to propagate from the launch node to the capture node in the specified time. A *slow-to-fall* (stf) fault is detected if the high-to-low transition fails to propagate between the launch and capture clocks (see Figure 4-1).

Using the context set up in Figure 4-1, Figure 4-2 shows the timing of the pattern application. Primary inputs and sequential elements have been sensitized to V1 prior to the launch event. V2 is also set up behind the sequential elements, with D1 being set to a 0. After the rising edge of CK1, the transition is launched down the path. The capture event on CK2 will detect a delay along the path if the propagation delay is larger than the time between the launch and capture events. Notice that clock edge placement and skew have an impact on t_{delay}. This and other timing issues are discussed later in Section 4.4.

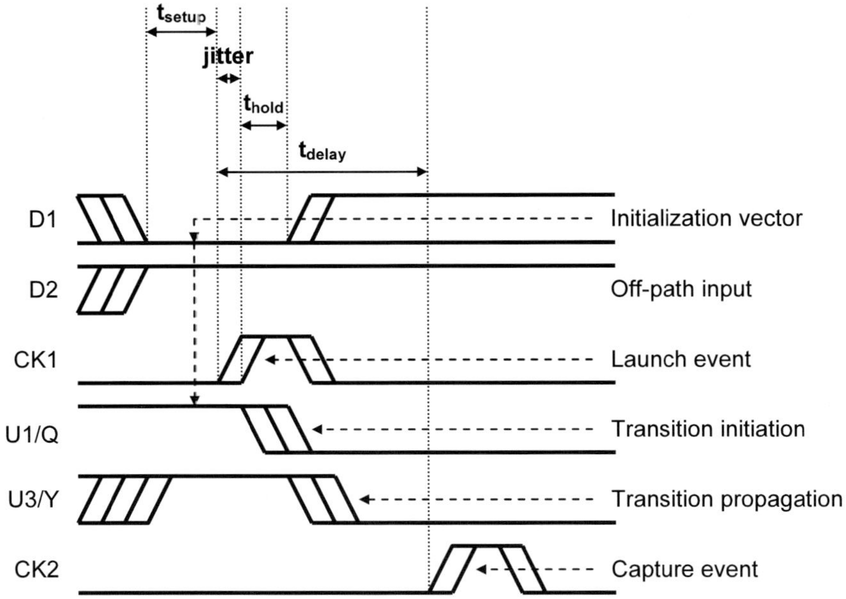

Figure 4-2: Delay test pattern waveform.

The path shown in Figure 4-1 is an example of a flop-to-flop path. There are many other path topologies to consider: input-to-flip-flop, input-to-output, flip-flop-to-output, flip-flop-to-memory, etc. In later sections, these topologies will be touched upon in an attempt to point out issues related to them.

Pure combinational paths are rare these days as higher levels of integration proliferate. On occasion, a path that begins at the inputs of a device and terminates at the outputs of the device will be encountered. Section 4.7 discusses issues arising while using Automatic Test Equipment (ATE). Specifically, relying on data signal edge placement or output strobe accuracy may be asking too much of the ATE, especially for high-speed delay measurements. A separate chapter of this book provides more details on this topic.

Still, there are valid reasons to use input and output pads during the application of delay tests. Device characterization and I/O characterization will typically involve paths that either begin or end with a device pin. These paths might begin with a transition on an input pin, as shown in Figure 4-3, and terminate at a core-level flip-flop. Likewise, a path transition might begin at a core-level flip-flop and terminate at a device-level pin.

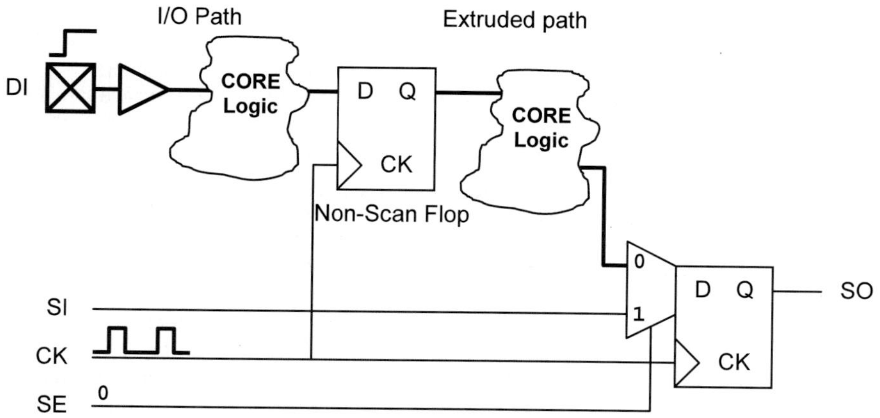

Figure 4-3: I/O path and extruded path.

Extruded paths begin or end at non-scan elements. These elements might be non-scan flip-flops, latches, or memories. For an Automatic Test Pattern Generation (ATPG) tool to craft such a delay test, it must be able to create sequential patterns, and it must have functional models of the non-scan elements. Effectively, the ATPG tool will pass the V1 and V2 pattern data from primary inputs and scan elements on to the non-scan elements for path sensitization. Then, after the launch and capture events, the results may need to be moved from non-scan elements to scan flops and primary outputs for extraction via normal scan techniques.

In the case depicted in Figure 4-3, the path begins at an I/O pad and ends at a non-scan element. This path is both an I/O path and an extruded path. The launch event, V1, will emanate from the ATE. In the Figure, this event is a zero-to-one transition on the `DI` input. After waiting the correct amount of time, the capture clock, `CK`, pulses to trap the result of the transition, V2, into the non-scan flop. In order to observe this response data, the result in the non-scan flop must be transferred to the scan flop. In this case, this is accomplished by another pulse of `CK`, but this second pulse need not be at speed. Once the captured response is safely in the scan flop, a regular shift cycle can extract the data to the ATE for validation.

There are many ways to craft a delay test. Table 4-1 lists some methods and fault models used to generate delay tests, along with some pros and cons. We will delve into two of these methods in the rest of this chapter: the transition delay fault model and the path delay fault model. It is important to remember that the emphasis of this chapter is a practical application of delay testing. To be practical, support must be available in production EDA tools to aid design synthesis, test generation, and defect diagnosis. Not all methods can claim this level of support. Yet research is available for many facets of delay testing. And, as the needs of the electronics industry change and evolve, new solutions to old and new problems will develop.

Method	Definition	Pros	Cons	Use
Functional	"Hand made" pattern set exercising the true function of a component	- Uses real function - May leverage existing simulation pattern sets	- Difficult to construct - Difficult to measure coverage - May not integrate well with ATE	- Speed binning - Minimum specification determination and validation
Path Delay [22]	Pattern exercising a specific circuit path	- Path of transition well defined - Good for distributed / small defect detection	- Exponential fault list growth	- Speed binning - Process monitoring - I/O characterization
Transition (Gate, Line) Delay [22]	Pattern exercising a specific circuit node	- Good for gross defect detection - Easier test generation - Linear fault list growth - "Geographically" diverse coverage	- May not address subtle delay defects	- Broad defect coverage - Process monitoring
Segment Delay [16]	Pattern exercising a smaller segment of a path	- More linear growth of fault list compared to path - Easier test generation	- No tool automation support - Benefit not well studied	- Impractical solution without automation

Table 4-1: Delay test generation methods and fault models.

4.2.1 Transition Delay Basics

The transition delay fault model places the faults (str and stf) at inputs and outputs of cell primitives (leaf cells in the technology library) in the design. As defined earlier, the str and stf faults, then, become the targets of the pattern generation tool. As you might have noticed, these fault sites are the same sites used for stuck-at ATPG. If you consider a slow-to-fall fault as being stuck-at-1 for a period of time, then you can begin to imagine how pattern generation tools can build tests using the transition delay fault model quite easily. In practice, test engineers target as many faults as possible using the transition delay fault model. However, there are many issues regarding transition delay pattern generation and test quality that will be discussed in Section 4.4. It is also important to note that the transition delay fault model is very specific with respect to the fault sensitization and propagation: the initialization vector is only sufficient to sensitize the transition. The capture pattern need only be sufficient to propagate the transition. There are no requirements for test robustness, as we will see for the path delay fault model.

4.2.2 Path Delay Basics

Like transition delay faults, the path delay fault is based on a logic transition propagating along a circuit path. These faults are either slow-to-rise or slow-to-fall faults, too. But, unlike the transition fault, path delay faults include the entire path through which the transition will propagate.

Since the quantity of "paths" through a device tends to increase exponentially according to device size, it is impractical to try to test all paths. Instead, a selection of paths is typically tested. The trick, then, is choosing paths that will provide enough visibility to meet the targeted quality goals. Several papers [6], [28], [26], [30], [15] suggest that path delay testing is meant for either producing a *process quality metric*[3] or for *speed binning*[4].

There are different classifications of a path delay test. A good discussion of these can be found in [17], but some basic ideas and definitions are presented here. A *robust* test is one in which the captured value of the transition is independent of delays off the target path. Figure 4-4 shows the values of off-path inputs during the application of the delay test. For the case of a falling transition along the path to an AND gate, for example, the off-path input must be a `steady` (`S`) logic `one` (`1`). A *non-robust* test may be dependent on delays of the off-path inputs. Figure 4-4 shows an AND gate with a falling transition test. The off-path input, however, shows a rising transition. The delay being measured at the output, `Y`, of the AND gate, then, is dependent on the delay characteristics of the off-path input, `B`. If the off-path input on pin `B` is late, then it will mask off the transition being measured at input `A`, and a "passing" result will be measured. There are also different qualities of test. For example, a *hazard-free* test is one in which the possibility of glitches along off-path inputs is eliminated. Figure 4-4 shows the potential for a glitch from the AND gate feeding the off-path input, `B`, of the OR gate during a test. Tools are also available which can produce tests during which all off-path inputs are always static. These might be used for *device characterization*[5], *process characterization*[6], or *improved test diagnostic accuracy*[7].

---

[3] A measurable parameter indicating the relative goodness of a process when compared to a benchmark

[4] Speed binning is a technique used to separate manufactured parts into collections that run at different speeds. For example, a processor may fail delay tests run at 200MHz but pass at 175MHz. This processor could be sold from the 175MHz bin.

[5] Device characterization might include determining operational parameters of a specific chip such as propagation delays or device speed.

[6] Process characterization might be done on test wafers to determine technology parameters or library characteristics such as performance.

[7] In the general case, off-path inputs might change during any given test, increasing the ambiguity between the actual path being tested and those defined by changing off-path inputs during diagnosis of failing test patterns.

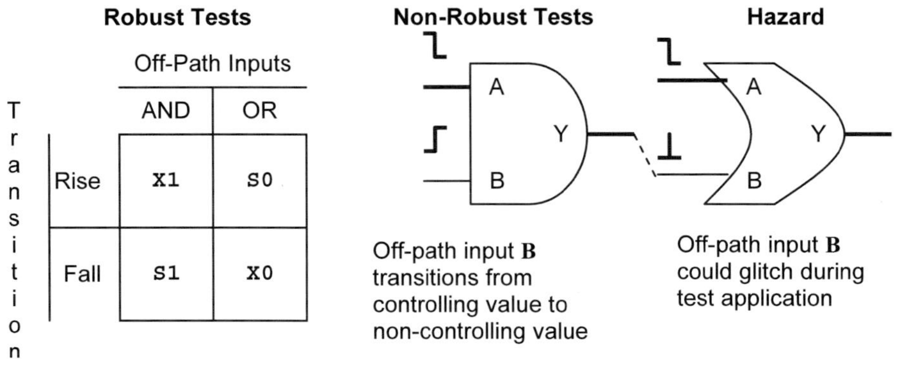

Figure 4-4: Path test classification and quality [8].

4.3 TEST APPLICATION

This section details how delay tests are actually applied to a design. Practical issues are discussed so the reader can make good judgments before proceeding down a particular design and ATPG course.

4.3.1 Scan Architectures

The Figures shown thus far have depicted sequential elements as simple flip-flops. In practice, however, designers will employ scan-based Design-for-Test (DFT) strategies to enable Automatic Test Pattern Generation (ATPG). The scan chain insertion will typically be based on either a multiplexed flip-flop design or a level-sensitive scan design (LSSD) latch. Also typical is to apply a "full scan" approach: all flip-flops are scanned. The controllability and observability that a full-scan application affords the ATPG tool far outweighs any extra area or speed penalty of this DFT style.

However, if a design is generated with only partial-scan (not full-scan), then tools are available to still make use of the facilities provided in the design. In this case, V1 (the initial vector), and V2 (the capture vector) are set up with sequential ATPG techniques. The delay test and its result, then, may be applied and extracted, respectively, with full-sequential techniques[9].

4.3.2 Last-Shift-Launch

As we have seen, delay tests require a transition in logic value to propagate across the circuit path of interest. This transition might initiate from a change on a primary input that the test equipment might toggle. Or, a launch clock might initiate the

---

[8] X0 means off-path input is don't care in V1 vector and 0 in V2 vector, X1 means it is don't care in V1 vector and 1 in V2 vector, while S0 and S1 mean stable 0 and stable 1, respectively.

[9] Multiple clock cycles may be required to apply V1 and V2 or extract the test results.

transition. A trick that is used sometimes is to actually use the last shift operation of the scan chain load to initiate the transition. As the last bit in the scan chains is loaded, the transition begins to launch through the circuit of interest. Then, the scan enable signal (SE) switches to enable the functional capture to occur. By the time the capture clock arrives, the scan enable signal must have propagated to all the scan elements involved in the capture operation. This feature is sometimes referred to as an "at-speed scan enable".

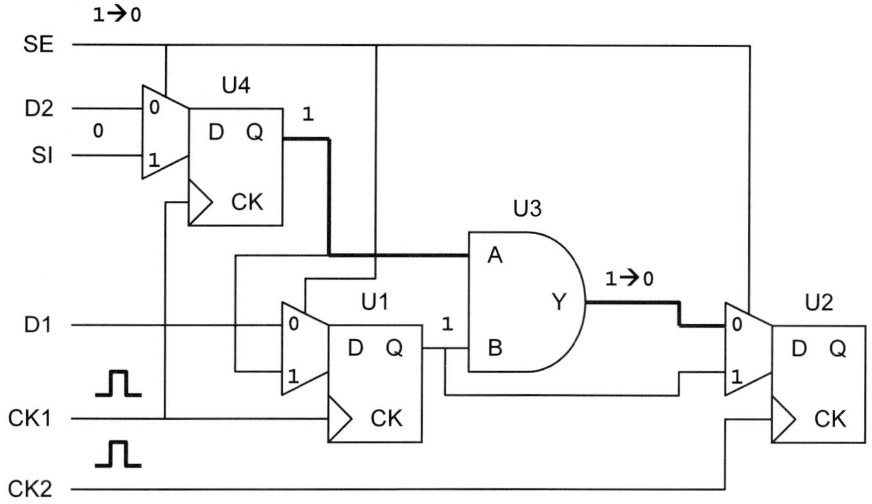

Figure 4-5: Last-shift-launch delay test.

Figure 4-5 depicts a *Last-Shift-Launch* (sometimes called *Skewed Load*) scenario. In this case, U1 and U4 have 1s loaded during shift. The last shift will load a 0 into U4. This 0 is the last shift of the scan chain applied when the scan enable signal, SE, is active (1). Upon shifting the last bit, the SE signal must transition to its inactive state (0) as fast as the functional logic before the capture clock, CK2, is applied. And here-in lies the major design challenge for chip designers: the scan enable signal must transition at-speed.

The scan enable signal is typically a heavily loaded signal. For simple stuck-at testing, "dummy" vectors can be inserted before an actual test vector to allow this signal time to transition. In delay testing, if this signal does not transition in time for the capture event, then the wrong side of the mux will be captured. But for a Last-Shift-Launch delay test, no "dummy" vectors may be inserted because they will take away from the at-speed nature of the delay test. Several solutions permeate the industry. The most prevalent is to layout the scan enable signal using classic clock-tree methodologies. As long as the scan enable signal can transition within the at-speed constraints, and still allow sufficient set-up time for the functional path to settle at the capturing end of the path, then a Last-Shift-Launch test may be applied. Another solution is to add pipelining stages to the scan enable signal. These pipes produce the same effect as a clock tree, but their effect is delayed synchronously by one or more clock cycles rather than variably with buffers. In practice, one or more pipeline stages are added to the scan enable signal such that all circuit flip-flops receive equally piped scan enable signals. Many times, the loads on the pipes are balanced and their placement physically optimized in a manner similar to that used in clock tree synthesis. Figure 4-6 shows a fairly complete two-stage pipeline structure for scan enable that can be used to feed positive and negative edge-triggered flip-flops. Note that for Last-Shift-Launch scenarios, the OR gates cause the core-bound flip-flop scan enable signal to switch immediately, without having to wait for the pipeline stages, in order to move from capture to shift mode, again.

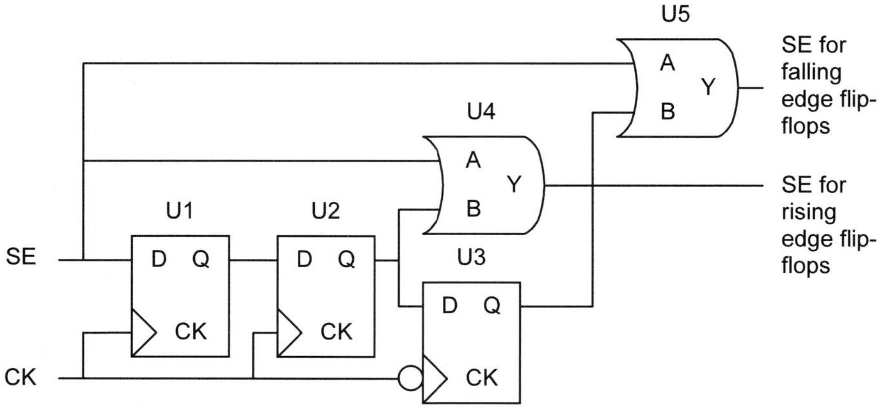

Figure 4-6: Pipelined scan enable signal.

Before such an elaborate solution as pipelined scan enable signals is implemented, it is wise for a design team to validate the need for such a feature. Whereas many designs do get higher fault coverage and lower pattern counts when using a Last-Shift-Launch test application strategy, not all designs benefit in this way. Sometimes the benefit is too small to matter, or not worth the extra design trouble to implement such a design feature. Physical tools are just beginning to automate the placement of such features. Also, ATPG tools must be able to understand this design style for all fault models.

Another interesting side effect of using the Last-Shift-Launch test application methodology is the following: any delay defect on the scan enable signal will result in a defect detection and the part failing on the tester. In other words, a yield penalty will be taken (functionally correct chips will be rejected) for defects in the scan enable circuitry. Clearly, this logic has no bearing on the actual functional speed of the device. To reiterate, there are serious ramifications to choosing the Last-Shift-Launch mechanism for applying delay tests.

4.3.3 System-Clock-Launch

System-Clock-Launch (or *Clock-Launch*, or *System-Launch* or *Broadside*) delay test application comes with few of the design constraints imposed by a Last-Shift-Launch test application strategy. After V1 is scanned into the design (at a slow speed, typically), the scan enable signal can switch at a slow pace. V2 is set up by functional justification. That is, the signal values are created by values rippling through the functional logic from primary inputs and scan flip-flops.

However, this test application methodology is not a panacea. Designs with cross-clock-domain communications, multi-cycle, and false path design characteristics will need special consideration when creating delay tests. Section 4.4.1 details solutions to these issues. In fact, all methods of test generation and application techniques will need to carefully consider timing exceptions before being applied to a design.

4.3.4 Hybrid Launch

A recent novel proposal [18] splits the scan enable drive between those flip-flops that receive the scan enable signal directly and those that receive a re-timed version of it through pipeline registration. This gives some flip-flops better controllability due to their ability to perform a Last-Shift-Launch, while not having to design this capability for all flip-flops in the design. An analysis of controllability similar to SCOAP[10] is performed to determine which flip-flops should receive the direct signal, and which get the re-timed signal. This architecture is not supported directly by EDA tools, but could be supported by custom flow development.

4.3.5 BIST and Delay Testing

A recent surge in the application of logic Built-In Self-Test (BIST) architectures demands a look at the pros and cons of these techniques. BIST applications popular today can be divided into two basic application types: Random Logic BIST (LBIST) and Deterministic Logic BIST (DBIST).

With LBIST architectures, a large number of clock cycles are required to generate patterns with sufficient coverage for the fault models of choice after a single initialization. Advantages of this type of application are the near-zero vector data required to run the test. DBIST, on the other hand, relies on a small amount of externally applied data that is expanded on-chip, and compressed prior to exiting the device under test. Tester time for DBIST should be smaller and coverages higher, when compared to LBIST applications.

Three interesting notes should be raised, however, when discussing BISTed designs: data compaction, fortuitous (or unmodeled) defect detection, and “at-speed” LBIST.

One of the most fundamental reasons to adopt a BIST architecture is for pattern compaction and tester time reduction. Certainly, coverage should not be sacrificed when using these techniques. When comparing data volumes of stuck-at test pattern sets with those of transition delay tests, transition delay pattern sets might be 2-6 times the size for top coverage goals. So, a means to compress this data and lower test application times are required to enable a steady or lower test cost structure. This might come from the ability to reuse existing ATE infrastructures or by simply lowering test times. LBIST tests, however, might not run in a shorter amount of time when compared to a regular scan test. This is due to the random nature of the data

---

[10] Sandia Controllability/Observability Analysis Program.

applied. In fact, test times might be longer for random BIST architectures. Coverage can also suffer without near-exhaustive test pattern sets or proper application of test points (control and observe nodes) within the circuit. One goal of synthesis, then, might be to generate circuit structures that are random pattern testable, if LBIST architectures are to be used.

BIST applications also typically require a higher number of patterns to be applied when compared with traditional scan test application approaches. With these larger numbers of vectors come the benefits of multiple sensitizations of faults. It has been well documented that simply the application of more patterns can result in fortuitous defect detections. This concept also applies to using multiple fault models on every design. Again, the more patterns applied, the more likely a defect is detected [8].

An often heard, but not well understood notion is that "LBIST applied at speed is the same as delay testing." As we have seen in the waveforms of Figure 4-2 and discussed in previous sections, it is required that a *transition* be launched and captured within the cycle time of the path. Assuming that the LBIST is relying on a Last-Shift-Launch mechanism (which may or may not be true), in order for a transition to occur, the launching cell must have contained a value that the last shift inverts. A complete fault simulation must be run to guarantee that this is the case.

4.3.6 Philosophy and Delay Test Application

Now that the predominant test application methodologies of Last-Shift-Launch and System-Clock-Launch have been presented, a short discussion about the philosophies of using each should be noted. In particular, ease of design and ease of pattern generation have been contrasted. But the implications of using scan as a method for delay test application has its own detractors.

The main issue facing test application with scan-based architectures revolves around functional verification, defect detection and product yield. To illustrate the point, the question is posed: should a defect in a piece of sequentially redundant logic indicate a failure if detected with a scan-based test? If the defect will never be sensitizable during normal system operation, should we allow it to affect product yield by failing the device during manufacturing test? And here-in lies the problem [19].

Some say that, due to scan architecture's inherently excellent controllability and observability, test generation can proceed and apply tests to paths normally unable to be sensitized in the design. On the other hand, some say that a defect is simply an indicator of a bad manufacturing process, and should be detected and filtered from the customer at all costs; or at least used as an input to process quality improvement. Such an over-application of test as a factory manufacturing filter is sometimes called over-kill (or over-testing). The philosophy one chooses to adapt may depend on many things. For example, if a chip is headed for a high-volume consumer application destined to be obsolete in one year, then failing parts due to over-kill may be too expensive. On the other hand, a team developing a part that will be deployed in a space exploration application may accept the notion that a failure of non-functional paths is an indicator of worse things, so may be more accepting of the yield loss due to over-kill.

4.4 DELAY TEST DETAILS

Delay test generation and application are fraught with issues involving clocks and signal timing. This section covers the issues that seem to crop up in the industry, today, and some of the useful features in pattern generation and other EDA tools that attempt to resolve these issues.

4.4.1 Clock Domain Issues

Different product types and design styles interact to cause varying troublesome environments for delay testing. The root of many of these problems stems from clocks and their control in a design. This section illustrates some of these issues.

The first set of issues is a simple ramification of design style. Some designers allow the use of falling edge-triggered flip-flops (negative edge domains) to be synthesized into the design. Regardless of their design necessity, they create an environment where-by delay test application is not carried out across two tester periods, but between the rising and falling edges of a single clock applied in a single tester period. Sometimes, functional lock-up latches are used between clock domains so that the delay across or between two clock domains is guaranteed by design. Other design styles use latches and a time-borrowing or cycle-stealing scenario using both levels of the clock to achieve timing goals that complicate timing analysis and delay path extraction. The reader is cautioned that these "rising/falling" environments are designed to work within a particular pulse-width of the clock. Due to the way many ATE timing protocols are generated, these clock waveform requirements might not be maintained between the design and test EDA environments. So, if delay tests are to be applied to circuits of this nature, the test engineer should check the clock waveforms specified in the ATE protocol and pattern files to be sure they meet the input specifications required of these waveforms.

The next set of issues involves single or multiple clock design environments. In particular, some areas of a design are timing critical and some are not. For this reason, sometimes designers specify to design synthesis tools that a part or section of the design is not timing critical, so that the tool will not waste undue amounts of time optimizing that part to meet some specific timing. These sections of a design are called *false paths*. Similarly, a designer might realize that the time it takes for a signal to propagate from one sequential element to the next will take more than one system clock cycle. Therefore, a designer might specify that these signal paths are *multi-cycle paths*. Both of these signal or design topologies present challenges for a delay test. Figure 4-7 illustrates multi-cycle paths, false paths as well as true paths[11].

[11] A true path is a timing critical path.

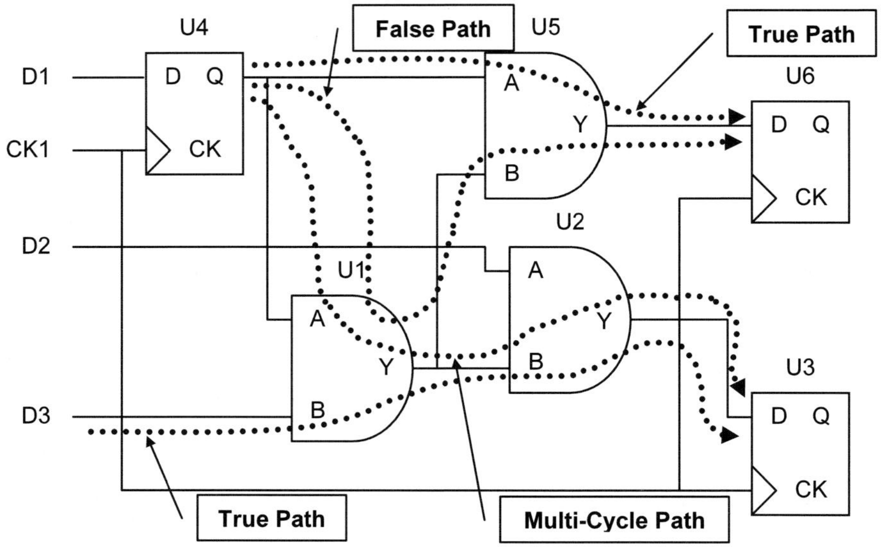

Figure 4-7: Multi-cycle and false paths.

Figure 4-7 depicts a few different paths that the designer thought should be handled in different ways. In a synthesis environment, the designer might tell the tool to treat the path from `U4` through `U1` and `U2` to `U6` as a false path. That is, the designer feels that the synthesis tool should not "waste" its time trying to close timing on this path. Likewise, the designer might feel that the path from `U4` to `U3` is a multi-cycle path. That is, the signal propagation might take more than one clock cycle to traverse this path. In these cases, delay tests need to be crafted which do not try to test these paths at the highest clock speeds, but rather at a speed corresponding to that of the path (or two clock cycles long, for example).

Recall that delay tests launch a transition in one cycle and capture a result in the next. If a path is not designed to run at this speed, then the transition is not guaranteed to arrive by the time the capture event arrives. If the pattern generator is not aware of this "design deficiency", then many improper delay tests will be generated causing failure on the tester. This issue effects both transition and path delay tests. While generating the test for one fault, all the other flip-flops in a design are simulated so that the capture results for these can be loaded into the pattern at the same time. This simulation, by default, will assume that the paths through which all these circuit signal transitions propagate are designed to run at the speed of the applied launch and capture clocks. To compensate for this problem, ATPG tools have several mechanisms that may be applied to resolve these timing-driven issues.

For transition delay ATPG, one feature is the ability to mask[12] the capture value of a flip-flop. The second is the ability to define a flip-flop as driving the head of a

[12] Masking out a bit in a test pattern is achieved by applying an `X` to the expected value in the test pattern.

slow path set. Unfortunately, there is typically no way to stipulate the slow path, in particular. That is, all of these specifications apply to the flip-flops all the time, and not to the enumerated elements in the path. As such, they are typically a pessimistic solution to the problem, and will affect overall test coverage. But it is one of the only supported solutions in existence today. Specification of the slow flip-flops and masked flip-flops falls on a quality static timing analysis tool to determine these constraints. A static timing analysis tool can be used to analyze the design in the context of the test mode being used to apply delay tests, including test mode constraints and clock waveforms. Once the timing analysis is complete, the constraints can be delivered to the ATPG engine. The solution's pessimism and resulting coverage loss is illustrated by looking at Figure 4-7 and considering a mask on the false path. This same mask will also mask the true path to `U6`.

For path delay ATPG, there are ATPG tool features that allow the masking of all flip-flops not participating in the collection of valid path delay fault results. So, any flip-flop that is not at the tail of a path delay fault specification can be masked on a pattern-by-pattern basis. Such a solution will not cause a degradation in test quality.

Both of the solutions just defined are designed to minimize false failures on the tester and save yield. These solutions revolve around known issues in the design. But some issues are not known in advance, but may still effect the quality and reliability of the test. For example, paths that begin in one clock domain and end in another can cause trouble when delay tests are applied from the ATE. Such traversals of signals from one clock domain to another are called *inter-clock domain communications*, or *domain crossings*. There are several reasons that inter-clock domain communications can be problematic. First, the clock waveforms applied by the tester may not truly mimic the timing applied in the system. As such, the clock timing may not allow for proper signal propagation times. Also, ATE hardware limitations may add skew to the clock timing that will adversely effect the propagation delay of signals traversing between the launch and capture clock domains. Likewise, ATE hardware limitations or pattern waveform formatting might not allow the exact clock waveforms required to be composed on the tester.

Some pattern generation tools provide solutions for inter-clock domain communications issues during delay test pattern generation in several ways. Path delay ATPG is the simpler problem to solve, since the head and tail of the path are known. One solution is to mask out all but the scanned out bits required for the delay test. For example, if a test pattern has been created for a single path, then the scanned out data checked by the ATE will include only one bit of unmasked data for this path delay fault. All other bits will be masked during scan out. Transition delay ATPG also has solutions for inter-clock domain communications. One solution is to not allow any two clocks to capture that will result in a domain crossing. Table 4-2 shows some data for 5 chips that indicates a low level of coverage loss when all clocks are constrained not to pulse at once. Although clearly design dependent, the inter-domain fault universe is small in these examples. Another solution is to mask out any captured data that has crossed between clock domains. Instead of masking all the data that crosses clock domains, a pattern generator could justify the data that crosses the clock domains such that it appears static. Either way, the idea is to preserve coverage as much as possible, yet not create an environment that is ripe for generating false tester failures that capture too much product yield. Again, Table 4-2 indicates that the coverage loss from trying to create a safe capture environment is typically small.

	Size (M gates)	# CLK Dom	System-Clock-Launch				Last-Shift-Launch			
			1 Clock Pulse/Pattern		All Clocks Pulse/Pattern		1 Clock Pulse/Pattern		All Clocks Pulse/Pattern	
			TC %	Patts	TC %	Patts	TC %	Patts	TC %	Patts
A	0.76	2	62.44	2514	62.45	2515	89.07	1224	89.07	1180
B	0.86	3	75.47	16744	75.78	16533	80.98	7086	83.65	6845
C	1.07	3	74.37	4468	78.71	2691	79.97	1231	79.97	931
D	3.11	2	72.24	31670	72.28	26690	64.94	528	69.83	532
E	4.17	9	88.89	23541	89.11	22145	89.12	2019	89.12	1909

Table 4-2: Data comparison between last-shift-launch and system launch[13].

The illustration shown in Figure 4-7 combined with the solutions mentioned above lead to some interesting ramifications when applied to the transition fault model. In particular, coverage and pattern quality may be difficult to maximize as the circuit qualities and solutions pointed out earlier collide in practice. Referring to Figure 4-7, masking `U3` altogether due to the multi-cycle path shown will also mask out the ability to detect the true path launched from the `D3` input. So there will be a coverage loss for the transition fault model.

Likewise, `U4` can't always generate an `X`[14] as that would cause many paths to become untestable. With current ATPG tools, there is no way to tell a transition ATPG engine that only certain paths are multi-cycle while others are false or true. The best solution to these problems is to not design using multi-cycle and false path directives. Use good synchronous design practices, as these tend to lead to the simpler ATPG environments: requiring no fancy tricks.

4.4.2 I/O Issues

As with transition delay test generation and application, path delay testing also makes a distinction between core-based paths (flop-to-flop) and paths containing I/O pads. However, I/O characterization[15] is a good application of the path delay ATPG technology. For paths beginning at input pads, the transition is launched from the ATE, itself. These paths typically terminate at core-level flip-flops. Paths terminating at output pins typically launch their transitions from flip-flops, but

[13] Data supplied by Vinay Jayaram, Ken Butler of Texas Instruments.

[14] Like masking, "generating an `X`" is a tool the pattern generator (ATPG tool) can use by applying an `X` to the stimulus data of the pattern.

[15] I/O characterization is required for some high-performance busses or other applications that require that certain timing measurements be met in order for the chip to be successful in its system application. For example, certain communications subsystems require data to be moved on and off chip within specific timing windows.

terminate at the output pad, thus requiring an accurately placed strobe. Figure 4-8 shows example waveforms.

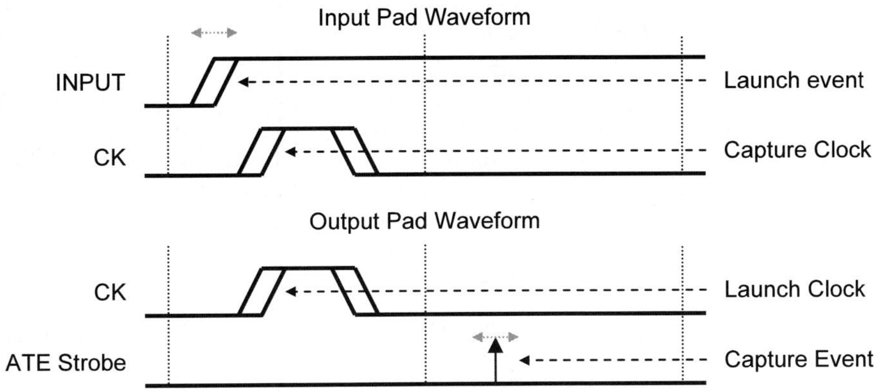

Figure 4-8: I/O pad characterization waveforms.

I/O characterization must be done with a complete understanding of the environment in which the test is being performed. For example, load characteristics of the test environment might be very different from those of the target application. Similarly, ATE constraints must be understood in order to set realistic timing targets for the test such that false failures are kept to a minimum, yet maintain meaningful testing criteria. Spice-level simulations should be made to help quantify these effects and help generate a meaningful test.

Even state-of-the-art test systems have finite performance metrics. Clock jitter may be 10ps. Input edge accuracy may be 50ps. Input edge definition may be in increments of 10ps (resolution). The load board[16] itself may add 350ps of delay. These are aggressive numbers. Even with this kind of performance, 60ps to 110ps of margin would need to be built into any test.

Test systems may also have a finite number of different input, clock, and output edges they can place at all. If each output pin, for example, has a different delay to be measured, all output strobes on each pin would need to be adjusted to a custom setting. Tester resources may be exhausted. To compensate, the schedule of tests may need to be adjusted to stay within the confines of ATE resources.

4.5 VECTOR GENERATION

Most ATPG tools do not use specific timing to aid in pattern generation. Stimulus and response event order is used, instead, in a zero-delay environment. However, timing information is used to select paths for path delay tests, and it is also used to constrain the pattern generation to avoid certain timing issues in a design such as

[16] A load board is a circuit card that plugs into the ATE allowing chip I/O and power connections to be routed to ATE resources.

multi-cycle and false paths. Recent efforts have been made to comprehend timing within the ATPG tools, however, the net effect is no different than methodologies that read constraints or rely on other sources for data. Having a good understanding of the defect types and their sources along with a *physical* database connection to ATPG will likely prove more useful in the future.

This fact is clear when looking at some of the more recent work done in attempting to consider crosstalk effects [20] and other signal integrity effects [21] during the generation of delay tests. The probability of a crosstalk-induced defect could be predicted using physical information. Delay test generation tools could be developed which consider the sensitization requirements and the physical circuit topologies when building the delay test pattern set.

4.5.1 Last-Shift-Launch

Last-Shift-Launch patterns can be generated using algorithms similar to combinational stuck-at patterns. The capture pattern (V2) contains all the data necessary to sensitize the fault and propagate it to a capture node. By shifting the pattern back by one bit, and making sure that the launch node will transition, the initial vector (V1) can easily be generated. Modified combinational test generators are used to create these patterns.

The easiest way for the Last-Shift-Launch pattern to be applied by the tester is to shift at-speed. That is, all tester cycles are at the system clock rate. In this case, chip power dissipation and the ability of the tester to supply such a power requirement must be considered. Creating patterns with two different periods is also an option: the shift occurring at a slow speed and the last bit shift and capture happening at a faster speed.

An interesting idea emerged a while back [22] which suggested using cell re-ordering or the insertion of dummy latches in order to change scan chain and circuit dependencies to improve Last-Shift-Launch transition test coverage. These ideas, however, are counter to other design requirements such as low area overhead and design routability. Given the high test coverages already achievable with full-scan designs and ATPG tools, and the fact that these design feature enhancements are not automated, scan chain re-ordering based on coverage seems to be an unnecessary enhancement.

4.5.2 System-Clock-Launch

Delay test patterns that require a launch and capture clock are usually generated with sequential pattern generators. This is because the initialization pattern (V1) needs to functionally justify a transition across one level of sequential elements. At the same time, it might need to also justify fault sensitization and propagation paths through sequential levels.

A relatively new idea in use today is to create a virtual combinational logic cloud out of the logic cones between the two levels of sequential cells [23], [24]. By transforming the sequentially deep circuit topology to one that is no longer sequential, a combinational ATPG engine can be used to generate the delay test pattern in a System-Clock-Launch environment. This can speed pattern generation and allow tricks currently employed by combinational pattern generators to be used in the transition delay pattern generation process. Figure 4-9 shows an example generated from [24] of how a circuit is transformed from a sequentially deep environment to a simple combinational environment for ATPG purposes.

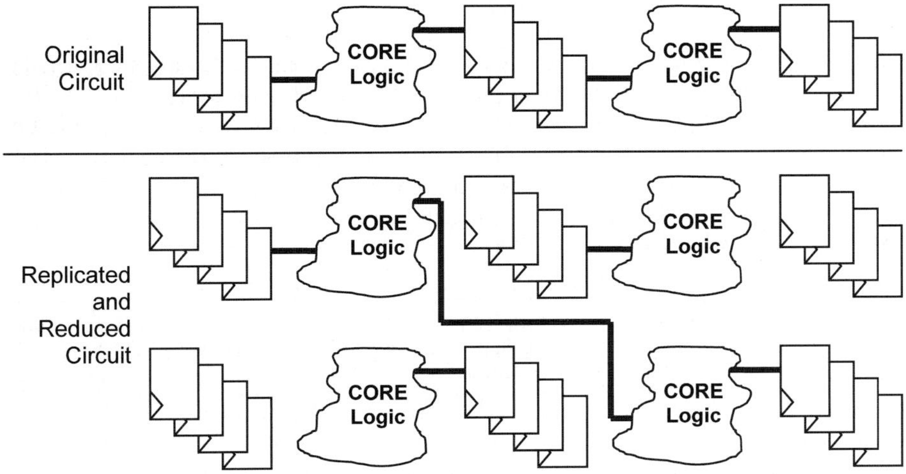

Figure 4-9: Replicate and reduce transform picture From [24].

4.5.3 Fault Model Tweaks

Recently, work has been done to compare the results of treating the str and stf faults as equivalent in the transition delay fault model [25]. In this investigation, it was proposed that the bulk of the delay was in the wire, and not the gate. The defect being detected was proposed to be the in-line resistive open. The results proved to be very interesting for the technology process in which the experimentation was done (0.18μm). In particular, by treating these two faults as equivalent, pattern count is reduced substantially, and effective fault coverage goes up, as well. This experimentation showed that the actual fallout from the tester closely matched the expected levels. In other words, by performing what amounted to slightly "less" testing, actual yield is expected to be better while test escapes remain about the same.

It is expected that with the continuous rampage of technology, new fault models and new uses of old fault models will be proposed and validated. Experimentation such as the one reported in [25] will continue. It is hoped that publication of these new findings will continue to be reported, too, so that all can benefit.

4.5.4 Selecting Faults

Before deciding which faults to address during manufacturing test, several factors should be considered. First, consider how much ATE memory is available for these tests. Second, consider test application time. Of course the faults themselves are important, and the stuck-at fault model continues to detect a wide range of device defects. Although this chapter does not specifically cover memory defects, it is recommended that memories be surrounded by a BIST structure that allows testing to proceed at speed. This will insure that data access speeds and memory integrity is as required. Recalling that the typical target of path delay ATPG is process variation, speed binning, or device characterization, selection of path delay faults will be our primary focus.

A typical flow to produce a regimen of tests, shown in Figure 4-10, would include generating delay tests that detect the targeted selection of path delay faults, first. Next, a transition delay fault set would be chosen. The path delay pattern set could then be fault simulated across the transition delay fault set and topped off with transition delay patterns. Lastly, the entire pattern set could be fault simulated against the entire stuck-at fault universe, and these would then be topped off to get the highest coverage possible.

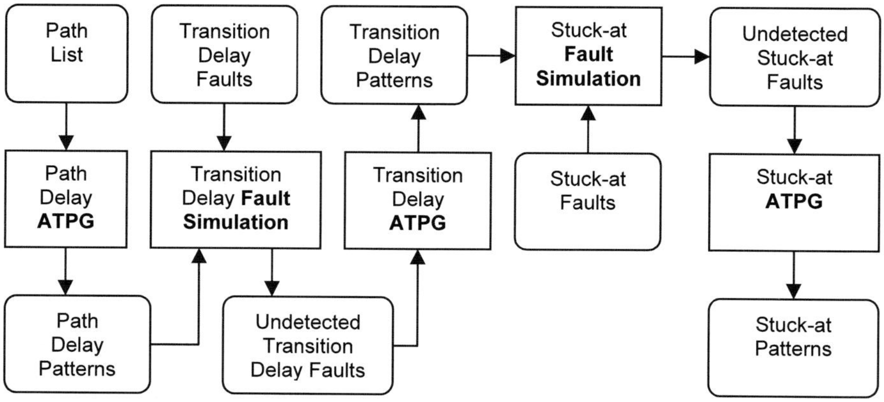

Figure 4-10: Minimum pattern count for maximum coverage.

As discussed earlier, fault list specification can have a lot to do with the success of the pattern set at both detecting defects and limiting over-kill or yield loss. Most ATPG tools have switches or other mechanisms allowing the selection of transition delay faults within a clock domain, between domains, or other fault list filter functions. Some can even automatically exclude certain types of faults such as scan enable faults.

Although methods have been proposed for creating a metric of goodness for path selection, most design teams use ad hoc path selection mechanisms. Path delay faults are typically selected by their criticality. Tools can select only those paths with a slack less than 1ns, for example. However, since the list of path delay faults is kept small, a more sophisticated method should be chosen to make the best use of the path delay tests. For example, [26] suggests selecting paths from geographically diverse regions of the die. This paper also suggests removing paths with too much timing variation. The work in [27] presents a novel approach wherein paths with high levels of overlap are removed, while keeping the path with the least slack. This could be an important technique as synthesis drives more and more paths to the edges of criticality [28].

Another selection process [29] suggests selecting paths based not just on criticality, but also on operating conditions. For example, it is possible that a path is critical at high voltage but not at low voltage. Other paths may become more critical at low voltage [14]. Selecting a variety of paths across the chip operating range helps guarantee that the process filter being developed is more useful.

Knowing the defect mechanisms at work, it may be possible to construct a path selection based not only on slack, but also on circuit topology. For example, knowing that resistive vias are a likely defect to be detected, selecting paths that traverse a lot of vias would create a useful set of targets. In fact, unpublished results of some experiments looking at 130nm technology showed that paths through many levels of logic did not necessarily lead to long delays. This result supports using an actual static timing analysis tool to build the path list, and using some other device features to augment path selection.

4.6 CHIP DESIGN CONSTRUCTS

Some specific design constructs which both help and potentially hurt delay test solutions are discussed in this section.

4.6.1 Phase-Locked Loops (PLLs)

The internal Phase Locked Loop (PLL) is used on many chips to multiply slower external clock references to faster speeds. For example, an external 10MHz clock might be used to generate an internal 800MHz clock. For ATE-applied pattern sets, synchronization must still exist between the application of test data (like scan chain values) and the collection of the response data. The actual at-speed test clocks, however, can be applied via the internal PLL. To achieve this, a glitch-free multiplexer needs to be designed which can switch between the ATE clock (or clocks) and the internally generated launch and capture clocks at the appropriate times.

Figure 4-11 shows waveforms pertinent to this discussion. Typically, the delay test is set up via scan path control, as usual. In this case, the ATE supplies the shift clock, `ATE_CK`, and scan enable signal, `ATE_SE`. Once the initial vector has been scanned in, the clock switching circuit is allowed to switch to its at-speed "program". This will cause a sequence of events to occur. First, the internal scan enable signal, `INT_SE`, will switch to its inactive state (for System-Clock-Launched delay tests, this can be done at the chips leisure). Sophisticated PLL control circuits might cause a delay that will allow `INT_SE` to switch to its inactive state allowing for the high fanout of this signal before proceeding on to the next step. Next, the PLL controller will allow two clocks to be ejected from the PLL onto the internal clock tree, driven by `INT_CK` in this case. These are the launch and capture clocks. The switch between the externally generated shift clocks (`ATE_CK`) and the internally generated PLL clocks (`PLL_CK`) must be glitch free. Once the last PLL clock occurs, the internal scan enable can become active again, and the ATE-driven clocks can extract the test results from the scan chains. Once again, the switch between the internal

PLL clocks and the external ATE clocks must be glitch free so as not to upset the contents of the scan cells.

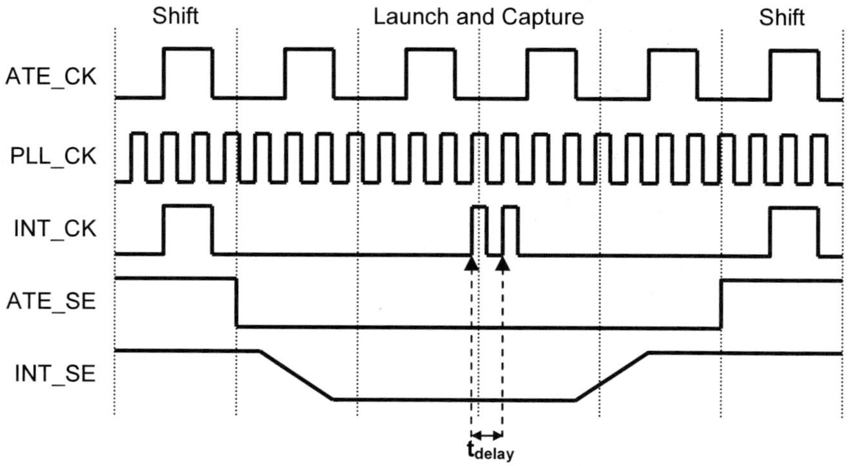

Figure 4-11: PLL-generated at-speed test waveforms.

In [30], the authors allude to some of the device features enabling at-speed testing to be successful when applied from a slow-speed test environment using an on-chip PLL. It also shows the block diagram of that controller. Most companies consider this controller and switching function a trade secret or specialized intellectual property, and are reluctant to share its design with the public. The keys to remember when designing such a clock selection circuit are that the two clocks (ATE and PLL, for example) should be considered completely asynchronous. Typical rules to follow include synchronizing control signals with the active clock; gating off clocks on their falling edges; and enabling one clock only after the other clock has been completely disabled.

4.6.2 Core Test Support

Many corporations are turning to third parties to provide specialized cores or Intellectual Property (IP). Many of these cores are "wrapped" with circuitry enabling a pre-processed test pattern set to be applied to the core, while allowing the logic outside of the core to be tested without the user being required to have knowledge about the logic content of the core. Figure 4-12 shows an example of a core wrapper added to an input pin. The bold signal path is the path taken during normal system operation. Specifications of core wrappers are given by the IEEE 1500 Standard for Embedded Core Testing (SECT)[17].

[17] http://grouper.ieee.org/groups/1500

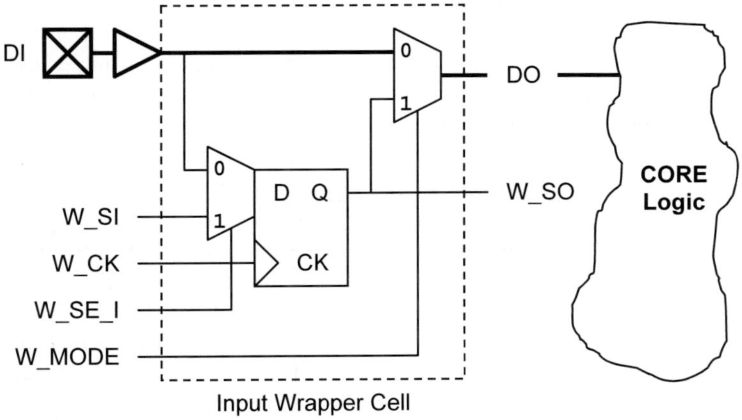

Figure 4-12: Core wrapper circuitry on an input.

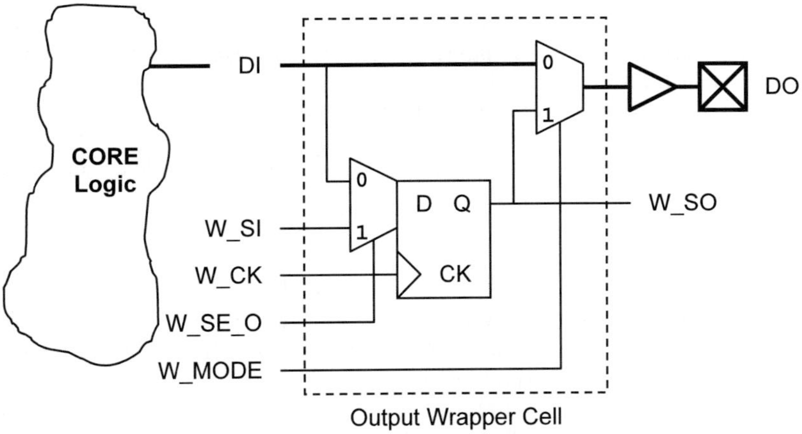

Figure 4-13: Core wrapper circuitry on an output.

Likewise, Figure 4-13 shows an output wrapper cell with its functional path highlighted. Since delay tests require a transition launch and subsequent capture, an examination of these wrapper cells and their impact on delay test generation is required. While studying the effect these cells have on the core, we can also examine the effects they have on delay test generation outside the core. Delay tests external to the core will still make use of the wrapper cells to either capture the termination of a delay test transition, or launch the transition into logic outside the core.

First, focusing on delay tests launched from the input side of the core into the core logic, one can see that the logic transition must be formed out of the data stored in the wrapper cell itself, and some input value to the wrapper cell. This input value (of opposite value from that stored in the wrapper cell flip-flop) might come from the `DI` pin or the `W_SI` pin, depending on the value of `W_SE_I` (see Figure 4-12). But, in order for the patterns to remain portable, the `DI` pin would not be a valid choice because the state of this pin is indeterminate with respect to the core and its wrapper. Therefore, the W_SI pin is the only valid pin one can set to cause a transition to occur on the launch clock of the delay test. Once this point is made, it is clear that the W_SE_I signal must be set in the "shift" state of the wrapper so that the preceding cell in the wrapper chain can supply the data required for the transition.

For System-Clock-Launched delay tests, then, there must be different system-level signals driving the scan enable signals for the core versus those driving the wrapper cells (proven, above, for the input wrapper cells): while the core is set to launch and capture through the functional paths of the core, the wrapper is set to shift such that only the wrapper cells need be set for portable pattern re-use. Next, let's examine paths terminating at output wrapper cells. These tests require that the wrapper cells capture the transition in the second phase of the test. During the launch phase, the state of this cell is inconsequential. So, the output wrapper cells can stay in functional capture mode while the delay test is being applied.

Similar logic is used to determine whether the wrapper cells require separate scan enable signals for delay test application to circuits outside the core. And indeed, it is required that the scan enable signal value for output wrapper cells is different from that of the logic outside the core: output wrapper cells should remain in shift state while they launch transitions into the surrounding logic. This is because the logic inside the core may not have a suitable model for ATPG.

An obvious alternative to using separate scan enable signals is to use a wrapper cell architecture with two flip-flop elements in it, but this solution can get very expensive in silicon area, data volume, and test time.

4.6.3 I/O Loopback

Similar in broad concept to using the device' PLL for driving high-speed clocks for delay test purposes, I/O loopback has grown in popularity. This technique uses on-chip resources to both deliver a stimulus and measure a response. For I/O loopback, the stimulus might pass out of the I/O pad and back into the device via the same pad (for a bi-directional buffer) or a different pad via an off-chip loop located on the load board. A simple pass or fail notification can then be delivered to the ATE. This type of test architecture [31] has been popularized by communications chips containing Serializers/Deserializers (SERDES) that have data-rates well into the Gigahertz range. But microprocessors and other devices can also operate at speeds that outrun the clock and other resources of the test equipment available on the factory floor.

4.7 ATE REQUIREMENTS

This section concentrates on issues that are generic to any tester. Low-cost test systems (detailed in a separate chapter of this book) are becoming quite popular. Likewise, making use of older ATE on the factory floor puts off expensive capital acquisitions. These older and slower machines or the under-resourced low-cost testers are still viable candidates for test application provided the power supplied to the device is still sufficient to affect a useful test.

Specifically, shifting in and out scan data can still be done at slow speeds. PLLs or free-running off-chip oscillators can actually be used to drive the much faster launch and capture clocks, as mentioned earlier. BIST or other pattern compression mechanism can be used to compress the quantity of data to be pushed into and pulled out of the device during test. So, whereas resources for test application are very important to the successful application of delay tests, not all these resources need to come directly from the ATE. Some (or most) can be on-chip.

4.7.1 I/O Requirements

Section 4.4.2 detailed some ATE constraints that will affect the ability to apply any delay test. Tester load and skew need to be considered when setting up the timing for delay tests. This is why most users of delay test limit their fault universe to areas between chip-internal flip-flops. With this limitation, the only signal that needs to change at-speed is the clock, which will be used for launch and capture. Also, limiting the coverage to areas solely within and not across clock domains further limits exposure to tester-related timing violations.

Once the at-speed signal is limited to just the clock, even insertion delays are removed as a source of timing errors, as this is accounted for in both the launch pulse and capture pulse, and should be nearly equivalent, pulse to pulse. All other delays are truly design related.

4.7.2 Speed Requirements

Other chapters in this book provide details regarding ATE functionality and limitations. This section examines a few issues that should be kept in mind when performing delay testing, and some solutions to these issues. Achieving the desired clocking speed to perform the delay test is the first hurdle one might come to when trying to match the device clock speed requirements to the tester speed limitations. Certainly Section 4.6.1's PLL usage solves any of the speed issues associated with high-speed launch and capture. However, for moderate speeds, the ATE clock might be sufficient to develop a successful delay test.

Some testers have the ability to multiplex two pins to derive a clock. Sometimes called “mux-clock”, this technique effectively increases the available frequency that can be applied for a launch and capture. The two clocks are ORed together, sometimes on the load board. A similar technique is to build in a second clock pin on the device which, during at speed test, will enter an on-chip OR gate along with the regular system clock pin. Then two tester channels can be used to generate the chip clock during test modes. Still other test systems can create a double pulse in a single tester pattern. Using these tester tricks, it may be possible to achieve the at-speed test goals without having to manipulate the source of the internal clock using an internal clock-switching circuit.

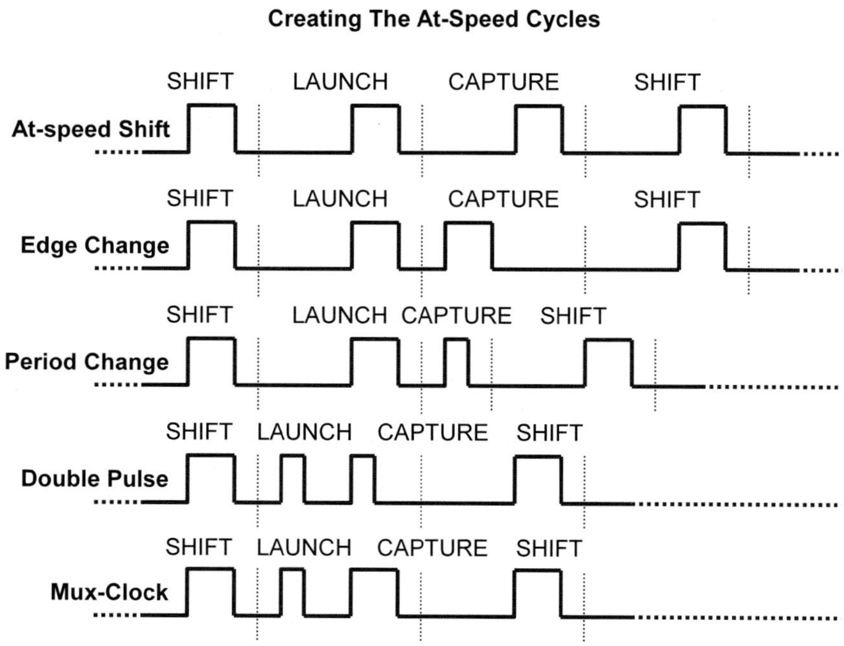

Figure 4-14: Alternate methods to generate delay test clocks.

Figure 4-14 shows waveforms for the preceding examples of how to use existing (slow) tester resources to produce fast clocks. Consulting the ATE user manuals or vendor technicians can guide you to successful implementations. But clock speed may not be the entire story: defect coverage is. Figure 4-15 shows data [10] for 189 chips that failed System-Clock-Launched transition delay test patterns.

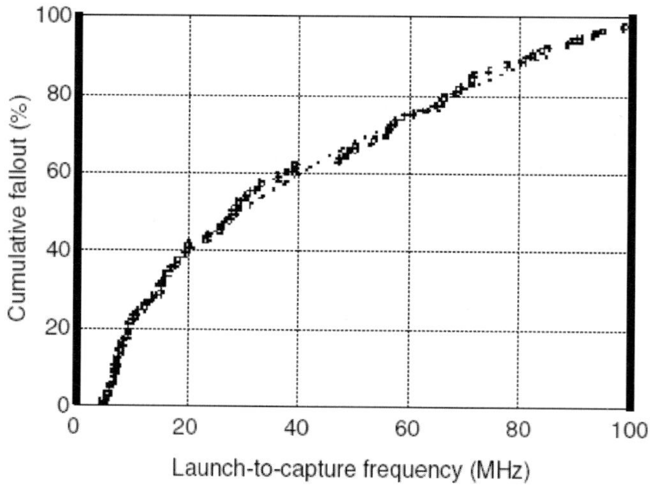

Figure 4-15: Fallout versus frequency [10].

This Figure shows that some defects were detected at very slow frequencies, too. The formula that fits this data can be represented as:

$$\text{Defect Coverage} = (f_t / f_0)^{1/2} \times \text{Fault Coverage}$$

where f_t is the frequency at which the test was run, and f_0 is the normal operating frequency of the device. Although this data is presented as is in [10], and should be validated for new processes and technologies (the manufacturing process of [10] is 0.18μm from a particular silicon vendor), it shows that there is a relationship between test frequency and defect coverage. For this experiment, parts failed at 5MHz all the way up to the 100MHz operating frequency.

4.7.3 Power Requirements

Power issues manifest themselves in many ways. The responsibility for power fortification lies with many sources, as well: chip design, synthesis, layout as well as ATE supplies, load board layout, etc. The power issues illustrated here with ATE applied resources is equally well suited to being illustrated with the chip-internal resources being used to route power from one side of a device to the other: the power grid or structure must be stoic and able to maintain the requirements of the system. This is especially true with scan-based systems that shift synchronously, typically generating 3 to 10 times more system activity than is used under normal operating modes.

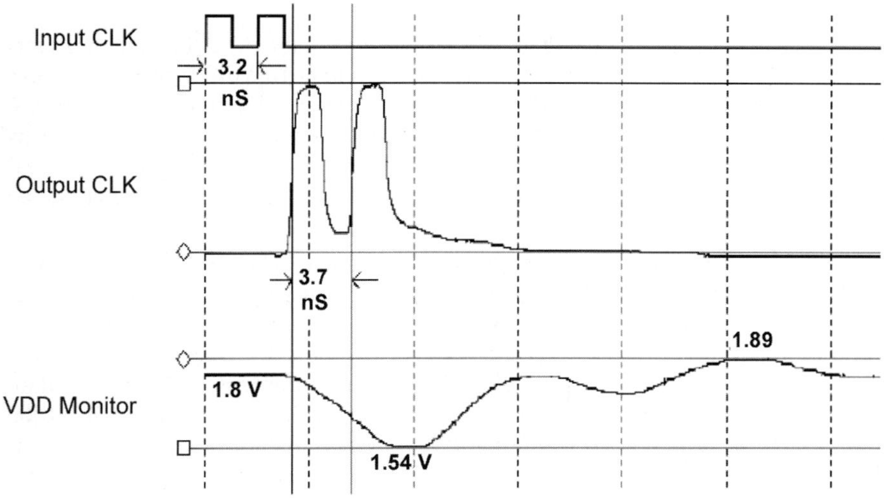

Figure 4-16: ATE waveforms showing clock stretch and V_{DD} droop[18].

Figure 4-16 shows the effects on a device's power supply connected to the chip being tested when two high-speed clocks are applied to the device. In an attempt to

[18] Waveform image furnished by Jeff Rearick, Agilent Technologies.

satisfy the chip's hunger for more energy, the ATE supply lines dip, unable to keep up with the demands of the device. This Figure defines another reason to apply delay tests using the System-Clock-Launch mechanism instead of the Last-Shift-Launch format. Using this pattern style, the supply source can rejuvenate after shifting, but before applying the two at-speed clocks.

Figure 4-16 also shows the *output* clock after passing through the device and output pads. Note that the clock has actually slowed down by about 15%. Although perhaps not a completely accurate picture due to the output drive coming from a power-starved I/O buffer, the point is nonetheless clear: it may be possible that the actual at-speed path being measured inside the device is also receiving a slower version of the input clock. In other words, in this example, there is actually 15% more timing margin between the launch and capture clocks, so more devices will pass this test than actually should. It is hoped that at these speeds, an internally generated PLL-driven clock source would have more success of performing at its rated speed, and therefore result in a valid test.

Clock architectures and power management tricks used in chips, today, could also be used under test mode conditions. For example, clock-gating cells with gating logic driven by scannable registers are well understood by ATPG engines. By maintaining the viability of these gating structures during delay test pattern generation and application, actual functional coverage should increase while power expenditures are decreased.

4.8 CONCLUSIONS: TESTS VS. DEFECTS

In [32] the author suggests that delay testing is a very viable filter for silicon defects in deep and very deep sub-micron processes. Testing at two different temperatures will likely be required to detect both resistive vias and resistive shorts. Resistive vias seem to make worse (more resistive) contact when cold, but other delay mechanisms are active when the device is hot. In fact, [33] seems to suggest that there is no real correlation between delay and temperature, but that a temperature versus delay characterization might be able to differentiate between different defect types.

As shown in the references made in the Introduction, delay tests in fact do detect many defects. As the investigations unfold, and the pattern generation tools evolve, many more useful procedures will be revealed which improve on the already very useful infrastructure. Copper voids and resistive vias in copper seem to be dominant defect types. However, the dissemination of factual information from the failure analysis labs at the foundries is still to come. Much of this "information" about the prominent defect types is from informal discussions with engineers and researchers "in the trenches" [34]. So, test engineers should build up a relationship with chip fabricators. Knowledge about which defect types to generate patterns against is very important when driving test pattern fault selection.

When applying delay tests to a manufacturing process, test engineers should not lose sight of the goal: to detect silicon defects before customers do. Therefore, other test methods and pattern generation features such as bridging fault ATPG should not be ignored. These tests are crafted to detect specific defect types. New models and new algorithms and tool integrations are sure to improve the usefulness even further. But, don't delay: defects await your scrutiny.

ACKNOWLEDGEMENTS

Many folks contributed to the content of this chapter. Certainly all the references made below belie the work entailed. A special thanks is extended to Bill Underwood, Cy Hay, Tom Williams, Roger Hsu, Al Crouch, Teresa McLaurin, Phil Nigh, Rob Aitken, Jeff Rearick, and Ken Butler for their inputs or reviews.

REFERENCES

The following references were used directly to pull research data for this chapter. However, most of the papers, in turn, have reference lists that are interesting, as well. Please avail yourself of these other references to obtain a good grounding in the topic of delay testing.

[1] G. Moore, "Cramming More Components Onto Integrated Circuits", Electronics, April 19, 1965, ftp://download.intel.com/research/silicon/moorespaper.pdf.

[2] "International Technology Roadmap For Semiconductors", 2003 Edition, Semiconductor International Association, http://public.itrs.net/.

[3] A. Krstic, J. Liou, Y. Jiang, K. Cheng, "Delay Testing Considering Crosstalk-Induced Effects", Proceedings International Test Conference, pp. 558-567, Oct 2001.

[4] S. Chakravarty, "On The Capability Of Delay Tests To Detect Bridges And Opens", Proceedings 6th Asian Test Symposium, pp. 314-319, Nov 1997.

[5] A. Pierzynska, S. Pilarski, "Pitfalls In Delay Fault Testing", IEEE Transactions on Computer-Aided Design of Integrated Circuits and Systems, v. 16, i. 3, pp. 321-329, March 1997.

[6] S. C. Ma, P. Franco, E. J. McCluskey, "An Experimental Chip To Evaluate Test Techniques Experiment Results", International Proceedings Test Conference, pp. 663-672, Oct 1995.

[7] J. Saxena, K. M. Butler, J. Gatt, R. Raghuraman, S. P. Kumar, S. Basu, D. J. Campbell, J. Berech, "Scan-Based Transition Fault Testing - Implementation And Low Cost Test Challenges", Proceedings International Test Conference, pp. 1120-1129, Oct 2002.

[8] P. C. Maxwell, R. C. Aitken, K. R. Kollitz, A. C. Brown, "IDDQ And AC Scan: The War Against Unmodelled Defects", Proceedings International Test Conference, pp. 250-258, Oct 1996.

[9] M. Renovell, J. M. Galliere, F. Azais, Y. Bertrand, "Delay Testing Of MOS Transistor With Gate Oxide Short", Proceedings 12th Asian Test Symposium, pp. 168-173, Nov 2003.

[10] R. Madge, B. R. Benware, W. R. Daasch, "Obtaining High Defect Coverage For Frequency-Dependent Defects In Complex ASICs", IEEE Design & Test of Computers, v. 20, i. 5, pp. 46-53, Sept 2003.

[11] J. Liou, A. Krstic, Y. Jiang, K. Cheng, "Modeling, Testing, And Analysis For Delay Defects And Noise Effects In Deep Submicron Devices", IEEE Transactions on Computer-Aided Design of Integrated Circuits and Systems, v. 22, i. 6, pp. 756-769, June 2003.

[12] S. F. Midkiff, W. Y. Koe, "Test Effectiveness Metrics For CMOS Faults", Proceedings International Test Conference, pp. 653-659, Aug 1989.

[13] R. David, S. Rahal, J. L. Rainard, "Some Relationships Between Delay Testing And Stuck-Open Testing In CMOS Circuits", Proceedings of the European Design Automation Conference, pp. 339-343, March 1990.

[14] P. Nigh, D. Vallett, A. Patel, J. Wright, F. Motika, D. Forlenza, R. Kurtulik, W. Chong, "Failure Analysis Of Timing And IDDq-Only Failures From The SEMATECH Test Methods Experiment", Proceedings International Test Conference, pp. 1152-1161, Sept 1999.

[15] K. Baker, G. Gronthoud, M. Lousberg, I. Schanstra, C. Hawkins, "Defect-Based Delay Testing Of Resistive Vias-Contacts A Critical Evaluation", Proceedings International Test Conference, pp. 467-476, Sept 1999.

[16] M. Sharma, J. H. Patel, "Enhanced Delay Defect Coverage With Path-Segments", Proceedings International Test Conference, pp. 385-392, Oct 2000.

[17] A. Krstic, K. Cheng, Delay Fault Testing for VLSI Circuits, Kluwer Academic Publishers, Norwell, MA, 1998, ISBN: 0792382951.

[18] S. Wang, X. Liu, S. T. Chakradhar, "Hybrid delay Scan: A Low Hardware Overhead Scan-based Delay Test Technique For High Fault Coverage And Compact Test Sets", Proceedings Design, Automation and Test in Europe Conference and Exhibition, v. 2, pp. 1296-1301, Feb 2004.

[19] J. Rearick, "Too Much Delay Fault Coverage Is A Bad Thing", Proceedings International Test Conference, pp. 624-633, Oct 2001.

[20] H. Li, Y. Zhang, X. Li, "Delay Test Pattern Generation Considering Crosstalk-Induced Effects", Proceedings 12th Asian Test Symposium, pp. 178-183, Nov 2003.

[21] A. Attarha, M. Nourani, "Test Pattern Generation for Signal Integrity Faults on Long Interconnects", Proceedings 20th IEEE VLSI Test Symposium, pp. 336-341, April 2002.

[22] J. Savir, S. Patil, "Scan-Based Transition Test", IEEE Transactions on Computer-Aided Design of Integrated Circuits and Systems, v. 12, i. 8, pp. 1232-1241, Aug 1993.

[23] S. Ohtake, K. Ohtani, H. Fujiwara, "A Method Of Test Generation For Path Delay Faults Using Stuck-at Fault Test Generation Algorithms", Proceedings Design, Automation and Test in Europe Conference and Exhibition, pp. 310-315, 2003.

[24] M. Abadir, J. Zhu, "Transition Test Generation Using Replicate-and-Reduce Transform For Scan-based Designs", Proceedings 21st IEEE VLSI Test Symposium, pp. 22-27, 2003.

[25] B. R. Benware, R. Madge, C. Lu, R. Daasch, "Effectiveness Comparisons Of Outlier Screening Methods For Frequency Dependent Defects On Complex ASICs", Proceedings 21st VLSI Test Symposium, pp. 39-46, April 2003.

[26] B. D. Cory, R. Kapur, B. Underwood, "Speed Binning With Path Delay Test In 150-nm Technology", IEEE Design & Test of Computers, v. 20, i. 5, pp. 41-45, Sept 2003.

[27] T. McLauren, "Debugging And Diagnosing Delay Defects In Deep Submicron Designs", Silicon Debug and Diagnosis Conference, 2004.

[28] E. S. Park, B. Underwood, T. W. Williams, M. R. Mercer, "Delay Testing Quality In Timing-Optimized Designs" Proceedings International Test Conference, pp. 897, Oct 1991.

[29] B. Seshadri, I. Pomeranz, S. M. Reddy, S. Kundu, "On Path Selection For Delay Fault Testing Considering Operating Conditions", Proceedings 8th IEEE European Test Workshop, pp. 141-146, May 2003.

[30] N. Tendolkar, R. Molyneaux, C. Pyron, R. Raina, "At-Speed Testing Of Delay Faults For Motorola's MPC7400, A PowerPCTM Microprocessor", Proceedings 18th IEEE VLSI Test Symposium, pp. 3-8, April 2000.

[31] B. Provost, T. Huang, C. H. Lim, K. Tian, M. Bashir, M. Atha, A. Muhtaroglu, C. Zhao, H. Muljono, "AC IO Loopback Design for High Speed uProcessor IO Test", Proceedings International Test Conference, pp. 23-30, Oct 2004.

[32] R. Aitken, "New Defect Behavior At 130nm And Beyond; Emerging Ideas Contribution, Extended Abstract", Proceedings 9th European Test Symposium, pp. 279-284, May 2004.

[33] P. Nigh, A. Gattiker, "Test Method Evaluation Experiments And Data", Proceedings International Test Conference, pp. 454-463, Oct 2000.

[34] B. Kruseman, A. Majhi, C. Hora, S. Eichenberger, J. Meirlevede, "Systematic Defects in Deep Sub-Micron Technologies", Proceedings International Test Conference, pp. 290-299, Oct 2004.

Chapter 5

High-Speed Digital Test Interfaces

Wolfgang Maichen

5.1 NEW CONCEPTS

5.1.1 Introduction

Historically, the interface between the tester's pin electronics (driver, comparator, DC circuits) and the device under test (DUT) has not received too much attention, apart from mechanical considerations[1]. Figure 5-1 shows a block diagram of a typical test setup. We will go into more details on many of the components shown later in this chapter.

But as usual any chain is only as strong as its weakest link [1], [2]. In this case it means that even the best performing, highest bandwidth, most accurate tester will

[1] For production test, electrical performance has so far usually taken a back seat compared to life time considerations (number of device insertions before damage to the interface occurs), compatibility with a certain handler, ruggedness, and material cost.

D. Gizopoulos (ed.), Advances in Electronic Testing: Challenges and Methodologies, 141-178.

fail to reliably sort good devices from bad ones or give accurate characterization results if the connection between tester and device – i.e. the interface – does not perform equally well: the tester has no direct knowledge about the "real" device itself – it only "sees" it through the interface. Thus if the interface degrades the signals sent to or coming back from the device, the device may fail the test even though it is actually behaving perfectly fine – resulting in costly yield loss in production test, or unnecessary and time consuming design respins because of inaccurate characterization data. Or an interface with insufficient bandwidth could hide sudden glitches in the device output so a faulty device escapes detection, increasing the failure rate at the customer. Because it increases test accuracy, solid and clean high-performance interface design also helps to achieve cross-platform correlation of test results, important for large device manufacturers and test houses that aim to retain second-source capability with regard to the test platforms used.

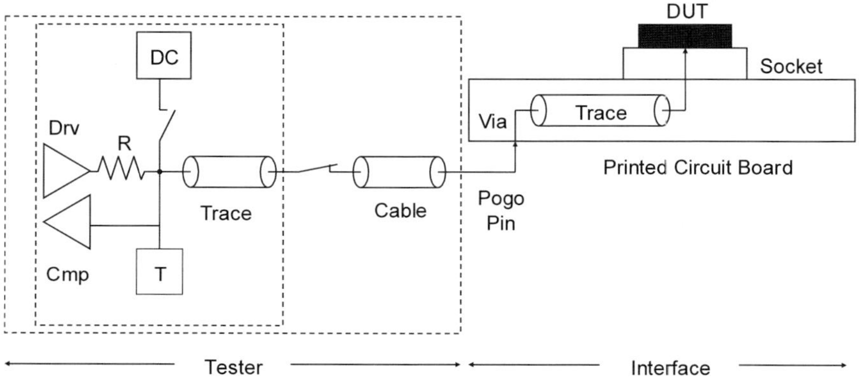

Figure 5-1: Simplified typical block diagram of a single channel in an automated test setup: Drv = driver, Cmp = comparator (receiver), R = driver impedance (matched to line impedance), T = termination circuitry, DC = DC measurement unit. Details are explained throughout this chapter.

Often the tester including the connection (printed circuit board, cables, relays, receiver, see Figure 5-1) was considered to be a simple capacitive load on the DUT's output. This still lingers for example when one sees printed circuit board (PCB) vendors specify the "total capacitance" of each of their traces on a PCB [3]. Unfortunately this simple picture completely ignores the way an electric signal travels along the transmission path between DUT and tester electronics (see Figure 5-1), and it cannot account for any effects that occur within the order of (or less than) one or a few times the propagation delay of the path. Since in a typical dielectric ($\varepsilon_r \approx 4$)[2] the signal propagation speed is (c is the speed of light):

[2] The dielectric constant ε_r describes the polarizability of a material (dielectric, insulator) in an electric field. For a capacitor it gives the ratio of the capacitance with the dielectric compared to the case where the material is replaced by vacuum (or, for practical purposes, air). It also determines the propagation speed of electromagnetic waves in this material, as shown in formula (1).

$$v = \frac{c}{\sqrt{\varepsilon_r}} = \frac{3 \times 10^8 \text{ m/s}}{\sqrt{4}} = 15 \text{ cm/ns}, \tag{1}$$

one can see that for an interface (including PCB, cables) that is 1 meter long this limit is reached (at the latest) for times smaller than about 40 ns (assuming $\varepsilon_r \approx 4$), corresponding to frequencies exceeding 25 MHz. What's more, the figure of merit determining a signal's bandwidth requirement is not given by the clock frequency or data rate, but rather by its knee frequency f_{knee} (assuming a trapezoidal waveform):

$$f_{knee} \approx \frac{0.5}{T_r} \tag{2}$$

where T_r is the rise time.

This required bandwidth always exceeds the clock frequency, which means that even devices running at clock rates of less than 25 MHz clock rate will be affected.

To deal with this situation, and correctly describe and understand the operation of the interface, one has to view the interface as a transmission line and apply its characteristics to the propagation of the signals sent through it.

Figure 5-1 makes clear that it is next to impossible to reduce the length of the path to less that maybe 50 cm, simply because there are too many components that each have a certain minimum space requirement: driver and comparator electronics, termination circuitry, DC measurement unit (today all these elements are often integrated into a single chip), relays to disconnect the sensitive electronics from the outside during device insertions or interface board changes, cables from the board to the mechanical tester-to-board interface (often pogo pins), the PCB, and finally the socket. Since a typical tester contains hundreds, if not thousands, of channels like the one shown in Figure 5-1, many of them will have to be rather far from the DUT and require long traces on the PCB.

5.1.2 Transmission Lines

In fact, transmission line theory has been around for almost as long as electronics technology, and it is extensively described in literature (very practice-oriented treatments are found in [4], [5] and [6]). But up to now for most test engineers it was not much more than some theoretical knowledge acquired during their studies, with seemingly little practical use. The basic concept is the *ideal* (*loss-less, homogeneous*) *transmission line* (*TL*), which takes into account that any conductor configuration has not only some capacitance, but also some inductance, and those properties are distributed along the length of the path. The ideal TL further assumes that the distribution is homogeneous (which in practice is achieved by a well-defined, homogeneous geometry of the conductors; a coaxial cable is a good example); any deviations from this ideal behavior are considered to be *parasitics* (which can be either capacitive or inductive), and they cause partial *reflections* of the signal. As a second addition come *losses* (resulting in a *lossy TL*) which will degrade the signal by reducing its amplitude and/or deforming the shape of the signal's transition[3].

The Ideal Transmission Line

Figure 5-2 shows a discrete picture of an ideal TL, consisting of a long chain of capacitors and inductors. This is of course only an approximation to any real system, in reality one would have an infinite number of infinitely small elements; but this discrete picture is well suited for actual modeling and simulation. L_u and C_u are the inductance and the capacitance per unit length, respectively. Since they are only used as a ratio in the equations below, the exact choice for the unit length is irrelevant.

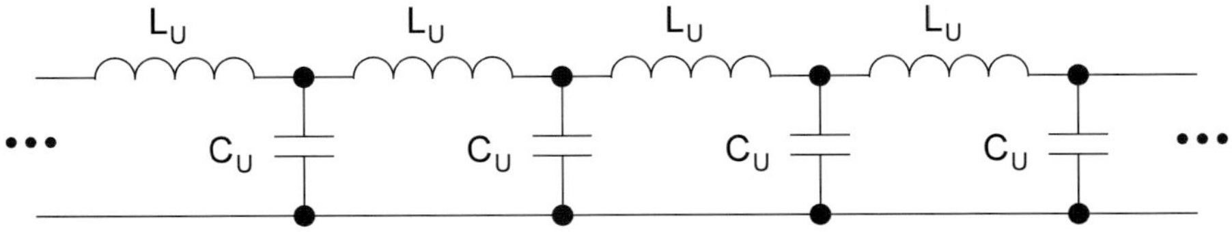

Figure 5-2: An ideal, lossless transmission line has uniform inductance (LU) and capacitance (CU) per unit length. The inductance is caused by the loop area enclosed between signal and return current. The capacitance is the capacitance between signal conductor and return conductor.

Such and ideal line is completely described by two values, the *characteristic impedance* Z_0 and the *propagation delay* T_{pd} [4], [5]:

$$Z_0 = \sqrt{\frac{L_u}{C_u}} = \sqrt{\frac{L_{tot}}{C_{tot}}}, \text{ and}$$

$$T_{pd} = \frac{c}{length \times \sqrt{\varepsilon_r}} = Z_0 \times C_{tot} = L_{tot} / Z_0, \qquad (3)$$

$$(L_{tot} = L_u \times length, C_{tot} = C_u \times length).$$

The total line capacitance C_{tot} is what one would measure with a capacitance meter using a signal frequency with a period much larger than the propagation time through the TL, but this would completely neglect the inductance. A somewhat surprising fact is that even though the line does not contain any ohmic element (only reactive components), it behaves like a perfectly ohmic resistance for any incoming signal. For example, when applying a 1 Volt *change* to one end of a line with

3 An ideal TL without any parasitics has the desirable property of transmitting signals without any distortion. It is a good idea to stress the difference between reflections (caused by parasitics or by impedance mismatches) and losses: Reflections mean that only part of the signal sent out continues towards the receiver, and the rest of the signal is reflected back to the sender. The total electromagnetic energy is conserved. On the other hand, in the case of losses, some part of the electromagnetic signal energy is "lost" (more precisely, converted into heat).

$Z_0 = 50\ \Omega$ (the usual choice for digital TLs), a current *change* of 20 mA will result. However, there are some important differences between a simple ohmic resistor and a TL. First, the signal will arrive at the other end of the line only after the delay T_{pd} (while a resistor does not have any significant delay). Second, no energy is dissipated, but it is stored in the electric and magnetic fields along the line. Third, at the end of the line a *reflection* of the signal will occur; more precisely, reflections occur whenever there is a change in the impedance along the line (and the end of the line can be seen as a change to infinite impedance), as shown in Figure 5-3.

The relative portions of the signal that gets reflected and transmitted, respectively, at a transition from the line impedance Z_0 to a load impedance of Z_L are given by the reflection coefficient ρ and the transmission coefficient τ:

$$\rho = \frac{Z_L - Z_0}{Z_L + Z_0}; \qquad \tau = \frac{2 \cdot Z_L}{Z_L + Z_0} \tag{4}$$

The load impedance does not have to be an ohmic resistor (i.e. real-valued and time-invariant), but can also be a capacitor, some network, or even another transition line with different characteristic impedance. Three important cases can be derived from the above equation. First, if the load impedance is equal to the line impedance, we speak of *matched termination*. The reflection coefficient disappears, so no reflection occurs. This is the ideal case since it avoids any reflections that could interfere with our data signal. If the load impedance is infinite (meaning no termination at all), the coefficient is +1, i.e. the full signal is reflected back and will interfere with our transmitted data. The last case would be a load impedance equal to zero, i.e. a short to ground. The coefficient becomes –1, meaning again the full signal is reflected, but with opposite polarity.

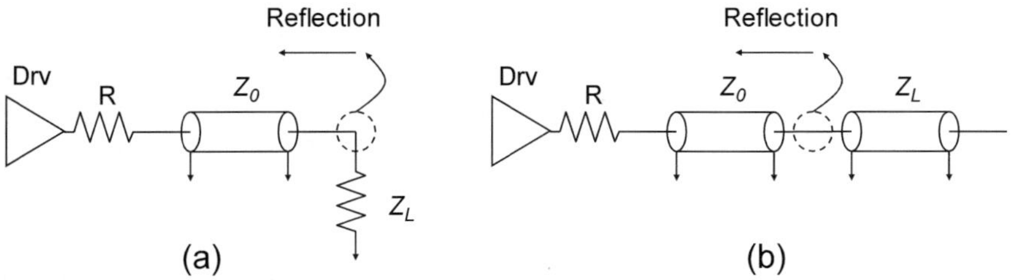

Figure 5-3: A discontinuity (change in characteristic impedance) in the path causes part of the incoming signal to be reflected back to the source. It is of no importance if the impedance is (a) some lumped network (symbolized by a resistor, but it can be any network) or (b) another transmission line.

Parasitics

Localized deviations from a constant ratio between inductance and capacitance per unit length (which determines the line impedance) can be regarded as either excessive shunt[4] capacitance or excessive series inductance at the particular location. (Note that each section of a TL has of course a certain capacitance and inductance, but only the *excess* capacitance or inductance—the part that exceeds what is needed to obtain an impedance of Z_0, typically 50 Ω—will affect signal fidelity[5]). If the deviation extends only over a length (propagation time) smaller than a fraction of the rise time of the signal traveling down the line, we can treat it as lumped at a point and we usually talk of inductive or capacitive "*parasitics*". Such parasitics are caused by any imperfection in the path geometry, like connectors, vias, narrow bends, etc., so minimizing those is one of the main tasks in good high-speed test engineering.

Parasitics are unwanted guests since *they cause reflections, waveform distortion, and limit the path bandwidth* [7]. The latter effect is easily understood if we recall that a TL acts as an ohmic load Z_0 to any transition, so if there is a capacitance C (or an inductance L) somewhere along the line, it forms an R-C or R-L filter with a time constant of:

$$T_{R-C} = R \times C = \frac{Z_0}{2} \times C, \text{and}\, T_{R-L} = \frac{L}{R} = \frac{L}{2 \times Z_0}. \quad (5)$$

The factor ½ for the R-C filter comes about because the line leading to and the line leading from the capacitance act in parallel, giving a Thevenin equivalent source impedance of half the line impedance. On the other hand for the R-L filter they act as two lines in series, effectively doubling the source impedance. The rise time (10% to 90%) $T_{10/90}$ and the 3 dB bandwidth[6] BW_{-3dB} of such a filter are given by:

$$T_{10/90} \approx 2.2 \times T_{R-C\,(\text{or}\, L-C)}, \text{ and } BW_{-3dB} \approx \frac{0.35}{2.2 \times T_{R-C\,(\text{or}\, L-C)}}. \quad (6)$$

The maximum reflection amplitude is given by the following formula, assuming a linear ramp as input signal (the rise time T_R *in this formula only* is the 0% to 100% rise time, T_C is the time constant of the filter and V_{inc} is the incoming voltage):

$$\frac{V_{refl,peak}}{V_{inc}} = \frac{T_C}{T_R} \times \left[1 - \exp\left(-\frac{T_R}{T_C}\right)\right] \approx \frac{T_C}{T_R}. \quad (7)$$

---

[4] i.e. between signal conductor and return conductor.

[5] "Signal fidelity" denotes the ability of a system to transmit a signal without distortion, i.e. incoming signal and outgoing signal look the same. Overshoot, ringing, and rise time degradation are examples for signal fidelity limitations.

[6] In the case of a low pass filter one is usually interested into the highest frequency component it can transmit without excessive attenuation. While the definition of the exact cutoff is always somewhat arbitrary because the attenuation increases gradually with frequency, a commonly used value is the point where the attenuation reaches 3 dB, i.e. the signal is reduced to roughly 70% of its initial amplitude. For most filters the attenuation increases rapidly for frequencies above this limit.

The approximation in formula (7) is valid when the filter time constant is small against the signal rise time, which is virtually always true for practically usable interfaces (otherwise the reflected amplitude would be large enough to cause false triggering in the receiver). Figure 5-4 illustrates the influence of a parasitic capacitance. The transmitted waveform has increased rise time, and there is a signal spike reflected back to the source, reaching some fraction of the incident amplitude given by formula (7).

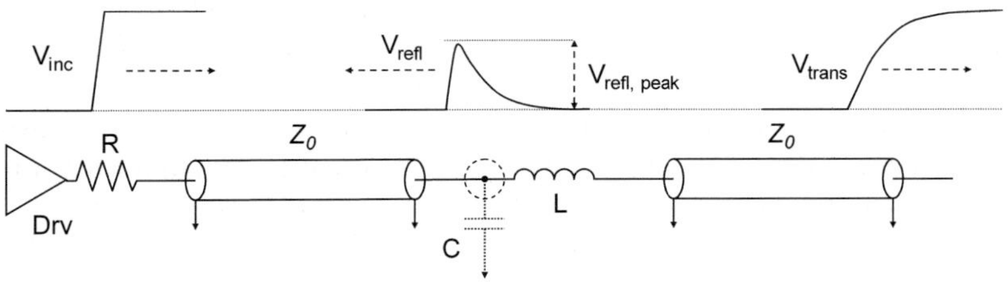

Figure 5-4: A parasitic series inductance causes a spike V_{refl} to be reflected back to the source, and it degrades (increases) the rise time of the transmitted signal V_{trans} and distorts its waveform (a shunt capacitance *instead* of the series inductance – as indicated in the figure – would give a similar reflection, but with negative polarity).

Why parasitics cause reflections can be understood with the following simple consideration. If a transition hits the initially uncharged capacitance, the capacitance will "soak up" any charge it can get, effectively acting as a short to ground (i.e. $Z_L = 0$ in Figure 5-3), thus with a reflection coefficient of –1. When it charges up more and more, the current into it decreases, until it is finally completely charged up and no longer has any effect on the signal (until the next transition arrives), i.e. a reflection coefficient of zero. In other words, there is a strong initial reflection that then decays over time. The exact height, shape and duration depend on both the size of the parasitic capacitance as well as the rise time of the incoming transition. Longer rise times smear out the response, so the reflection becomes shallower (see formula (7) for an approximation) but longer. For parasitic series inductances it works very similar, except that it initially acts like an open (no current can pass, $Z_L = \infty$, and we get a positive reflection spike) and then gradually opens up.

Losses

Apart from reflections, the second enemy of signal fidelity is signal losses [4], [5], [6], [3], [8]. The most important contributors here are:

- *Ohmic DC resistance*: It comes about by the ohmic resistance of the conducting material, and it causes a reduction in signal amplitude, but no distortion of the waveform.
- *Skin effect*: Caused by the fact that for high frequencies – for typical geometries this means anything higher than a few MHz – the current flows only in a thin layer on the conductor surface, effectively reducing the current-carrying cross section and thus increasing the ohmic resistance. Its resistance increase goes roughly with the square root of the frequency, so

seen in the time domain, after a transition it only disappears with $1/\sqrt{time}$, causing an increase in signal rise time and a long, very slowly vanishing voltage droop [5]. It is one of the major causes for *pattern dependent timing errors*[7].

- *Dielectric losses*: In a simplistic picture, the molecular dipoles in the dielectric surrounding the conductor have to follow the fast changing electric field, but have only a limited response time and thus exhibit some time lag, which partially counteracts the external field. As the end result, electric energy is lost from the signal and converted into heat (the same effect heats food – acting as the dielectric – in a microwave oven!). Dielectric losses degrade (increase) the rise time of the transitions, but disappear quickly with increasing time distance from the transition [5]. Dielectric losses increase approximately linearly with frequency.
- *Radiation losses*: While very important for EMI (electromagnetic interference) compliance [9], radiation losses only have negligible effect on signal fidelity [5], even for frequencies as high as 10 GHz. This may change in the future when clock frequencies will exceed those speeds by a wide margin[8], but at the moment radiation losses are only of very limited concern for the digital test engineer.

Losses are modeled in the discrete TL picture by introducing a per-unit-length series resistance R and shunt conductance G, as shown in Figure 5-5. Both parameters are of course strongly frequency dependent, R modeling the ohmic DC resistance as well as the skin effect, G taking care of dielectric losses and radiation losses.

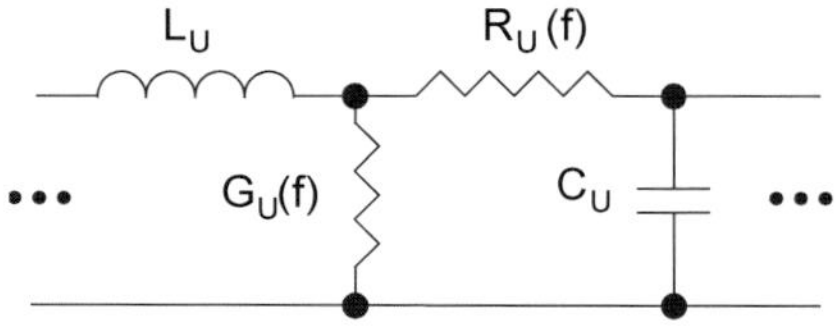

Figure 5-5: Discrete model element of a lossy transmission line: the series resistance Ru(f) models DC resistance and skin effect losses, while the shunt conductance Gu(f) takes into account dielectric and radiation losses. Both parameters are strongly frequency dependent.

Figure 5-6 displays a schematic view of how the total loss of a transmission path changes with frequency (the frequency scale should not be taken too literally as the exact ranges depend on design and dimensions of the path; it is only meant to give a

[7] "Pattern dependent" means that the timing of a particular transition is influenced by the specific bit stream that preceded it – the timing becomes dependent on the "data history" of the transmission path.

[8] As a rule of thumb radiation becomes significant when the wavelength of the highest frequency component in the signal gets within an order of magnitude of the transmission line's conductor spacing [5]. For typical printed circuit boards with a signal-trace-to-ground separation of around 0.25 mm (10 mil) this means radiation won't be much of a concern – at least for signal integrity! – below 100 GHz (corresponding to rise times of 3.3 ps).

general idea). At designs for a few 100 MHz, skin effect losses are still dominant, but because the skin effect only increases with the square root of the frequency, while dielectric losses go linearly with the frequency, above some frequency the dielectric loss mechanism becomes dominant [5]. However, the transition region is very broad[9], so in today's designs we have always a mixture of both.

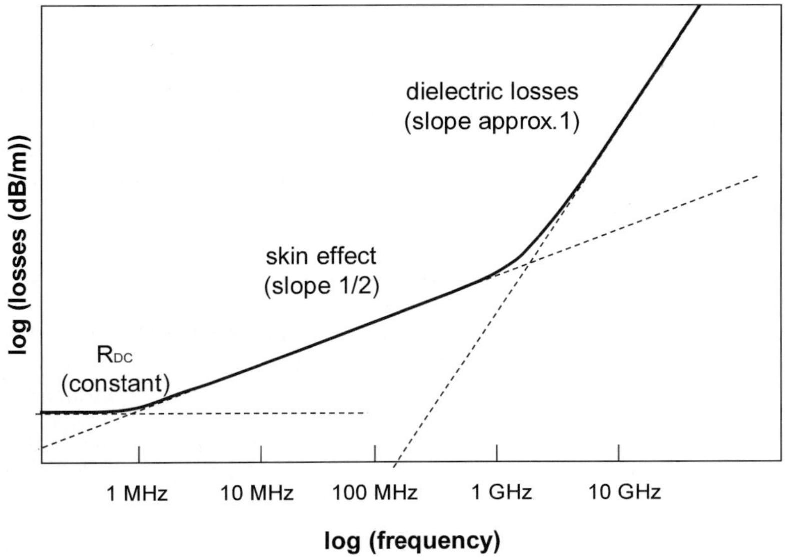

Figure 5-6: Schematic view of transmission losses (ohmic, skin effect and dielectric) vs. frequency. The frequency scale is only meant to indicate approximate ranges and should not be taken literally.

Effects on Signal Fidelity

When a signal transition (an "edge") travels down the transmission path, losses, parasitics and reflections modify and degrade its shape, each type in a different way [5]:

- Ohmic DC losses merely reduce the signal amplitude, but leave the edge shape intact.
- Dielectric losses increase the signal rise time, but disappear quickly after a transition; for a homogeneous TL the increase is linearly proportional to the line length.

[9] Since dielectric losses increase proportional to frequency and skin effect increases proportional to the square root of frequency, their ratio changes only with the square root of frequency. Thus if at some frequency both effects are equally strong, one needs to increase the frequency by a factor of 100 – i.e. a lot! – in order to reduce the skin effect contribution to no more than 10%.

- Skin effect on the other hand not only increases the rise time, but also it settles so slowly with time that it causes an additional voltage drop for a very long time after the transition. The rise time increase here is proportional to the *square* of the line length.
- Parasitics form low-pass filters, again increasing the rise time, but their rise times add as root-mean-square[10]. In addition, they cause reflections, which can interfere with subsequent transitions.

Since rise time and bandwidth are inversely proportional, one can always view a rise time increase as a reduction in effective path bandwidth.

Figure 5-7 shows a transition on the output of an imperfect, lossy transmission path, degraded by reflections and different loss types, compared to the clean input signal sent into the path.

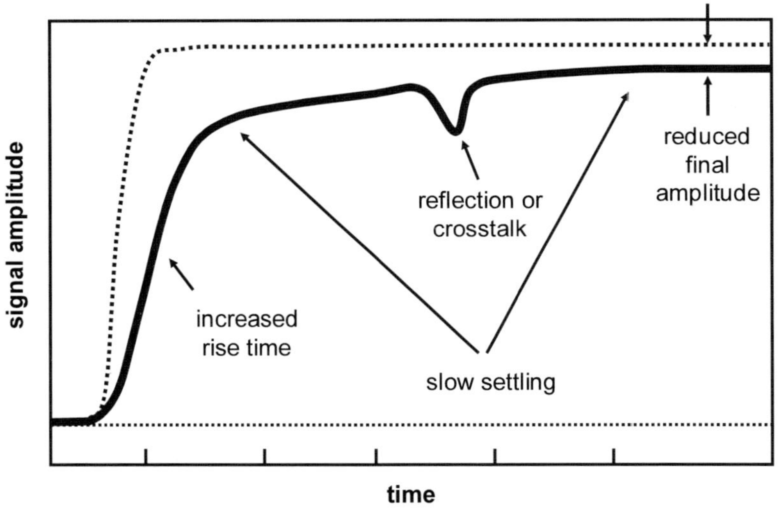

Figure 5-7: Signal degradation caused by an imperfect transmission path: Shown are increases rise time (caused by parasitics, skin effect, dielectric and radiation losses), slow settling (due to skin effect), reduced final amplitude (caused by ohmic resistance), and spikes from reflections (at impedance mismatches or parasitics) or crosstalk. The dotted curve shows the clean input signal into the path.

[10] Root-mean-square (RMS) means that if one has a series of low-pass filters with rise times of $T_1, T_2, T_3, \ldots$, the aggregate rise time of this combination is $\sqrt{T_1^2 + T_2^2 + T_3^2 + \ldots}$.

5.2 TECHNOLOGY AND DESIGN TECHNIQUES

While the tester hardware is quite often given and not easily modified, the interface between the tester and the DUT, usually a combination of printed circuit board, discrete elements like resistors and capacitors, socket/probe head, and sometimes cables, is an area where the test engineer has a lot of influence on the final performance of the whole test setup [10].

5.2.1 Parasitics Minimization

To reduce the number and size of the parasitics along the transmission path between the tester electronic and the DUT, every single component of the path must undergo thorough scrutiny and, if necessary, be optimized. Inside and outside the tester, this amounts to good PCB design and layout, proper choice of connectors, cables, sockets and/or probe heads. We discuss these aspects in the following subsections.

Cables and Traces

All cables and all PCB traces should have tight tolerances on their characteristic impedance, as a 10% mismatch between two sections of the transmission path produces already a 5% reflection of the signal. To assess the severity of this, keep in mind that a reflection of ρ percent of the signal can cause a maximum timing error of approximately:

$$\Delta T \approx \frac{\rho\,(\text{in }\%)}{100} \times T_r \quad (T_r \text{ is the signal rise time}) \tag{8}$$

if the reflected signal coincides with a subsequent transition at the receiver. For example, a 5% reflection of a signal with 200 ps rise time means 10 ps of additional timing error, which eats into the test's accuracy budget.

As always, there is a tradeoff between performances and price; while a 10% tolerance on PCB traces is rather easily achievable for a good PCB vendor, the tighter 5% tolerance needed for high performance, high accuracy applications often results in painfully reduced yield, meaning the vendor has to cherry-pick good boards from a larger manufactured batch, which drives up the price fast.

Vias

On the PCB, vias are one of the main sources of parasitics. Their impedance is difficult to control precisely, and the stubs of simple through-hole vias as well as the via pads introduce capacitive parasitics. Thus, it is best to keep the number of vias per trace to the absolute minimum, and advanced manufacturing techniques like sequential lamination[11] (the best, but very costly) or reverse drilling[12] (not quite as

[11] For sequential lamination the manufacturer first laminates those layers together that the blind via shall traverse, then creates standard through-hole vias in this partial stack. Next he laminates the remaining layers together, and finally merges the two partial stack-ups. The required accuracy in lamination reduces yield and makes this scheme very expensive. What's more, each additional via depth increases the number of partial stack-ups by one.

high performance, but much less expensive and sufficient for data rates up to several Gb/s) can be used to reduce the parasitic stub capacitance [4], [9]. Examples for different via types are shown in Figure 5-8.

Via parasitics can also be reduced by making their diameter smaller, but this has limits because the maximum drilling aspect ratio (via depth to via diameter) cannot be made arbitrarily large, and high-pin-count devices need thick boards with many layers to route all the signals. Another option is to fill the vias so that contact pads can be placed on top of them (as opposed to on the side in a "dog-bone" like configuration); the reduced pad size results in reduced parasitic capacitance and also minimizes area requirements—an important advantage in the socket area where space is at a premium. Via capacitance is reduced by increasing the plane clearance around them, but too large a clearance around nearby vias (so the clearances touch) will result in large breaks in the power and ground planes, negatively impacting impedance control.

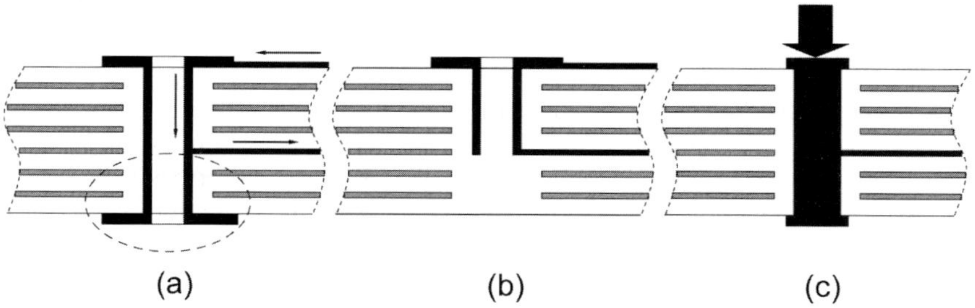

Figure 5-8: Different via types: (a) The standard through-hole via is easy and inexpensive to manufacture, but the stub hanging off (marked by the dashed circle) adds parasitic capacitance. (b) Sequential lamination or reverse drilling can avoid this stub, but increases manufacturing cost. (c) Filled vias allow reducing the total pad size, decreasing space requirements as well as parasitic capacitance.

Sockets and Probes

Just as vias, sockets and probe cards almost inevitably introduce parasitics, quite often including excessive crosstalk between channels. Short of doing away with them completely (which is rarely possible), the next best thing is to minimize their electrical length to a small fraction of the signal rise time. For sockets, this can mean using a thin pad of polymer with embedded metal particles as interconnect [11], [12], shown in Figure 5-9(b), instead of the common spring-loaded contacts (also widely called "pogo pins", Figure 5-9(a)). This can give amazing electrical performance, but

[12] Reverse drilling means that the via is first built up as a normal through-hole via. Then, using a drill slightly wider than the via diameter, the via stub is removed (drilled out) and the hole closed with a dielectric filler material. Manufacturing tolerances prevent complete stub removal, but on the upside the process is much less expensive than sequential lamination and there is virtually no incremental cost for doing more than one via depth.

usually for the price of greatly reduced life time; those specialty sockets can rarely withstand the constant strain put onto them by large-scale production test, but are rather geared towards application in bench-top characterization and analysis where accuracy and performance is a top priority and sample sizes (and thus insertions) are small.

For probe cards, the legacy needle probes are far from being matched impedance TLs and do not allow test speeds exceeding a few ten MHz, although their performance can be pushed a bit with some engineering ingenuity improving their impedance profile. But again, miniaturization is the key to performance improvement, like the vertical probing technique with microscopically small needles grown directly on a semiconductor substrate, or similar techniques [13], [14], which allow on-speed wafer level test of multi-GHz devices.

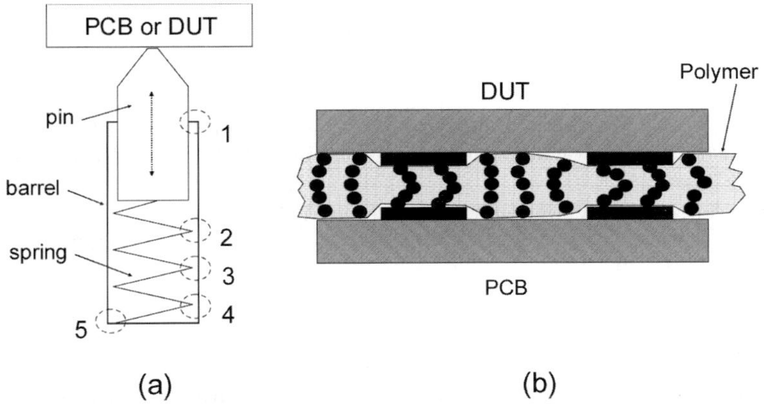

Figure 5-9: (a) Pogo pin assembly: Contact may occur randomly through any of the possible connections 1 – 5, making signal performance (parasitic inductance) variable for each insertion. (b) High-performance polymer pad with embedded metal particles provides minimum connection length and parasitics and thus maximum bandwidth. Since there are always several conducting paths parallel for each contact pad, parasitics variability is much lower.

Connectors

Connectors are a particularly difficult topic because as for sockets the tradeoff is usually between mechanical and electrical performance (and cost going up steeply with either of them). For large-pin-count production testers, where long life time with a large number of insertions is one of the top design goals, the connection of choice between the tester and the interface board is usually some pogo pin arrangement [15]. It offers small per-pin size and excellent compliance with interface board planarity (due to its spring-loaded contact mechanism), low cost and long lifetime with decent electrical bandwidth, especially when using small pogos. For best signal fidelity, a pogo arrangement with sufficient ground pogos to approximate a coaxial layout can yield some improvement. However, for speeds exceeding a few Gb/s, this concept reaches its limits as well. By design pogo pins incorporate a small needle pushed by a spring, so the contact point between this and the outer shell of the pogo pin is not very well defined and can change over time and with each operation[13]. If the high-speed pin count is small, some coaxial connection scheme like SMA[14] is a valid option. It can provide bandwidths into the far GHz region, but for the price of much larger space requirements and cost and often more difficult handling.

Another type of problem comes with very high pin count applications—the total force of the large number of compressed pogo pins required can exceed the mechanical stiffness of the interface board, making it warp and lose contact. To solve this, less stiff connection schemes are being developed [15], but they still must provide as good (or better) electrical characteristics and mechanical accuracy as the pogo interface.

5.2.2 Loss Mitigation

In Section 5.1.2 we saw a short overview about the losses that affect high-speed signals. Proper PCB layout has to strive to minimize the effect of those contributors. In the following sections we will see some general guidelines to help a test engineer when he has to specify his requirements to the PCB designer [16].

Ohmic Losses

Ohmic loss reduction is straightforward – those losses decrease with shorter trace length, larger cross section of the conductor, and higher-conductivity material [4], [5]. Since copper is the material of choice most of the time anyway, which is an excellent conductor, possible improvements are limited; highest-purity material (electrolytic copper) can reduce the specific resistance somewhat. At the same time, making the traces very wide soon runs into routing space constraints on the PCB (unless additional layers are added, which increases cost, complexity, and forces longer vias with more parasitic capacitance), so the best option is to make traces as short as possible and use the highest feasible plating thickness.

Skin Effect Losses

Since skin effect losses are nothing more than ohmic losses aggravated by a reduced effective cross section, some of the same considerations apply: use the highest conductance material available, and make the traces as short as possible. But there are differences as well: because the current flows only on the conductor's surface, increasing the thickness (which for a PCB trace is usually small compared to the width) has only negligible effect. Widening the trace (or, in the case of coaxial

---

[13] There is no way telling if already the needle itself makes contact with the housing - a high-bandwidth connection - or if the current has to go through all of the spring, experiencing a heavily inductive parasitic.

[14] SMA ("sub-miniature A") is one of the most widely used threaded connection systems for test instrumentation, providing a bandwidth of around 20 GHz.

cables, increasing the cable's diameter) is one possible solution, but again, space and routing constraints soon become overwhelming[15].

Keep in mind, also, that the ratio between trace width and dielectric layer thickness must be kept constant to obtain a constant characteristic impedance (50 Ω), i.e. wider traces mean thicker boards and longer—and as a consequence thicker—vias.

Also the transversal trace dimensions (width, distance to ground plane) must be kept small compared to the signal's wavelength, otherwise non-TEM (*transversal electrical modes*) will occur. Put into simpler terms, it means that the simple one-dimensional picture of the propagation breaks down, the propagation must be treated truly three-dimensionally, and the end result is dispersion and distorted edge shapes. Fortunately, below 10 GHz this is not yet of concern with typical PCB trace dimensions [5].

Since the resistance in the presence of skin effect becomes strongly dependent on the surface roughness (because the current has to follow the – longer! – path along all the surface features), providing a smoother surface can yield some improvement. Unfortunately, a certain amount of roughness is usually required to assure bonding between the dielectric and the copper. Elaborate bonding schemes that produce microscopic "stubs" of copper reaching into the dielectric, assuring bonding while keeping the rest of the surface flat, aim at improving this situation ("double-treat process", [5]).

Often the real concern in a test environment is less the absolute amount of skin effect loss, but its change with frequency (or, equivalently, time), because this is what introduces pattern dependent timing errors. Some cable manufacturers thus use a core of relatively high resistance material, plated with a thin, smooth layer of low-resistance material like silver [5]. The result of this is that even at DC (or low frequencies) most of the current is already flowing in the thin low-resistance layer on the surface, so the change in current density due to skin effect is much less pronounced. Of course this only works as long as the skin depth is larger than the thickness of the plating. Beyond that range, the usual square-root behavior with frequency takes hold again. The same plating method can be employed for traces in a PCB as well.

Dielectric Losses

The amount of dielectric losses per unit length in a given material is a constant. So the only choices to reduce total signal loss are to shorten the path (a cure-all good for any loss effect), and second, to use lower-loss materials [4], [5], [6], [9].

The theoretically best "material" would be of course vacuum, which has no losses at all (for signal integrity purposes, air is just as good). So how can we get as much of the electromagnetic fields to be in air, short of a free-flying cable in space? One tempting solution would be to use microstrip lines (i.e. traces on the surface of the PCB, with dielectric and a ground plane below, see Figure 5-10(a)) instead of

---

[15] Especially below the socket, where the traces have to wiggle through a dense jungle of vias leading to the device's solder balls, traces cannot be made arbitrarily wide.

striplines (traces sandwiched between two layers of dielectric and two ground planes, see Figure 5-10(b)). At least one side of the field around such microstrips is in free air, so the effective (average) dielectric constant and the losses are reduced, and as long as the path does not have to reach the other side of the board, there is no need for vias, further improving signal fidelity. But closer investigation shows that this also leads to dispersion because (in a highly simplified picture) the field in air propagates faster than the field in the dielectric (propagation is not true TEM), so edges get washed out over distance. In contrast to striplines, even a perfectly designed microstrip line will show rise time degradation and ringing. Also, especially for test purposes where there are many connections to make, routing space on the two surface layers will not be sufficient; on the other hand, for striplines we have an almost arbitrary number of layers available (sometimes up to 10 or 15 signal layers). What's more, since there is no shielding of the trace on the outside, propagation (impedance) is influenced by any dielectric close to the trace on the surface, be it the device socket, the test engineer's hand, or some dirt or dust.

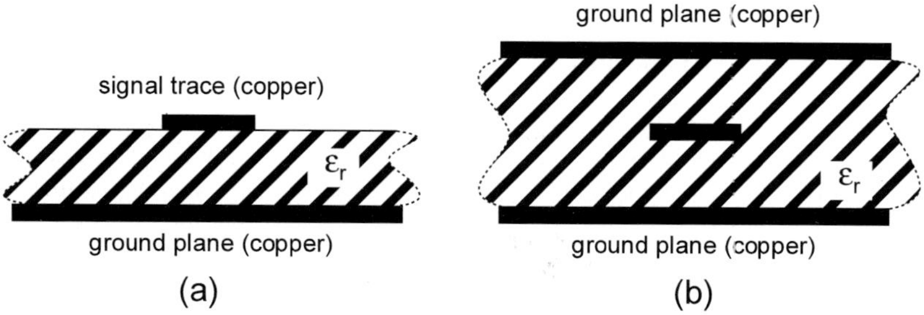

Figure 5-10: Signal traces in (a) microstrip and (b) stripline configuration.

Fortunately technology has made good progress in creating lower-loss materials [17]. Already the widely used and trusted FR-4, a fiberglass-epoxy composite, has pretty good electrical, mechanical, and loss characteristics up to several 100 MHz, combined with low price and wide availability. This makes it the material of choice for most low- to mid-performance PCB's, especially in consumer devices.

Materials with lower loss almost always make tradeoffs between cost, electrical and mechanical performance [18]. Teflon for example has one of the lowest losses, but is very difficult to bond to other materials (e.g. the copper of the traces). Typical problems of ultra-low-loss materials include: poor mechanical stability or stiffness; poor adhesion to copper or other dielectrics; bonding temperature requirements incompatible with other materials used for the PCB; coarse grain (for composites) which makes it difficult to drill the small-diameter vias preferred for highest performance. Last but not least those specialty materials are often produced by just a single supplier, so one lacks alternate sources in case this supplier has manufacturing problems or the demand exceeds his production capacity.

Finally, some cable manufacturers took up the idea that air is an almost loss-less medium. Instead of a solid layer of dielectric, they produce a foamed dielectric (Teflon is a good choice) that consists to a large part of tiny air bubbles (with a bubble size much smaller than the signal's wavelength). This reduces the effective loss factor and also increases the propagation speed. The latter is an additional advantage for I/O channels where the round trip time is of concern because it must be smaller than the channel turnaround time (the switching time from drive mode to receive mode).

5.2.3 Differential Signaling

Differential Signal Paths

In recent years *differential signaling*[16] [5], [6], [19] has made large inroads in high-speed transmission schemes. Estimates are that in a few years almost 100% of all PCB's will have at least some differential signal paths on them. There are several reasons for this. On one hand having a symmetrical pair of lines carrying opposite signals close to each other greatly reduces electromagnetic emissions because the electromagnetic far-fields of the two lines largely cancel. Second, since in such a transmission scheme the total signal current over the two lines of a differential pair is constant (at least as long as no differential skew is present), the current spikes drawn by the drivers are reduced by an order of magnitude, reducing power supply noise and ground bounce (see the section below about power decoupling). Third, differential transmission is much less sensitive to residual ground bounce or external influences than single ended signaling because the influences on the two lines of a pair are of similar size (as long as the two lines are close together) and so largely cancel out since the receiver is only sensitive to their difference.

For testing, this means that we have to concern ourselves with that topic as well when designing the interface, but the emphasis here is on accurate test results rather than optimizing the performance in an end application.

In contrast to wide-spread belief, there is nothing really special required *per se* for two lines to be "differential" - the only distinctive feature is that the signals the two lines carry are not independent, but are always complementary to each other. Things like "coupling" and "differential impedance" (see below) are the result of specific design techniques associated with differential signaling rather than prerequisites for it.

When two TLs come very close to each other, their electric and magnetic fields start to overlap and induce voltages and currents into each other, interfering with the original signals on the lines. In "normal" (single ended) signaling this is referred to as capacitive and inductive crosstalk, and it is an unwanted feature there. Another way to look at it is that the two lines have some mutual capacitance as well as mutual inductance which, depending on if there is a transition on the other line, adds to or subtracts from the self inductance and the capacitance against the ground plane [5], [6], [19].

However, if those two lines form a differential pair, then the transitions on them are no longer independent, and the crosstalk has always the same effect for the same

[16] When one line is high, the other one is low, and when one line transitions, the other one transitions in the opposite direction.

transition. In this case we talk of "coupling", but it really is just the same physical phenomenon. Which way the signals are influenced depends on the relative polarity of the transitions. If they are of opposite polarity ("odd mode"), the effective capacitance is increased, the effective inductance reduced, and thus the effective "odd mode impedance" Z_{odd} is reduced as well. For same-polarity transitions ("even mode"), the case is exactly the other way around, causing the "even mode impedance" Z_{even} to be higher than the impedance of an isolated line. These two impedances are related to differential impedance Z_{diff} (the total impedance seen by the differential signal, for which the two lines are effectively in series) and common impedance Z_{common} (where the two lines act in parallel) by simple formulas:

$$Z_{diff} = 2 \times Z_{odd} ,\ Z_{common} = \frac{Z_{even}}{2} ,\ Z_0 = \sqrt{Z_{even} \times Z_{odd}} ,\ Z_{even} \leq Z_0 \leq Z_{odd} \quad (9)$$

The equality between Z_{even}, Z_{odd} and Z_0 happens when the lines are completely uncoupled[17]. The beauty of this concept is that the transmission equations for differential and common signals stay the same as for single ended signals as long as one replaces the impedance with the proper value for each case.

In real-world applications of differential signaling close line spacing (resulting in considerable coupling) is used out of several reasons. First, it reduces emitted radiation—very important for a device in order to be compliant to EMI rules and regulations. Second, it makes differential lines less susceptible to external fields because those influences will cause the same disturbance in both lines of the pair, so they cancel out in the differential receiver. Third, routing both lines close together automatically means their propagation times will be well matched, so the differential signal fidelity at the receiver is conserved. Fourth, since crosstalk (coupling) between the two lines is of no concern, routing the lines close together conserves board space, allowing either for smaller boards or less layers, thus decreasing board cost and/or size. In summary, we see that close coupling is merely a side effect of other considerations in connection with differential signaling, but otherwise not an inherent necessity.

On the other hand, during test it can create a host of problems because it makes the impedances for common, differential, and single ended signals different from each other. If coupling is different for some sections of the path (e.g. because the first part consists of (uncoupled!) coaxial cables within the tester, and the second part is routed through coupled traces on a PCB), a path designed for differential signals will have impedance mismatches for single ended signals. This can wreak havoc with test accuracy if—what is often the case—path delay (deskew) calibration is done with single-ended signals. Moreover, minimizing electromagnetic emissions or assuring minimal susceptibility normally takes a back seat in testing compared to maximum accuracy and clean signaling. Therefore a better practice *for test purposes* is to route signals sufficiently far apart, so there is no coupling and thus no dependency of the impedance on the exact trace separation.

---

17 Note that Z_0 here is *not* the impedance of an isolated line, rather it is the impedance when one of the two lines of the pair is driven by a signal and the other one kept silent. Like Z_{odd} its value decreases with increasing coupling between the lines, but much less strongly than Z_{odd}.

Differential Path Routing Rules and Issues

So what are the real additional PCB routing requirements, compared to single ended signals [5], [6], [9], [20], [21]. A common misconception is that, unlike single ended signals, differential lines don't really need a continuous ground return path because the current flowing into one line already returns through the other line of the pair, so there is no current through the ground plane. This would be true if the signals were really perfectly differential, and - even more important - the traces were so close to each other that the coupling between them would far exceed the coupling to the ground plane (in a different view, this corresponds to the demand that the return currents below the traces completely overlap and thus cancel, leaving current only in the traces but not in the ground plane(s)). But in reality, due to rise/fall time mismatches, skew caused by path length differences or driver mismatches, etc. there is always some amount of common mode signal present, too, that has to return to its source though the ground plane. And the trace-to-trace coupling in realistic designs is always much weaker (or almost non-existent if following the trace-separation recommendations made before) than the coupling to a massive ground plane [6]. For the closest achievable spacing, where the trace distance of two 50 Ω lines is equal to the trace width, return current overlap (and cancellation) in the ground plane is only about 10% [6]. So there is no difference to single ended signals here - ground planes are indispensable. (One notable exception are *twisted-pair lines*[18] [5], [6].)

Still, for differential signal integrity purposes it is good practice to keep the two lines of a pair similar in performance (losses). Routing them similarly shaped and on the same PCB layer(s) (to avoid variations caused by dielectric material tolerances) helps achieve this goal, but does not put very stringent requirements on the matching tolerances. If the tester can deskew the timing on the two lines independently (and many testers can because their pin electronics drivers are still inherently single ended), it can account for any delay mismatch. The opposite direction (DUT driving and tester receiving) usually does not have this luxury because only in rare cases the DUT has this deskew possibility, and the tester's differential comparator usually lacks it as well, so in this case the matching must be within a fraction of the signal rise time. This puts high demands on the routing of the PCB. For example, if the signal has 50 ps rise time (this corresponds to a propagation of about 1 cm using a low-loss dielectric), the matching should be better than 10 ps or about 2 mm. While this may sound easily feasible, keep in mind that the exact delays of discontinuities like vias or bends tend to be very difficult to calculate and have considerable manufacturing variations, especially for the thick multi-layer boards common in high-performance test applications. Routing the two PCB traces of a differential pair far enough apart from each other (and of course from all other traces as well) to avoid coupling makes that keeping the distance exactly constant all along the line is of no importance, which eases routing restrictions.

---

[18] In twisted pair lines, where two wires are snugly wound around each other, the spacing between these two signal wires is usually small against the spacing to the shield, and thus the two return currents in the shield nearly cancel – in other words, the shield is hardly carrying any total current, and virtually all the current is flowing in one signal line and returning through the other. In this case one can even remove the shield without affecting the line impedance much (but immunity against external fields will be somewhat reduced).

What's more, skew between the two signals of a differential pair causes a rise time increase of the resulting *differential* signal (simply plot two slightly skewed edges of opposite polarity and add them up to their differential signal to see the effect). In other words, in addition to parasitics and losses, for differential signals path delay mismatches are an additional source of bandwidth limitations.

Remember that the change of effective differential impedance is caused by crosstalk (even when one calls it "coupling"), so it will only be there when the transitions on both lines occur at the same time at the same place. Skew anywhere along the lines - even when compensated further down - causes the two transitions (on the two lines) to arrive at the same location at different times. As a consequence they don't "see" each other anymore, thus there is no crosstalk/coupling affecting the transitions, and the effective impedance on this section is Z_0 (while during board design we were assuming an impedance of $Z_{odd} \neq Z_0$). So any skew creates impedance mismatches and, as a consequence, reflections. The best way out of this danger is again to route the traces uncoupled because then Z_{diff} equals Z_0 at all times.

5.2.4 Termination

In Section 5.1 we saw that good impedance control and matching is extremely important. But while mismatches along the line are usually in the range of just a few percent, the worst offender is the end of the path where the receiver sits. By itself, receivers (or comparators, if it is the tester end) have high impedances, so the end of the line would be virtually unterminated (resulting in 100% reflection) if there weren't some additional termination.

Diode Clamps

For slow-speed test setups reflections and ringing on the transmission path was mitigated through the use of *diode clamps* (see Figure 5-11a). In an idealized picture those clamps would clip any overshoot that exceeds one diode drop above (or below) the clamp voltage. But real diodes open up rather gradually with increasing voltage, they have only finite switching time – which must be faster than the signal rise time to have any effect – and together with the diode drop this leaves residual reflections [22]. All put together, diode clamps cease to be an effective termination method when speeds exceed a few 10 MHz or when voltage swings are small[19].

---

[19] For low-cost, slow-speed functional testers diode clamps remain an acceptable solution as long as the voltage swings are sufficiently high and one keeps realistically low expectations regarding the obtainable signal fidelity.

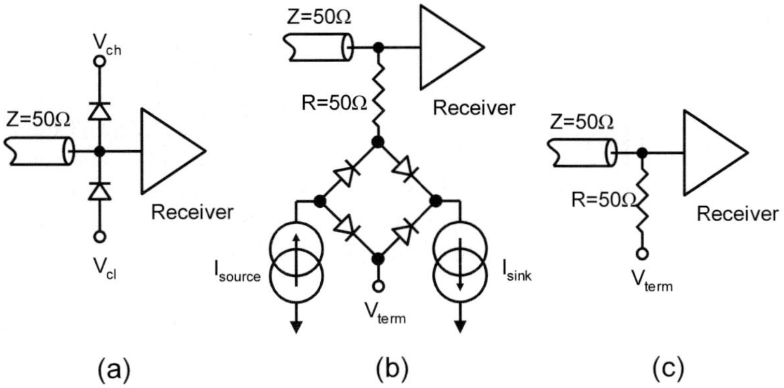

Figure 5-11: Common termination schemes: (a) diode clamps, (b) current load (I-load), (c) matched termination.

Current Loads (I-loads)

A more sophisticated termination scheme is the so-called I-load (Figure 5-11b). It consists of a 50 Ω termination resistor connected to a diode bridge, which is supplied by a current source and a current sink and referenced to the termination voltage. As long as the programmed source or sink current is not exceeded, the diode drops along the bridge keep the back end of the resistor on the same potential as the termination voltage. Thus, in this range the load acts just like a 50 Ω resistor to the termination voltage, and will provide very effective matched termination (assuming the current sources have switching times less than the rise time of the incoming signal). The behavior changes when the maximum current is reached - the effective termination impedance is no longer kept constant, and residual reflections occur. If the current limit is set to very small values, the I-load acts almost as an open. For large values (preventing it from ever reaching its current limits for the applied signal) it acts just like a matched 50 Ω termination all the time, suppressing all reflections. For current limits between those two extremes it provides partial termination. Some testers have active loads that work up into the low GHz range, but since for high-speed testing *all* reflections should be avoided, the "matched 50 Ω" mode of operation is usually the only one of interest, so the additional complexity compared to a simple 50 Ω resistor to the termination voltage is of little use here.

Matched Termination (R-load)

The principal function is almost trivial, as it consists of just a 50 Ω resistor connected to a DC voltage source (see Figure 5-11c), which as we have seen in the first section removes reflections completely, at least in theory. If available, this should be the termination of choice for high-speed testing. The challenge is to make the reaction time of the DC source smaller than the signal rise time, which is achieved through capacitive decoupling. It is imperative that the path (cable or PCB trace) has tightly controlled impedance as well so it really matches the termination impedance. A well-designed R-load will leave virtually no residual reflections even at data rates in the Gbit/s range.

Differential Termination

So far we have only considered termination of single ended signals. Differential signals add another layer of complexity to our considerations, because now we may have *two, unequal* characteristic impedances – Z_{odd} for the differential component and Z_{even} for the common component, so in this general case a single resistor will not suffice. Fortunately as we have seen in Section 5.2.3, in a test environment there is usually no compelling reason to couple true and complement line, and if one routes them uncoupled, even and odd mode impedance will be equal, and single ended termination is all that is needed.

For the general case (Z_{even} and Z_{odd} *are* different) several termination schemes are in use [6], [19], Figure 5-12 shows some of them, all based on resistive termination.

The most elaborate scheme (Figure 5-12a) uses a set of three resistors, and it can match both even and odd mode impedance. Note that if the lines are uncoupled (and thus Z_{even} is equal to Z_{odd}), the center tap resistor becomes infinite and we are back at two single ended lines with matched termination.

As far as test applications are concerned, this termination scheme can successfully remove reflections of the differential as well as of the common component of the signal. Its disadvantage lies in the necessary number of components (the total footprint requirements can create trouble). In addition, such a termination (if inside the tester) prevents any single ended usage of the two channels.

A single ended scheme (Figure 5-12b) can only perfectly terminate either the differential (the usual choice) or the common signal. If the termination resistance matches Z_{odd}, then the common signal will only be partially (but to a large extent) terminated, unless the lines are uncoupled so that Z_{even} and Z_{odd} are equal. A big advantage—not offered by any other differential termination scheme presented here—is that channels terminated this way can be used either in singled ended or in differential configuration, greatly improving the flexibility of the test platform.

On the other end of the complexity spectrum is the simple bridged termination (Figure 5-12c), which consists of just a single resistor between the two lines, matched to the differential impedance. It provides full termination for the differential signal component, but none whatsoever for any common component. Due to its simplicity and small footprint this is often the method of choice for on-die termination, but only to a lesser extent for the termination inside the tester, where space (and cost) constraints are not as pressing compared to accuracy and signal fidelity requirements[20].

One can improve the last setup by splitting the termination resistor in two and adding a buffer capacitor in the middle (Figure 5-12d). The capacitor acts as an AC ground that can terminate spikes of common mode (and spikes are what we care most about, since static common mode offsets do not add pattern dependent errors).

---

[20] While single ended termination will provide optimum signal fidelity in this case, sometimes one may instead prefer to use simple bridged termination in order to match the situation in the target application, or because the DUT driver requires a floating load. Thus single ended termination is the most common termination scheme in lower and medium speed multi-purpose testers, while solutions geared exclusively towards high-speed differential testing occasionally employ bridged termination.

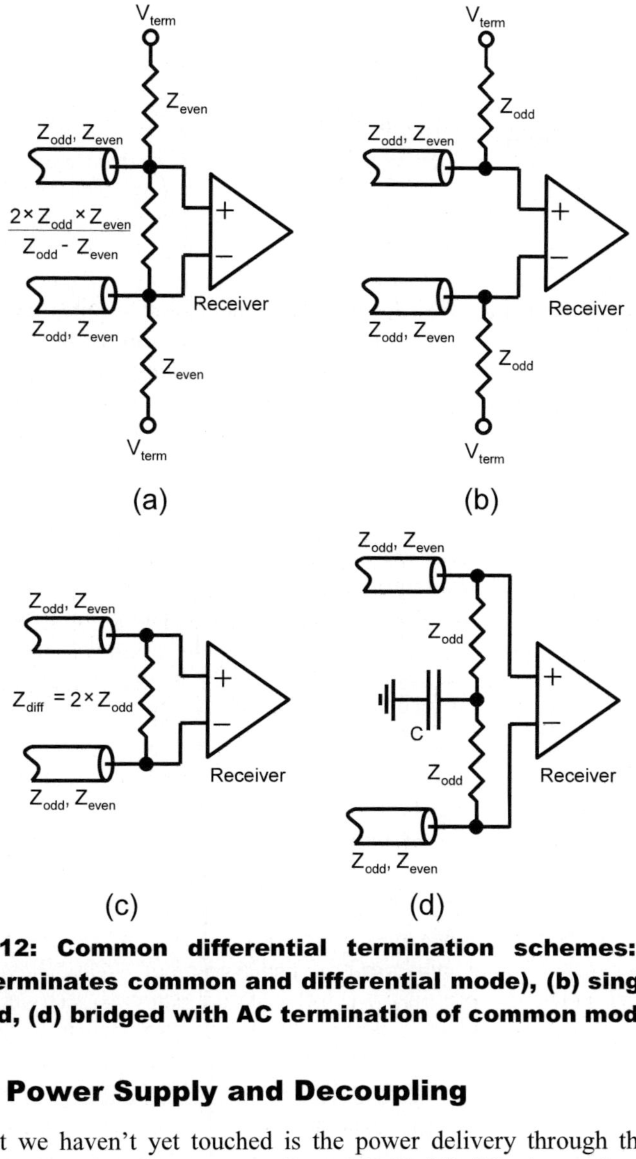

Figure 5-12: Common differential termination schemes: (a) ideal pi-type (terminates common and differential mode), (b) single ended, (c) bridged, (d) bridged with AC termination of common mode spikes.

5.2.5 Power Supply and Decoupling

A topic that we haven't yet touched is the power delivery through the interface, called the "*power distribution system*" or "PDS" [4], [6]. Apart from providing digital signals to the device under test, the tester must also supply it with electrical power. To obtain meaningful test results (and to avoid yield loss because of glitches in the device power supply during the test run) we have to make sure the supply levels remain stable independent of the device behavior. In the following section we will see how this can be achieved through proper decoupling (i.e. charge buffering).

At first glance one could be misled to believe that this is of no special interest because power is DC and we are only concerned with fast signals. However, this does not consider the fact that the power consumption of a device is not constant but can actually change extremely fast during a test run. For example, if a processor goes from sleep mode into full-blast activity, its power consumption will jump from near zero all the way to maximum in just a few clock cycles. Even worse, when the channel drivers on an output bus are switching, depending on the exact driver design large spikes of current can be drawn within just one signal rise time (this can easily be much less than 100 ps). Although bipolar circuit designs (used for highest-speed circuits instead of classical CMOS) in the case of core power consumption, and differential signaling in the case of the bus drivers mitigate those sudden surges in power consumption, they can't completely remove them. At the same time, as supply voltages decrease, the demands on rock-solid supply levels increase continuously. Typical specifications are a maximum ripple in the final application of only 5 to 10 percent (and less at test time if one wants accurate results), while supply voltages are at 2 Volts or below, which means only a few mV of margin for the supply voltage.

Quasi-static Decoupling

The classical approach for keeping the voltage at a variable current load constant is the well-known force-and-sense technique, shown in Figure 5-13 in addition to the power supply lines (force and return), two sense lines (sense and sense return) are attached to the device's power supply pins. Since only negligible current is flowing over the sense lines, there is no voltage drop across them and the power supply can measure the actual voltage at the load (as opposed to the voltage driven by the supply) and, using a regulator feedback loop, it adjusts the driven voltage to achieve the desired voltage level at the load.

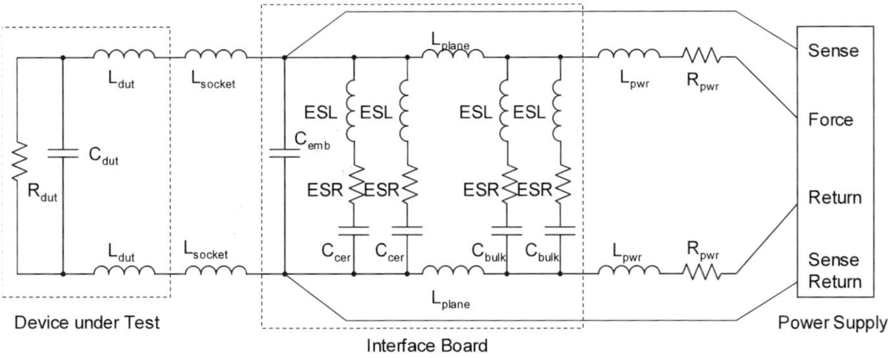

Figure 5-13: Simplified view of a test setup's power distribution system: The power supply regulates its driven voltage through a sense feedback loop. Faster transients are buffered through large bulk decoupling capacitors and smaller ceramic capacitors closer to the device. The main obstacle is the socket, pin, and bond wire inductance which limit the achievable slew rates, so as a first line of defense the device has some decoupling capacitance integrated on the die itself.

It is usually not possible to have the bulky power supply very close to the device (in actual production test setups the distance can be up to several meters), so there is a time delay before the power supply even notices a change at the device, and another delay before the changed levels make it back to the device (just as for any other TL, the propagation speed along the power supply lines is finite). What's more, since this is a feedback loop it will need several iterations before it settles to a final state, and the loop gain cannot be very large and the response not very fast, otherwise the whole setup will break into wild oscillations. So the sense loop feedback is a delicate tradeoff between reaction time (loop bandwidth) and stability (in addition, remember that higher bandwidth always means more total noise).

Practically achievable loop responses are therefore in the range between a few ten ns and several 100 μs, an eternity given device cycle periods of down to below 100 ps.

Capacitive Decoupling

Fortunately there is a good way out of this dilemma between poor reaction time or poor stability and the trouble with the distance between load (DUT) and supply [23]. Since the power supply itself is already taking care of basic supply and longer-term variations, all that is needed is a transient supply that has a very fast response and can cover for the short term variations. One can get this by adding capacitors close to the load (the device), as in Figure 5-13. They will charge up when the power supply is turned on. If now the load changes, they will supply (or sink) the current difference, slowly discharging until the power supply feedback kicks in. If the capacitance is large enough and the time short enough, the residual voltage drop will be negligible.

Several parameters affect the quality of this decoupling. Just as for the power supply itself, the distance between the capacitance and the device puts a lower limit on the fastest possible reaction time. With propagation speeds of around 15 cm/ns, and considering that a stable situation will only be achieved after several (typically around three) round trips, a distance of only 5 cm means changes faster than 2 ns cannot be buffered with such an external capacitor.

Another limitation of this method is the inductance of the path between the device and the decoupling capacitor. An inductance L withstands any sudden change in current and only slowly opens up (time constant L/R). Thus inductance reduction is paramount, while the trace or plane capacitance is of no importance or at best helps a bit (see next section). All this is in contrast to signal lines, where all we strive for is a constant, defined ratio between capacitance and inductance[21].

If the resistance between capacitance and load is not negligible, it will again introduce a voltage drop proportional to the current drawn. With typical requirements that the voltage drop be no larger than just a few mV, the maximum allowed resistance is in the order of only a few mΩ. Keep in mind that just like in

[21] Alternatively we could say that in the case of the power supply we strive for a minimum-impedance, minimum-delay TL.

signal lines, skin effect can increase the effective resistance to many times the DC value.

Finally, physical capacitors are not ideal capacitances, but have some amount of ohmic series resistance ("*effective series resistance*", ESR – see Figure 5-13) due to the finite conductivity of their leads and the plate material, as well as inductance ("*effective series inductance*", ESL – see Figure 5-13) caused by the leads and by the internal geometry, and shunt (leakage) resistance (especially in the case of electrolytic capacitors)[22]. All this limits the efficiency with which a given capacitor can supply current. The inductance makes the capacitor completely ineffective above a certain frequency[23] because its inductive impedance component becomes larger than the capacitive impedance, so in this range the "capacitor" looks like an inductor - its capacitance has no longer any effect.

Tiered Decoupling and Layout Considerations

To solve all those conflicting requirements - fast response, close to the device, large enough capacitance with minimum resistance and inductance, etc. - we have to employ a tiered approach for the decoupling (see Figure 5-13).

In fact, on every semiconductor device the designer has already placed a certain amount of on-chip capacitors as a first stage of decoupling. Though the maximum size is usually limited to a few nF, they are as close to the load as possible, and not hindered by the lead, package, or socket inductance, so they can successfully buffer very sudden spikes (above 100 MHz).

A second, also very effective capacitance comes more or less for free with the PCB. It is the inter-plane capacitance between power and ground plane, often called the "embedded capacitance". It can be maximized by using full power planes (as opposed to just wide traces) and choosing a dielectric with large ε_r and minimum thickness. A high loss tangent helps to dampen high-frequency ringing bouncing around on the ground planes, thus further improving the supply stability. In other words, the dielectric material requirements for power planes are the exact opposite to the ones for signal layers, and here for high-performance boards the inexpensive FR-4 is an excellent choice. When choosing the thickness, make sure the selected material can withstand the applied field and does not have reliability or manufacturing problems at small sheet thickness. The capacitance one can achieve is limited (in the order of just a few 10 nF even for a larger production test PCB), but has almost negligible inductance and resistance and thus the fastest possible reaction time.

One must be aware that depending on the time scale, only a limited fraction of the total plane is effective as embedded capacitance because the portions of the plane

---

[22] Values for ESR and ESL are commonly specified in the data sheet for the specific capacitor.

[23] A reasonable definition for the limit frequency f_{limit} is the point where the inductive impedance becomes equal to the capacitive impedance – the so-called self-resoance frequency of the capacitor. This point is given by $f_{\text{limit}} = 1/\sqrt{2\pi LC}$ (*L* being the parasitic inductance, and *C* being the capacitance).

that are too far away cannot respond fast enough - the propagation speed of the signal is again $v = c/\sqrt{\varepsilon_r}$.

Overall, the value of small plane thickness lies mostly in the fact that it provides a low-inductance path for the currents from the bulk capacitors described below, unless some special ultra-high ε_r material is used to increase the capacitance.

The second line of defense is an array of small ceramic capacitors (C_{cer}) as close to the device as possible (Figure 5-13). Choosing small (short, but wide) capacitor packages, putting them as close to the DUT as feasible, and connecting them to power and ground plane with as many vias in parallel as possible (all vias placed close to or even underneath of the capacitors) minimizes the loop inductance. Those capacitors still don't have excessively large capacitance, but will react fast. When choosing their size and number, keep in mind that it is always better to use two capacitors of half the size than a single larger one, because this - assuming same package and same via connections - reduces the series resistance as well as the series inductance by half.

Finally, a set of large electrolytic capacitors completes the decoupling buffer (also shown in Figure 5-13 – C_{bulk}). Since they only need to decouple rather long-term changes, they can be placed anywhere on the PCB (i.e. not necessarily very close to the device), but still inductance should be kept to a minimum - never use leaded (through-hole) package types, always use surface mount devices. The maximum amount of capacitance that can be put on the board is limited by the current drive capabilities of the power supply - too large a total capacitance will make it unstable and cause it to oscillate, or it may overload the supply during power-up. Since electrolytic capacitors also have large leakage currents, they negatively impact the accuracy of supply current measurements (especially I_{DDQ} and leakage tests) - this again demands that one puts on only the necessary minimum of those capacitors.

Beware of the often-used tantalum capacitors because they tend to have catastrophic failure modes where they suddenly break into hot flames, irreparably damaging the usually very expensive PCB. If you have to use them, make sure their rated maximum voltages as well as their maximum ripple current far exceed the conditions encountered during operation. Today there are good replacement types available (e.g. so-called aluminum capacitors) that fail less spectacularly.

5.3 CHARACTERIZATION AND MODELING

We have seen techniques and concepts applied to the design of quality high-speed interfaces viewed from the TL point of view. In order to design well-working interfaces, which always means to go through some learning to continuously improve one's designs, two more things are needed. First, measurements on a given interface that enable *characterization and quantification* of the signal path performance (bandwidth, losses, parasitics, impedance and impedance mismatches) and thus *identification of weaknesses* in the design. But that approach still requires building one or more physical prototypes of the interface and is therefore a high-cost practice. Second, instead of using simple trial and error, modeling and numerical simulations can *predict* performance and many parameters of a planned design (e.g. PCB trace impedance or parasitics) before anything is built. Very extensive treatments on modeling and simulations are provided in [24] and [25].

5.3.1 Characterization Techniques

Time Domain Reflectometry

One of the most frequently used tools in signal path characterization is *time domain reflectometry* (*TDR*). The basic concept relies on the fact that the reflection coefficient at a discontinuity depends on the relative impedances at this discontinuity (see Section 5.1.2, formula (4)). This means that time resolved observation of the reflected signal allows spatially resolved reconstruction of the impedance profile as well as calculation of propagation times, parasitic parameters (inductance, capacitance), and losses. Figure 5-14 shows a basic TDR setup. A signal source sends a voltage step into the path, and a sampler observes the reflected signal. This setup can easily be expanded to characterize differential paths - one simply doubles the setup and if needed inverts one of the drivers so the pair sends in a common signal or a differential signal, respectively.

TDR has a number of advantages over competing methods (especially *Vector Network Analyzer – VNA*, explained below) [24], [26], [27], [28]. First, it gives all results (e.g. delays) directly in time domain, which is the domain of interest for digital test anyway, and can be used directly as input into Spice or similar simulation tools (see below). This makes it very intuitive, and often a semi-quantitative analysis of the path is possible by just looking at the acquired TDR trace. Even with only approximate knowledge of the propagation speed along the path it is simple to pinpoint the location of parasitics and other weaknesses.

TDR is easy to set up and does not require any complex calibration routine. Common application of TDR includes:

- Path length (propagation delay) measurements: this is frequently the method of choice for internal channel-to-channel timing deskew in an automated test system (the tester has to account for the propagation delay which will be different for each channel).

- Locating and quantifying impedance mismatches: actually this has been around for a very long time – people have been using it e.g. to locate faults in underground telephone lines. A common application today is quality control of printed circuit board signal traces.

- Quantitative measurement of inductive and capacitive parasitics: it can resolve capacitances of fractions of 1 pF and inductances far below 1 nH, which is unrivalled by any other method. One example is assessing the severity of parasitics caused by vias or connectors.

- Determination of losses: the shape of the reflected edge (assuming the line is either open or shorted at the end) carries information about the amount of DC resistance, skin effect, and dielectric losses.
- Rise time/bandwidth measurements: again the shape of the reflected edge provides the information. This is especially useful when a direct through-measurement is not possible because one end of the line is not accessible or too fragile to be contacted (e.g. needles on a probe card).

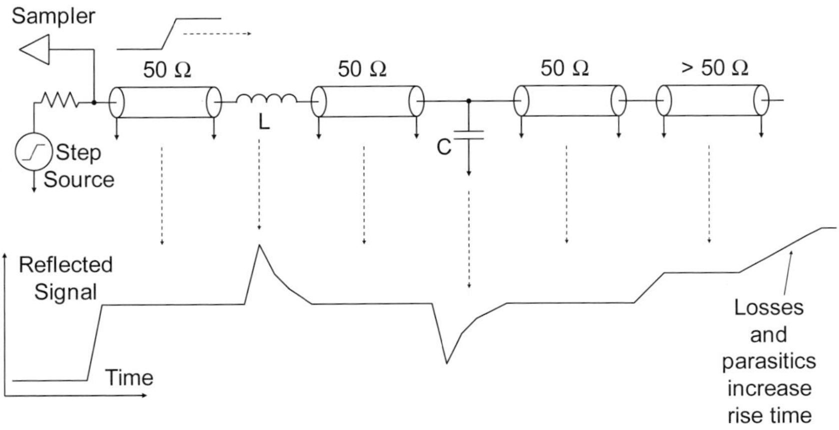

Figure 5-14: Schematic view of a TDR setup: A voltage source sends a fast-rising step into the line, a sampler is looking at the time-resolved reflected signal. Below are the typical signatures of a parasitic inductance, a parasitic capacitance, and an impedance change. The curve carries quantitative information about the size and location of those parasitics, losses along the path, and the bandwidth of the path.

TDR's weaknesses come into play mostly when increased accuracy is required [27], [28], [29]. First, to be able to easily judge the expected signal quality and the fitness of the path for a particular application, one should use a TDR signal rise time and edge shape similar to the one present in the final application. Otherwise conclusions are more difficult to make and rather inaccurate because the shape of the reflections is directly and strongly dependent on the exact signal edge parameters. What's more, cabling and probe behavior enter directly into the end result. Though it is possible to compensate mathematically for TDR signal effects caused by those elements, which commercial software packages do, this process involves numerical convolution and deconvolution of the signal used, which is an algorithm that is extremely sensitive to measurement noise, and thus not very accurate. The same goes for resolving features that are further down the line, which is rendered less and less accurate because of the multiple reflections and bandwidth degradations caused by the elements in front of them (this is comparable to a dense fog - the further away an object is, the less details one can make out). Again, deconvolution (often called "layer peeling" in this context) could save the day, but it depends on a very good signal-to-noise ratio of the signal as well as excellent linearity of the measurement system, which TDR has trouble providing[24].

The maximum frequency range covered by TDR is given by the signal rise time, and today is limited to less than 20 GHz for typical high-end sampling oscilloscopes with TDR option (this corresponds to rise times of only about 17 ps)[25]. It is not trivial to supply much faster rise times to the system under test (note that the stated rise time is the aggregate rise time of signal source plus receiver), especially since one must conserve a very clean edge shape with very little overshoot as well as fast settling. This will very soon become a bottleneck for the characterization of leading-edge interfaces.

Also the extraction of loss parameters (skin effect and dielectric) is rather cumbersome since those effects - while rather easily described in the frequency domain - have complex influence on the reflected edge shapes in the time domain, so their determination based on TDR responses always needs computer based numerical fitting. (For loss or bandwidth measurements one can also use the transmitted signal - in this case we speak of time domain transmission – TDT – measurements, which can yield more accurate results).

Finally, one of the most serious shortcomings is that there is no intuitive way to deduce the reaction of the whole path when given only the TDR responses of its components (again, it involves numerical convolution of the partial responses, which except for special cases requires the help of a computer).

Vector Network Analyzer

As an alternative to TDR/TDT, the *vector network analyzer* (*VNA*) (Figure 5-15) is also widely used, and it has been a very common tool for RF type (narrow band) applications already for a long time. It yields its results - matrices of S-parameters[26] - in the frequency domain instead of the time domain. Thus - with the exception of losses - those results are often much less intuitive to understand in terms of parasitics, reflections etc. for someone working on digital applications. However, the conversion from frequency to time domain can be made mathematically through Fast

---

[24] Even with averaging the signal-to-noise ratio for TDR with a good oscilloscope is limited to about 1:1000 (or 60 dB) at best. In contrast vector network analyzers achieve $1{:}10^5$ (100 dB) or more. On the other hand for basic signal integrity considerations for digital signals 1% accuracy (40 dB) is very often already perfectly acceptable.

[25] There have been some recent developments to reduce the TDR system rise time that push the bandwidth limit to around 30 GHz using nonlinear transmission lines as external pulse sources [24].

[26] For an N-port system (a system with N connections to the outside world) the S-parameters form a $N \times N$ matrix. Its elements $S_{xy}(f)$ give the (frequency dependent!) phase and magnitude of the signal coming out on port x when applying a sine wave signal of frequency f to port y, with all other ports terminated to the characteristic impedance. For example a simple transmission line can be seen as a two-port system, with S_{11} corresponding to the reflection coefficient, and S_{21} to the transmission coefficient. (Note that the stimulus here is a continuous sine wave, not a step as assumed in all previous sections, so the exact meaning of "reflection coefficient" and "transmission coefficient" is somewhat different in this case).

Fourier Transformation (FFT), and most high-end VNAs have this option built in; in this case, in addition to frequency-domain output, they are able to produce outputs that look exactly like what a TDR setup would produce[27].

The basic principle of operation is to send a continuous sine wave into the system and measure amplitude and phase of the reflected as well as the transmitted signals (see Figure 5-15); the VNA performs a sweep over a wide frequency range and records the measured values for each step. More expensive 4-port VNAs suitable for crosstalk and differential path analysis have only recently become widely available, whereas differential TDR and TDT have been around for much longer.

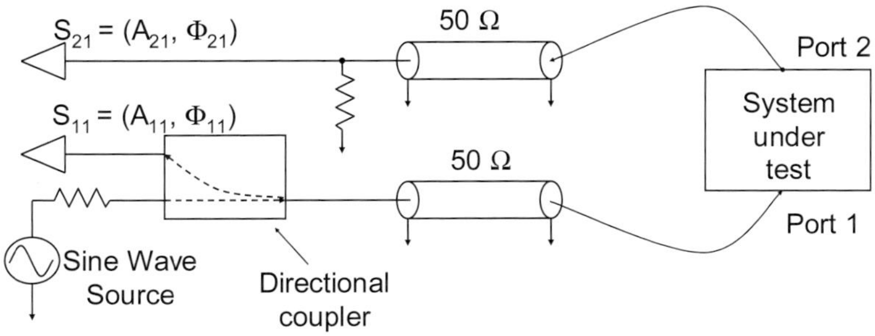

Figure 5-15: Schematic view of a single-ended (i.e. non-differential) VNA setup: The VNA sweeps the frequency of a sine wave signal and for each step measures amplitude and phase of both reflected and transmitted portion of the signal (a third part may be absorbed in the system under test because of losses).

We should stress that TDR and VNA really obtain exactly the same fundamental information about the path - they are merely two alternative ways to the same final goal. There is nothing that one method can do and the other can't, at least from a theoretical point of view. Of course this does not imply that one method cannot have certain practical advantages over the other [24], [26], [27], [28].

As one may already suspect from the discussion about the TDR, the strengths of the VNA method - and the reason why it is used in digital applications - lie in the achievable accuracy. It begins with the fact that it is easier to generate a highly clean high-frequency sine wave than a clean, fast rising isolated edge with no settling or ringing issues. Also VNAs can be calibrated to completely remove the effects of cables and probes used to connect to the system under test (this process is called "de-embedding"), and be trimmed for optimum performance at the specific impedance of the system (important if the system impedance is not standard 50 Ω). In addition, VNAs with upper frequency limits well beyond 100 GHz are readily available

[27] Note that due to mathematical reasons (necessity to apply windowing before performing FFT to avoid aliasing effects) a VNA needs approximately *double* the equivalent bandwidth of TDR to obtain comparable results in the time domain, e.g. around 30 GHz to compete with state-of-the-art TDR oscilloscopes (with 15 GHz equivalent bandwidth). The same is true in the opposite direction as well (obtaining frequency domain results from TDR data), but in the case of digital application the domain of interest is usually the time domain, which gives TDR a head start.

(although quite expensive), thus by far exceeding the range covered by TDR (as mentioned before, 30 GHz at best).

Loss parameters can be determined in a very straightforward way from the curves showing the transmitted signals - remember that skin effect is roughly proportional to the square root of frequency, and dielectric losses are directly proportional to frequency, so in a log-log display chart they are simple straight lines.

The major disadvantage of the VNA method is that it does require more elaborate preparation - the high accuracy and superior signal-to-noise ratio can only be achieved if the calibration is done correctly and accurately as well, and since the results are less intuitive, more experience is required to judge the reliability of the results obtained. Calibration must be done with calibration structures close to the impedance of the system to be measured (within 5%), lest the potential high accuracy be lost, which means several sets of such structures are necessary, driving up instrumentation cost. Furthermore, each new setup requires the user to perform a new calibration, in contrast to TDR where once a system is calibrated once, not much more need to be (or even can be) done, except for periodic recalibration every few weeks to account for instrument drift.

Table 5-1 gives an overview of the advantages (+) and disadvantages (–) of TDR vs. VNA for *digital* transmission path analysis. Overall TDR tends to be the better choice for everyday *digital* applications because of its ease-of-use, much lower cost, and the results being in time domain (which is the domain of interest for digital signals), while VNAs are better suited when highest accuracy is absolutely required, and of course for RF and microwave applications (where VNAs originated after all!).

5.3.2 Path Modeling

Measurement-based Equivalent Modeling

A very accessible road for any test engineer is modeling based on electrical measurement data (provided by TDR or VNA characterization of the path). The goal here is usually an "equivalent circuit model" that reproduces the behavior seen in the characterization. The TDR signal is analyzed and TL parameters (delays, parasitics) are extracted for each segment. It helps, but is not indispensable, to know the physical structure of the path (e.g. position of connectors, vias, etc.). There are software tools available to do this in a semi-automated way and to optimize fit parameters. The end result is an equivalent circuit[28] consisting of passive elements (capacitances, inductances, resistors, and ideal as well as lossy TLs) that can be used to run simulations in Spice or some other circuit simulator to determine the interface performance for specific conditions [7]. The achievable accuracy is limited, but on the other hand this method is able to handle complete, complex real-world interfaces in reasonable time and with reasonable effort.

---

28 "Equivalent" means that there isn't necessarily a one-to-one correspondence between the circuit elements in the model and physical structures in the actual interface; the requirement is merely that the two behave the same for the same stimulus within the required level of accuracy.

TDR	VNA
+ Results in time domain – shows size and timing of reflections directly.	– Results in frequency domain – no direct correspondence to digital data streams, edge shapes, jitter. Needs FFT to convert results into time domain, which reduces effective bandwidth.
+ Good for pinpointing problem areas; signal peaks and dips correspond directly to location of physical features, parasitics.	– Response lumps whole path together – difficult to find out which component causes reflection/loss.
+ Easy and intuitive to use.	– Requires expert knowledge to make use of superior accuracy and to assure validity of calibration.
– Not very accurate or straightforward for loss measurements (losses do not have easy time domain description).	+ Good choice for dielectric loss and skin effect characterization – simple trends in the frequency domain.
– Sampling scopes with TDR yield around 15 GHz of bandwidth. With external options up to 30 GHz.	+ Over 100 GHz VNAs are available (but necessary windowing for FFT to get time domain data reduces effective bandwidth to about half of the "spec value").
– Less accurate (50 – 60dB at best), but enough for digital levels.	+ Can be much more accurate, large dynamic range (> 100dB noise floor) *if and only if well calibrated.*
+ Needs at best only simple voltage gain/offset and timebase calibration (if at all). No need to recalibrate between setups. Even an uncalibrated setup can provide useful (though less accurate) data.	– Needs new calibration for each new setup and every few days, if not hours. Calibration is very tedious and time consuming (several minutes to several hours). Does not work at all (yields no useful data) if not calibrated. Quality of calibration (and data taken!) is difficult to judge from measurement data.
+ Differential TDR is almost as simple and fast to do as single ended TDR, and carries only a small price premium. It provides true differential stimulus.	– Most VNAs are single ended. Differential (4-port) VNAs are rather recent and very expensive, and in fact still provide only single ended stimulus and then create the differential result mathematically (a problem if there are any non-linear components in the path). Calibration time is even longer.
+ About half the price of a VNA with similar effective resolution in the time domain.	– About double the price of a TDR unit with similar effective resolution in the time domain. Moreover, FFT requires a true – expensive - Vector Network Analyzer (yields both phase and amplitude), not a simple S-parameter analyzer (amplitude only).
+ In many cases (moderate reflections), approximate windowing is trivial (only look at trace features of interest).	– Calibration has to be done to the end of the probe/cable/fixture, otherwise fixture effect cannot be taken out of the result. Major obstacle if the end of the line is not easily accessible or does not connect directly to the return line for calibration.
+ Modeling based on TDR data can provide topological interconnect models (direct correspondence between model elements and physical features in the path).	– Modeling based on frequency-domain VNA data yields behavioral models (no direct correspondence between model elements and physical features, "black-box models"). (If data converted to time domain data before modeling, there is no modeling advantage over TDR).
– Even with most recent software, modeling is semi-automated at best.	+ Modeling can be done automated, generated models can be very accurate.
– Predicting the response of a signal path based on measurements on single components is more complicated and much less accurate.	+ It is mathematically rather straightforward to accurately predict the total path response based on knowledge of the response of each component.

Table 5-1: Overview of advantages (+) and disadvantages (–) of TDR vs. VNA for digital transmission path analysis.

Two-dimensional Field Solvers

This class of simulation tools [5], [30] aims to calculate electrical path parameters based on known physical parameters (e.g. width, thickness, ε_r) of the interface structures, especially PCB traces or cables, by solving Maxwell's equations numerically. To keep complexity down (and handling user-friendly), and to achieve fast run times, they restrict themselves to two-dimensional cross-sections of the structure and make certain assumptions and approximations such as the following.

First, they assume the path extends uniformly and infinitely (or at least very far) in both directions perpendicular to this cross section, so it is sufficient to treat the structure as a two-dimensional problem only (hence the name for this type of tools). This means they neglect any fringe effects (e.g. end of the line) and discontinuities like connectors, vias, bends, tapered line sections, breaks in the ground plane, etc. This clearly limits the applicability to long, uniform structures such as cables and straight traces. In reality uniformity may not be given if the trace has significant manufacturing variations of its width or thickness.

Important numerical approximations usually made are the so-called "quasi-static assumption", which requires the wavelength of the signal to be much larger than conductor separation and trace width, further the assumption that the skin depth is small against the conductor thickness (so the inductance does not vary with frequency), and last but not least the simulator has to approximate a continuous field in a discrete grid of finite resolution ("finite-element method"), introducing quantization errors, especially at sharp corners.

The main application is the extraction of trace parameters like characteristic impedance, skin effect and dielectric losses, crosstalk and coupling, heating, and current distribution. In this case these simulators do a good job and the achievable accuracy of leading-edge tools is limited only by insufficient knowledge of the input parameters (due to measurement accuracy, material and manufacturing tolerances) rather than by the simulation accuracy as long as none of the assumptions above is violated. Clearly they can only give answers for certain parts of a real interface, namely trace and cable sections.

PCB designers routinely use those tools to determine the physical dimensions of their traces and planes in order to match desired impedances and crosstalk requirements. Their calculation speed and ease of use make them a valuable asset in anybody's toolbox.

Three-dimensional Field Solvers

The top end of software (also with regard to price and complexity) does away with many restrictions of two-dimensional field solvers. They take into account the full three-dimensional structure of the elements under investigation, again performing *ab-initio* finite-element calculations based on the physical properties and dimensions of the element, and Maxwell's equations (often enhanced by thermal and mechanical calculations). This thoroughness comes at a price, namely long run times and difficult, time-consuming model setup because it requires the creation of a full three-dimensional model of the structure. Apart from geometrical data depending on the type of analysis the user has to supply material properties (conductivity, electric susceptibility, magnetic permeability, specific thermal capacity, surface roughness, for each component of the structure, e.g. a connector launching onto a trace, or a via. Those properties can all be frequency or temperature dependent, and a thorough analysis will also include information about the tolerances for each property to assess the accuracy of the end results.

On the upside they can deal with arbitrary structures like connectors, vias, bends, and inhomogeneous, non-uniform lines - for such systems it is the only way to get good and accurate theoretical answers. Accuracy can be excellent, of course dependent on the quality of the input data (material constants, dimensions, etc.). Model complexity and simulation run time increases exponentially with model size, so this method is usually restricted to the investigation of small, critical subsections of the whole path. Run times of several hours are not uncommon even for smaller structures. Often the end result can be translated into an equivalent Spice model that can then be integrated into a Spice model of the complete path, greatly speeding up subsequent simulations that use those substructures.

Of course it is possible to combine any of the tools mentioned - e.g. use two-dimensional solvers for sections of straight TLs, calculate discontinuities three-dimensionally, and add equivalent models (based on TDR measurements) for drivers and receivers, in order to achieve a reasonable tradeoff between modeling complexity, effort, and accuracy.

5.3.3 Power Distribution System Modeling

Similar tools as the ones used for signal path modeling - from equivalent Spice circuits (here power planes are modeled as meshes of capacitances and inductances) to three-dimensional full-wave solvers - are employed for the PDS [31]. The trouble here is that - even more than for signals - the overall performance is dependent on the system as a whole (power supply, bulk decoupling, PCB design including vias, DUT) rather than just on its components, which makes such modeling very complex. On the other hand, often exact answers are not needed (more often than not, the power consumption profile of the DUT is not known very accurately to begin with), as long as the model can predict a sufficient safety margin. What's more, the maximum frequency of interest rarely exceeds 1 GHz[29], which corresponds to a rather coarse necessary spatial resolution of maybe several millimeters at most (compare this to typical signal trace widths of around a few tenths of a millimeter).

Analysis is mostly done in the frequency domain. Every good capacitor manufacturer supplies ESR and ESL data for his devices (see Section 5.2.5), so this is the easy part. More difficult is the performance prediction for connection vias, power planes, and DUT package leads and bonding wires, which again requires full three-dimensional modeling. On the upside, the main parameters of interest here are

---

[29] The inductances of the device package prevents supplying very high frequencies to the die at the low impedance (fractions of an Ohm) required for a power supply, so as discussed earlier the first line of defense has to be some decoupling capacitance integrated into the device itself, making the interface performance at those speeds irrelevant. (Data and clock signals can enter the device at higher speeds because the characteristic line impedance is much larger, typically 50 Ω).

just inductance and maybe approximate propagation times, which reduces the demands in the calculations. Currently however full simulation is still often replaced with very approximate modeling.

One large obstacle is that it is rather difficult to verify results experimentally because of the extremely low impedance of the structure; after all, low impedance over a wider frequency range is the design goal here. Current profiles for specific DUTs are usually much less well defined than their signal output, and in addition, since the PDS by definition has to be a very low impedance structure, standard 50 Ω TDR is ill suited to measure such a system; good VNAs stand a much better chance here since they can be calibrated to the low resistances, and also yield the results directly in frequency domain, which is of advantage here. A good discussion on suitable methods is found in [32], [33], [34].

Second, the numerical results can depend strongly on exact component parameters. As an example, capacitors all have resonant frequencies (resonances - the Q-factor - get stronger with reduced series resistance, which puts a limit to the reduction of this parameter), but this only wreaks havoc if all capacitors in a bank have their resonances at exactly the same frequency (which is a frequent assumption in simulation, using ideal capacitor parameters); so manufacturing variations can save the day, but at the same time this is a very fragile result that needs good knowledge of the actual distribution [35].

5.4 OUTLOOK

With ever increasing system speeds, the test engineer needs to become proficient in a large variety of subjects ranging from TL theory over PCB design and layout, to path characterization, modeling, and analysis [36]. Since performance and cost are almost always opposing factors, good command of these tools enables one not only to achieve the required performance needed to ensure accurate testing, but also helps in making educated choices as to where in the setup additional effort (and money) should be spent, and where shortcuts can be taken. All those challenges are only going to become more pressing with each new advance in speed and accuracy requirements. Apart from technological advances in the test hardware and PCB design, interface simulation and modeling will more and more become a pure necessity in order to understand, correlate, correct and guardband test results. But today many advanced modeling tools still require extensive expert knowledge to use them gainfully, so significant development effort will be needed to make those tools more accessible to the average engineer. As far as layout is concerned, high-performance PCB design needs a large portion of manual work – as opposed to auto-routing – to optimize trace layout for good trace matching (e.g. it's often more an art than science to take via delays into account correctly), crosstalk avoidance, loss minimization etc. Developments in this area should streamline and accelerate the manufacturing of well-performing interface boards. On the hardware side, sockets and needle probe heads are one of the main limitations in the test signal path – increasing their performance while at the same time keeping them rugged and low-cost enough for production applications will be a challenge. As copper based (electrical) interfaces start approaching their limitations due to increasing losses, crosstalk, and signal fidelity issues, optical transmission schemes are likely to gain a foothold in the interface area as well.

REFERENCES

[1] U. Schoettmer, C. Wagner, T. Bleakley, "Device Interfacing: The Weakest Link in the Chain to Break into the Giga Bit Domain?", *IEEE International Test Conference*, 2000, pp. 995–1004.

[2] S. Re, M. Li, "A Generic Test Path and DUT Model for Datacom ATE", *IEEE International Test Conference*, 2003, pp. 528–536.

[3] T. Sargent, "Physical Principles of Interface Design", *IEEE International Test Conference*, 2002, pp. 549–554.

[4] Howard Johnson, Martin Graham, *High-Speed Digital Design*, Prentice Hall, 1993.

[5] Howard Johnson, Martin Graham, *High-Speed Signal Propagation*, Prentice Hall, 2003.

[6] Eric Bogatin, *Signal Integrity - Simplified*, Prentice Hall, 2004.

[7] T. P. Warwick, "Mitigating the Effects of the DUT Interface Board and Test System Parasitics in Gigabit-plus Measurements", *IEEE International Test Conference*, 2003, pp. 537–544.

[8] W. Humann, "Compensation of Transmission Line Loss for Gbit/s Test on ATEs", *IEEE International Test Conference*, 2002, pp. 430–437.

[9] Mark I. Montrose, *Printed Circuit Board Design Techniques for EMC Compliance*, Wiley-IEEE Press, 2000.

[10] D. E. McFeely, "The Process and Challenges of a High-Speed DUT Board Project", *IEEE International Test Conference*, 2002, pp. 565–574.

[11] Roger E. Weiss, David M. Barnum, "High Speed Connections for 40 GHz and Beyond", *Chip Scale Review*, April 2003.

[12] Ron Iscoff, "Reach for the Sky! As Socket Choices Multiply in the GHz Range, Makers Face New Demands", *Chip Scale Review*, April 2003.

[13] Mark Brandemuehl, "Challenges in Advanced Wafer Level Probing", *SST*, September 2003.

[14] Ken Smith, Reed Gleason, "Membrane Probe Speeds Digital and RF IC Testing", *Microwaves & RF*, Jan. 1995.

[15] Tom Lecklider, "ATE Interfacing Outgrows 12" Probe Boards", *EE Evaluation Engineering*, Oct. 2001.

[16] Rick Hartley, "Design at the Speed of 6 Gb/s", *Printed Circuit Design & Manufacture*, March 2004.

[17] Mike Kuszaj, Art Aguayo, "A Substrate for All Seasons", *Printed Circuit Design & Manufacture*, March 2000.

[18] Lee W. Ritchey, "A Survey and Tutorial of Dielectric Materials Used in the Manufacture of Printed Circuit Boards", *Circuitree Magazine*, Nov. 1999 (also available for download from www.SpeedingEdge.com/press-articles.htm).

[19] Stephen H. Hall, Garrett W. Hall, James A. McCall, *High-Speed Digital System Design*, John Wiley & Sons, Inc., 2000.

[20] Douglas Brooks, "Differential Trace Design Rules: Truth vs. Fiction", UltraCAD Design, Inc., and Mentor Graphics Corp., 2002, downloadable from www.ultracad.com.

[21] Lee W. Ritchey, "Differential Signaling Doesn't Require Differential Impedance", *Printed Circuit Design Magazine*, March 1999.

[22] Howard Johnson, "Diode Terminations", Online Newsletter Vol. 2, Issue 19, July 1998, downloadable from www.signalintegrity.com.

[23] Jiayuan Fang, Jin Zhao, "Low Impedance Power Delivery over Broad Frequencies", *Printed Circuit Design & Manufacture*, Sept. 2003.

[24] Brian Young, *Digital Signal Integrity: Modeling and Simulation with Interconnects and Packages*, Prentice Hall PTR, 2001.

[25] D. G. Swanson, Jr., W. J. R. Hoefer, *Microwave Circuit Modeling Using Electromagnetic Field Simulation*, Artech House, 2003, ISBN 1-58053-308-6.

[26] Mark Alexander, "Power Distribution System (PDS) Design: Using Bypass/Decoupling Capacitors", Xilinx Application Note XAPP623 (V2.0), April 2004, download from www.xilinx.com.

[27] Dima Smolyansky, Steven Corey, Mike Resso, "Choosing the Right Signal Integrity Tools for InfiniBand Measurements", *DesignCon* 2002, paper HP7 (paper also available from www.tdasystems.com under the title "Choosing Signal Integrity Measurement Tools: Time or Frequency Domain?").

[28] Mike Resso, Jeff Tehan, Eric Bogatin, "TDR and VNA: The Right Tools for the Right Measurements", Archived web cast, transcript and slides on www.gigatest.com, Aug. 2002.

[29] James R. Andrews (Picosecond Pulse Labs), "Time Domain Spectrum Analysis & S-Parameter Vector Network Analysis", Application Note AN-16, May 2004, download from www.picosecond.com.

[30] Howard Johnson, "2-D Quasistatic Field Solvers", *EDN Magazine*, Sept. 27, 2001.

[31] W. Prasad Kodali, *Engineering Electromagnetic Compatibility*, 2nd edition, chapter 14, IEEE Press and John Wiley & Sons, Inc., 2001.

[32] TDA Systems, Applications Note PDNA-0703 "Simple Approach to Modeling of Power Delivery Networks and Components"), DesignCon 2003 (available for download from www.tdasystems.com).

[33] Howard Johnson, "Measuring Power Plane Resonance", *Online Newsletter* Vol. 2, Issue 27, at www.signalintegrity.com.

[34] Howard Johnson, "Measuring Power Ground Impedance", *Online Newsletter* Vol. 2, Issue 14, at www.signalintegrity.com.

[35] Douglas G. Brooks, "ESR and Bypass Capacitor Self Resonant Behavior: How to Select Bypass Caps", Ultracad Design, Inc., Feb. 2000, available for download from www.ultracad.com.

[36] Websites (in alphabetical order) that have a large number of freely downloadable literature (articles, book excerpts, application notes) about signal integrity, interfaces etc.: www.Agilent.com, www.Ansoft.com, www.BogatinEnterprises.com, www.EDNmag.com, www.Gigatest.com, www.Picosecond.com, www.Signalintegrity.com, www.Sigrity.com, www.SpeedingEdge.com, www.Tektronix.com, www.TMWorld.com, www.Ultracad.com

Chapter 6

DFT-Oriented, Low-Cost Testers

Al Crouch and Geir Eide

In recent times, the ever-evolving CMOS semiconductor design styles and the advance of manufacturing and fabrication process technologies have created a cost, schedule, yield, and throughput crisis that can no longer be efficiently handled with functional test vectors. The "Time-to-" schedule requirements coupled with modern volume requirements and nanometer defect types are eliminating the luxury of time involved in using behavioral vectors to evaluate manufacturing testing, to conduct first silicon bring-up and characterization, and to bring up yield during initial volume ramping. In addition, cost requirements are driving lower cost Automatic Test Equipment (ATE). A solution to this problem is the adoption of structural test and the use of *structural testers*.

This chapter introduces the concept of structural test and structural testers and delves into the adoption drivers for both and the requirements, design rules, and Design-for-Test (DFT) techniques that enable the use of the structural tester.

D. Gizopoulos (ed.), Advances in Electronic Testing: Challenges and Methodologies, 179-216.

6.1 INTRODUCTION

Test has always been viewed as a non-value added part of the semiconductor development process. The original focus of test being to ensure that the manufacturing process produced exactly what was submitted to the fabrication facility. If the manufacturing process was error free, then test would not be required – hence, test is viewed more as an expense or a tax. To buy back some of the non-value added, test has historically been applied as a mixture of verification of the design process and verification of the manufacturing process – by the use of behavioral or functional vectors[1].

As semiconductor science evolved, design sizes moved from small scale integration (SSI) up through very large scale integration (VLSI) and into the System-on-a-Chip (SOC) integration levels of tens of million of transistors on a single die; wafer dimensions moved from inches to what is now a foot (300 mm); feature sizes have passed through the micron, sub-micron and into the nanometer space; and packages have grown from just a few pins to hundreds of pins. Modern designs are so rich and complex in features and in application frequencies, that it is difficult to create the functional environment needed to verify them during the design phase.

All of these advances have made the reliance on functional vectors as a method to verify the manufacturing process, a costly disadvantage [1]. It takes longer to develop the vectors, they are more suitable for design verification, and they must be graded against fault and defect models to turn them into suitable manufacturing test vectors. Mostly though, functional vectors are costly in application on modern semiconductor ATE, also commonly known as testers. The semiconductor and the ATE are caught in an endless treadmill where the semiconductor advances, then the ATE must advance in order to test the semiconductor, then the semiconductor advances again and so on. During this advancement, as the cutting-edge semiconductors push the technological envelope, older versions branch off the relentlessly growing Moore curve to become their own markets and businesses, as shown in Figure 6-1. Devices above a certain point on the Moore curve are not driven by development costs as much as they are driven by Time-to-Market (TTM)[2].

At some point in time the cost function of test for both the cutting-edge parts and the parts in markets all up and down the Moore curve started to exceed the cost function of semiconductor development – making the 'cost of test' a dominating influence on the semiconductor production process[3]. This pushed the development of

---

[1] Behavioral, functional, or operational vectors are vectors that are used to exercise the development model in the software testbench to ensure that the design commits the correct actions in reference to the specification. These software simulation vectors are often converted from a software simulation format to a tester format to prove that the silicon repeats the same behaviors and operations as the software model.

[2] Time-to-Market (TTM) means that there is a market window where the profit potential of the semiconductor is maximized and missing that window can result in a minimal profit or even not making enough return to break even on development and manufacturing expenses.

[3] The point in time when the test cost exceeded the semiconductor manufacturing cost is different for different markets and relies on many variables, most of them manageable. The ability to conduct test cost management on the cost variables (test time, vector volume, tester

structural vector generation – deterministic vectors generated against a fault model and largely using scan technology. Scan vectors reduced the time involved with vector generation and produced vector sets that were more optimal and richer in coverage – fewer vectors for a given amount of coverage and coverage based on more fault models. Other structural methods were also adopted to solve various problems – for example, memory built-in self-test (MBIST) and logic built-in self-test (LBIST) were adopted when the cost of providing routes or signal pins to get access to internal logic favored on-chip vector generation and response assessment versus tester bandwidth; current and leakage measurements were adopted when it was discovered that non-ideal or un-modeled faults and some latent faults resulted in excess current flow.

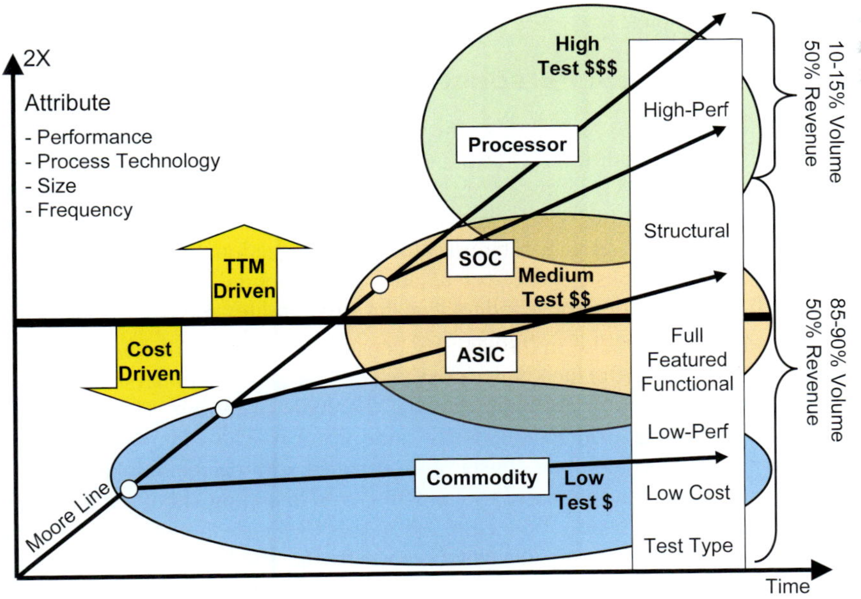

Figure 6-1: The Semiconductor markets, Moore's Law, and testing.

Scan vectors and BIST made the design-side vector generation task easier, but were ill matched for the ATE of the time largely because the ATE did not have enough memory to deal with the input vector volume or to capture memory fail data and scan results. This led to the myth that structural vectors were more expensive than functional vectors in application[4]. Eventually, the adoption of scan and memory BIST became a mature technology, but the mismatch between vectors and ATE still

cost, multi-die testing, yield, and others) has proven to take longer to implement than the rate of reduction in manufacturing cost factors (smaller die, more die per wafer, larger wafers). In some markets, the cost of test can be 50% to 60% of the total manufacturing, packaging, and assembly costs.

[4] Not all scan vector sets that exceeded tester memory were due to large vector sets, but many were because the scan architecture was implemented badly with too few scan pins and long, unbalanced, scan bit-depths. This resulted in trying to place all of the vectors in just a few tester channels and this usually easily exceeded the functional test memory of the testers of the time.

existed. This opened the door for a paradigm shift in ATE technology – the Low Cost Structural Tester.

It must be noted that there is not any specific name adopted by the industry for the low cost structural tester. In many cases the term structural and design-for-test (DFT) are used interchangeably, and the associated optimization of cost, reduced functionality, or low performance is added to the name. So, structural testers have been referred to under many names and acronyms such as: Very Low Cost Tester (VLCT); DFT-Optimized Structural Tester (DOT, DOST); Low Cost Structural Tester (LCT, LCST); Low Cost DFT Tester (LCDT); Reduced Functionality Structural/DFT Tester (RFT, RFST, RFDT); and Low Performance Structural/DFT tester (LPT, LPST, LPDT). We use the term Low Cost Structural Tester (LCST) throughout this chapter.

6.1.1 Historical Perspective on Structural Test

The LCST is a tester that is optimized to a feature set that lines up better with the application of structural tests – scan, logic BIST, memory BIST, and current measurement. The LCST is not meant to completely replace a high-performance functional tester – the high-performance functional tester still has its place (and conversely, a structural tester can be used for limited functional test). To be able to fully investigate the cause and effect of adopting structural test and a structural tester, the first step is to examine and understand structural test.

Although it is referred to in the industry constantly, there is still some confusion in some areas on the difference between structural test and functional test. Much of this confusion comes from not understanding the goals of testing. The majority of design and development organizations are quite familiar with running simulations, generating test-benches, and are constantly concerned with design validation. Many organizations even have automated systems to create operating code to apply to the software or emulation model of the chip. All of these vectors, though, are functional or behavioral vectors and are designed to verify the specified behaviors of the eventual chip against a golden specification or to assess a design coverage metric (code coverage). It is a common practice for manufacturing organizations to ask the design organization for manufacturing test vectors – and for the design organization to deliver a wealth of these functional "design verification" vectors.

For simplistic designs where a known few operations are all that the chip will be required to perform, the design verification vectors may be adequate for determining manufacturing correctness. However, modern designs have richness in their feature set, have many programmable operations, and are eventually destined for many different possible uses. The number of design verification vectors needed to verify the manufacturing process for this type of chip may be economically infeasible.

The manufacturing process is more concerned with making sure that the mask layers were applied in the correct order; that the masks are correct; that the process mix was correct; that processing steps do not result in missing or malformed wires, vias, dielectrics, and transistor elements; and that random defects don't alter the electrical characteristics or Boolean behavior of the final device. In addition, another concern is to not only find those parts with errors and defects, but to identify those parts with weaknesses and latent defects – they weren't broke when they came out of the fab, but they will break soon.

In the past, the "design verification" vectors were "fault graded" to assess their structural content – and only those vectors necessary to get the structural coverage up to some stated or defined quality level were kept. This means that the whole wealth of vectors that the design organization would like to see applied to the part after fabrication could not and would not be applied due to cost constraints – only those vectors needed for quality assessment would be kept. The problem with this methodology is that grading the wealth of vectors delivered by the design organization is a time-consuming proposition and may take months to develop a minimal-quality test program. Generating a high quality test program usually involves engaging the key architects and designers to make vectors for the end-cases and hard to exercise logic – and this process is also measured in months.

A more efficient methodology was developed over time to provide the manufacturing vectors. Electronic Design Automation (EDA) tools were created to generate vectors that were based on fault models. The vector set that resulted verified internal gate truth tables and wire routing connections – and with minimal overlap (faults tested multiple times). These vectors could then be used for the purposes of manufacturing test since they evaluated the structure of the device, not the behaviors and functions – and with vector management as a side effect (the number of vectors needed to achieve a certain quality level).

First attempts at these vector generation tools were against sequential circuits, which was time consuming, computationally complex, and fraught with inefficiencies. Over time, research, optimizations, and adoption proved that combinational vector generation was a much more optimal solution and this selected "mostly scan" (or "partial scan") and "full scan" as the most efficient methods to create high fault coverage vectors in a timely manner. As designs became more complex with other applied limitations, these same algorithms and optimizations were applied to the generation of logic BIST vectors – and similar evaluations drove memory BIST[5].

Ultimately, scan and BIST test methodologies became common for the digital logic arena, and BIST for memories became a mature technology – although it is still not universally adopted by all organizations. Since many of the structural fault models – stuck-at, transition delay, path delay, bridging, opens, leakage, etc. – are applied using DFT in conjunction with Automatic Test Pattern Generation (ATPG) tools, then structural test is generally associated with or is considered synonymous to DFT-based testing techniques, such as scan or BIST. A structural tester, by the

---

[5] Sequential vector generation dominated in the late 1980's and the early 1990s since no-scan and partial-scan test methodologies were used to augment functional vectors to raise fault coverage. A critical juncture in chip sizing and vector generation time led to full-scan being accepted in the mid-1990s which resulted in combinational vector generation becoming a dominating methodology. The SOC methodology began to dominate in the late 1990's and led to full-scan plus scan test wrappers and embedded testing which brought BIST to the forefront. In the early 2000's the vector volume crisis (more vectors required for coverage than testers could hold) led to embedded deterministic vector generation and compression – the use of BIST-like structures used in conjunction with ATPG.

extension of this thought process, is a tester that is optimized in its feature set to provide the ability to conduct scan, BIST, and leakage testing fairly easily. Since these structural techniques have lesser requirements than functional operations, then structural testers are simpler and less expensive.

6.2 TEST COST – THE CHICKEN AND THE LOW COST TESTER

Cost is always hard to quantify and it means different things to different people [2], [3], [19]. However, there is consensus in the industry today that cost of IC test is of growing concern. For instance, the International Technology Roadmap for Semiconductors (ITRS) identifies manufacturing test cost as one of the difficult challenges the industry will have to deal with over the next 5-10 years [4]. Higher gate count, smaller feature sizes, higher speeds, increasing complexity, increasing gate-to-pin ratio – you get the idea. EDA and ATE vendors are touting a variety of technologies as the ultimate solutions to the test cost problem. With most ATE bearing price tags in the million dollar range, it makes perfect sense to examine what it would take to develop less expensive systems.

Reducing test cost is more than building a cheaper box. As we will explain in more detail later in this chapter, there are perfectly good reasons why ATE systems are as expensive as they are. A cheaper system simply means fewer capabilities. However, by utilizing specific test methodologies, it is possible to reduce cost by using lower cost test systems without compromising test quality. Such test methodologies will rely heavily on DFT. Furthermore, the structural tester is cost managed by making intelligent bandwidth decisions and placing certain capabilities within the chip or on the loadboard as opposed to within the tester.

While the original motivations for DFT were productivity and quality requirements, the interesting side effect of DFT test is that when used *the right way*, DFT enables use of less expensive systems. DFT test patterns have different test system requirements than functional patterns. And this is what makes test cost reduction a story about poultry – about DFT-chicken and tester-eggs, and who will be first. True DFT testers will not exist until the DFT usage has reached a maturity level that would justify their existence. And DFT designers won't use DFT *the right way* until there is a justification for it (i.e. the lower cost DFT tester). Today, too many designers organize and implement their DFT to fit ill-matched functional ATE systems [5].

There are, of course, other motivating factors to LCST, but cost is what many people (product managers included) consider the most important one. With that in mind, let's take a look at what test cost means. We will start by dividing test cost into two categories: factors related to schedule (time-to-market and time-to-volume) and factors related to manufacturing test cost.

6.2.1 Schedule, Work Product, and Time-to-Market

DFT has been widely adopted by the industry because this methodology makes it possible to automate the generation of production test patterns through structural fault models and ATPG. At the same time, relying on traditional functional test patterns alone is virtually impossible for today's complex devices. Overall, the test development includes tasks prior to silicon manufacturing (pre-silicon) and tasks after manufacturing (post-silicon).

In a DFT based test flow, pre-silicon test development tasks include test structure insertion, test pattern development, and test pattern verification through fault simulation, as shown in Figure 6-2. These are tasks that can be (more or less) fully automated through commercially available DFT insertion, pattern generation, and fault simulation tools. In terms of work product, the more of the test program that relies on this approach rather than functional tests, the lower the cost. This all depends on the availability of methodologies, tools, and fault models that support new and complex design structures[6].

On the post-silicon side, when the patterns start hitting the tester, the amount of memory required for scan test might be an issue for certain ATEs. This depends on the scan configuration and patterns required for a particular design. Other than that, a DFT based test program can in most cases be executed perfectly well on a traditional ATE system, even though such a system might be overkill. Most likely, only a small fraction of the ATE's capabilities will be utilized.

During first silicon validation, it is critical to understand the failures, or one cannot proceed to full device manufacturing. At this point, one often does not debug defects, but inconsistencies between the model of the design used during test pattern development and real silicon. When the test patterns are up and running, the next step is most likely characterization[7] of the device and the process. One wants to understand how well the device will perform under different conditions.

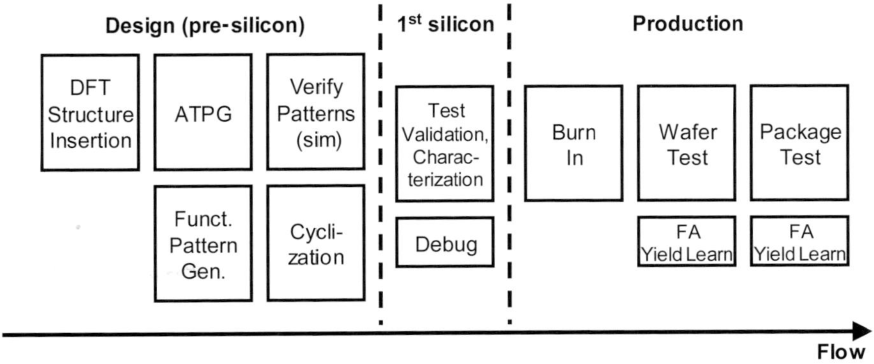

Figure 6-2: Test flow based on DFT and functional patterns.

6 For example, deep submicron leads to new fault types that need new test methods and new test tools.

7 The goals of the characterization process are generally to understand power, timing, and clocking behavior of the device and the effect of temperature or voltage stress.

When production is ramping up, it is normal to take a closer look at defects in the failing devices to help understand what can be done to improve yield. While not explicitly shown in Figure 6-2, the initial ramp-up of production, is critical to reach what one might define as "Time-to-Volume" (TTV)[8]. Both during this ramp-up period, and later, it is desirable to increase yield and that indirectly lowers the cost per device. An important part of yield learning is *failure analysis* (FA) – understanding what causes systemic defects and where they are located within the process, device or mask. While doing failure analysis on functional patterns is a near impossible task, DFT based failure analysis has proven itself to be a very effective methodology that can contribute to the yield learning process. DFT oriented test systems are typically designed specifically to fit in a DFT based diagnosis and FA flow. Large amounts of capture memory help improve the accuracy of the diagnosis process. These systems also deliver failure information in the exact format required by the diagnosis tool. This all makes the process more effective and illustrates that when a DFT based testing strategy is followed, yield-related costs can be reduced.

6.2.2 Manufacturing Test Cost

Manufacturing test cost is typically calculated by multiplying *test time* with *cost per second*. The per-second cost[9] is defined by equipment cost (price of testers, handlers, etc.) and operational expenses (maintenance, electricity, etc.). Test is typically done in multiple insertions[10] [6]. In a traditional test flow, the same test program may be used in all insertions. In a distributed test flow, the test patterns and the test equipment used in each insertion may vary, as shown in Figure 6-3. In the Legacy Test Flow, an expensive, functional test system is used for both wafer sort and final test. In a distributed flow, a functional system is still used for final test, but a lower-cost structural system is used for wafer sort. This way, a distributed flow can be used to reduce test cost without increasing the defect level (number of defective parts shipped to customer) after the final test insertion (usually measured as devices per million or DPM).

[8] TTV is Time-to-Volume production, which represents the schedule time required to go from first silicon to high volume manufacturing production and includes the steps of identifying and fixing yield problems; developing a smoothly-operating fabrication process; developing a high-quality test program; and organizing and assembling the test floor and equipment required to conduct wafer probe and final test.

[9] Typically 5 to 10 cents per second.

[10] Each time a device is tested during the manufacturing and assembly process is referred to as a test "insertion." For instance, Figure 6-3 shows test flows with three insertions: Wafer sort, burn-in, and final package test.

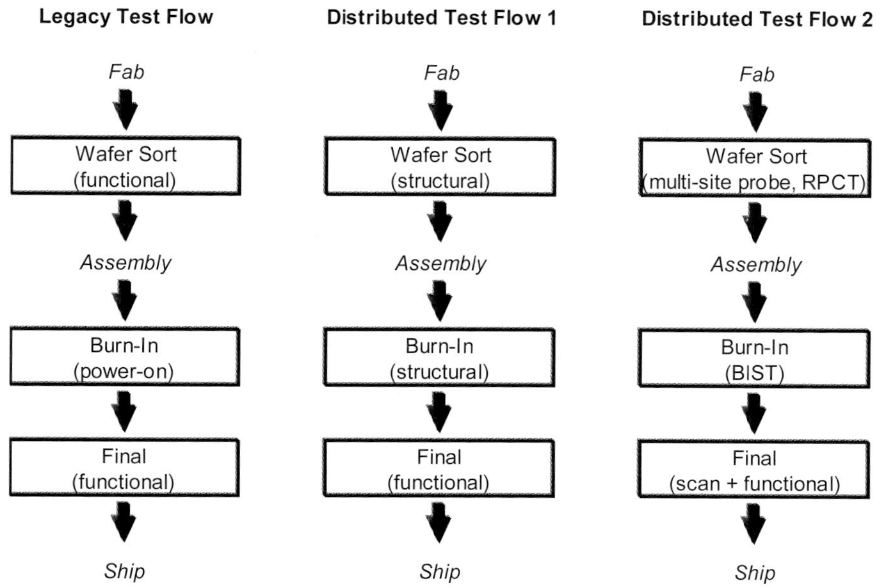

Figure 6-3: Legacy vs. distributed test flow.

As an example, assume that for a particular device, most of the test program is based on DFT tests that require a 100-channel, 50 MHz system, while 1% of the defects observed can only be detected by functional patterns that can only be executed on a high-performance, 1GHz, 1500-channel functional tester. In that case, it may be cost effective to skip this test step during wafer probe. Such a flow will cause defective devices to be packaged (but weeded out during final test). Depending on the cost associated with packaging these devices, this may or may not be the right flow. For both of these insertions, the purpose of the test is simply to detect whether the devices fail or not. For failure analysis, a third system might be the best alternative. For this application, the ability to capture a large number of results is a key factor and more important than for instance throughput. While some high-end production ATE systems can only capture less than 1K failing cycles, lower cost LCST systems can capture several million failing cycles and the more failures that can be captured, the higher the chance that the automated failure analysis tool can produce useful results.

Tester cost and test time are not independent. A test based on BIST might allow a significantly less expensive tester compared to a traditional test, but at the same time, the BIST test is likely to require longer test time because of the application of long pseudorandom test sequences. There are several commercially available solutions today that through compression of scan test data can reduce test time by a factor of 10X and beyond. This is all well and good, but the true effectiveness of such tools depends on how much of the total test program is DFT based. The better of two worlds can be achieved by using a DFT-only test program. This allows for the highest effectiveness of DFT-based compression tools, and usage of lower cost structural test systems.

The bottom line is that the DFT and the tester(s) must match. One has to understand that the tester requirements for wafer test, package test, and failure analysis may differ. Doing DFT correctly can reduce the cost for any and all test insertions.

6.3 TESTER USE MODELS

There are multiple uses for the structural tester – not just the common uses of a functional ATE. When asked, the handy-dandy structural test salesman will generally provide a listing that covers burn-in, characterization, test chip evaluation, first silicon bring-up, failure analysis, yield learning, vector validation, test program development and assembly, production probe and some insertions for final package test. From the customer point of view, if some item on that list matches a need then adoption is only a case of evaluation and matching the current vector and test environment to the structural tester. From the tester provider point of view, though, this makes for a schizophrenic sales model – having to support the features, software, and development and loadboard types for each different environment and conducting expensive evaluations with the result being selling only one tester to each customer. In the long run, this would be an unsuccessful business model.

The key to successful application of the structural test model is in understanding the differences in the target organizations, and in understanding the difference between the EDA software sales model and the ATE hardware sales model. A low cost structural tester has more uses in the design space than it does in the test space, so the 'use model' lines up more like an EDA software model than it does an ATE model. However, the structural tester is still useful in the production probe and production test space, so in that case, the 'use model' is more like that of traditional ATE. Then there is still the failure analysis, debug, and yield learning aspects that are related to production test, but are more easily and cost effectively accomplished on a machine that is not sitting on a production test floor or in a clean room – the use model for this area lines up better with that of an engineering lab tester. So how are these differences in perception, implementation, and use resolved into a coherent use model that makes sense?

In truth, there is not one use model, there are at least three. As shown in Figure 6-4, the three basic spaces that must be considered are: the *design-only space*; the *correlation-space* (we also use the term *design-to-test space* with the same meaning); and the *production-only* space.

The design-only space is the use of a structural tester by design, validation, debug-diagnosis, and design-for-test engineers – and with no thought of the part being tested going to production or onto a production tester. The goals in this space are to conduct experiments, to characterize the various electrical aspects of design, to bring-up first silicon (for understanding, not for production), and to evaluate test chips. The tools used in conjunction with the tester, in this case, are the design-side tools that are used to generate and evaluate the design: ATPG, static timing analysis, simulation engines and testbenches, schematic viewers, waveform viewers and layout extraction engines. The structural tester is treated much like EDA software licenses, in that multiple different users such as validation or DFT engineers can make use of it. If one word can be used to describe what is most important about this space, it would be "capability" – the ability to conduct the analyses and to investigate the part under test to the extent necessary.

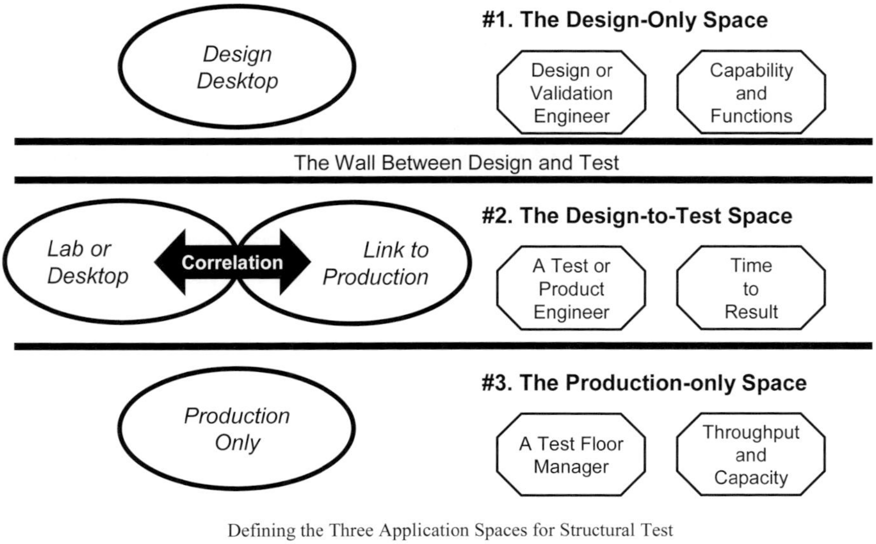

Figure 6-4: Defining the three application spaces for structural test.

The production-only space is the use of a structural tester on a test floor that is managed by a test floor manager. The goals of this space are flexibility, capacity, and uptime. The tools used in this space are those of the test floor – test cell controllers. The concerns of the test floor manager are hardware interfaces to probers, handlers, and other test floor equipment. The one word description of what is important about this space is "throughput."

The correlation space is the use of a structural tester by test, failure and yield analysis engineers. The goals in this space are to validate vectors that are to be included in the test program, to assist with the development of the test program, and to conduct correlated debug and diagnosis (finding the reason or location of the failure as it pertains to production ramp-up or yield analysis). The goals of this space are to do vector development, test program development, yield-analysis and debug-diagnosis on a more efficient platform that has a lower cost impact than that of a big-iron tester on a production test floor. The tools used in this space are the traditional tools of a tester – the tester interface, vector editors, and shmoo and margining tools. The one word description of what is important about this space is "schedule" – the schedule being defined as the "Time-to-Volume" (TTV) and "Time-to-Yield" (TTY)[11].

[11] The term TTY is Time-to-Yield and is a subset of TTV in that it involves the portion of the schedule related with identifying, diagnosing, and fixing systematic yield problems. Systematic yield problems are yield problems that source from repairable factors such as mask errors,

Note that the correlation-space and the production-only space are what can be termed as the "test-side" since they are driven by product and test engineers and managers, whereas, the design-only space is what can be termed as the "design-side." So, the use of the LCST crosses what is traditionally referred to as the wall between design and test.

What this assessment of the use-spaces or use models of the structural tester points out is that successful adoption has different drivers and metrics in the various different spaces. What makes a structural tester successful for test program development will not necessarily make it successful for test chip evaluation. This fact points out that the structural test and tester revolution is not necessarily about the tester hardware, but that the adoption of structural test is about the overall packaging, tools, capabilities, and interfaces in each space – in short, the space specific methodology. Each different space has a different mix of goals, requirements, tools, and solutions that will be used by different parts of the organization.

6.4 WHY AND WHEN IS DFT LOW COST?

Test cost reduction is a major motivator behind LCST, and can only be achieved when such a system is used together with DFT in a DFT based test flow. In this section, we will take a closer look at how DFT can be implemented correctly to reduce test cost, both in terms of enabling LCST systems, as well as simplifying tasks that go beyond pass/fail testing such as debug, characterization, and failure analysis for yield learning.

6.4.1 Functional vs. Structural Test

Traditionally, manufacturing test has been done using functional test. Functional test patterns verify that the model or logic behaves as it was intended. If the function is an adder, the test will be written to see if for instance 2+2 is 4 and 9+12 is 21. Functional testing is measured by the logic committing the correct action to the applied stimuli. 100% functional correctness is the standard expectation and this should be verified at the behavioral (RTL) or gate level of the design with a simulation process.

Structural test, on the other hand, is used to verify the topology of the manufactured chip. Given a good circuit before the manufacturing process, structural testing can be used to verify that all connections are intact, and that all gate-level truth tables are correct after the manufacturing process. Such testing relies on fault models, which assume that the physical defect will represent itself in a certain way, such as "stuck-at-1" (short to V_{DD}). Structural testing is measured using the fault coverage metric, with respect to the targeted fault model(s). It is important to notice that 100% fault coverage would only cover 100% of the defects in a perfect world where all defects are modeled in the fault model.

---

process mix errors, equipment problems, etc., as opposed to yield problems that source from random factors such as errant dust and chemicals in the clean room.

There is an ongoing trend in the industry transitioning from functional to structural test. Why is this the case? The key issues being addressed or solved by the adoption of structural test over traditional functional test are:

- Minimizing the time it takes to develop vectors for the production test program (because of the extensive automation employed in structural test).
- Minimizing the time to characterize the chip and understand the process.
- Evaluating items that are inefficient or difficult to test with functional vectors.
- Achieving the required fault coverage metric for chip qualification.
- Managing test program application time.
- Minimizing the test interface to support tester limitations or enable multi-site.
- In many SOCs, the internal cores have more interface signals than the SOC has pins. Therefore, traditional functional test of the cores becomes virtually impossible. DFT lends itself as a test methodology for SOC design styles.

6.4.2 Structural Test, DFT, and Cost

Literally, Design-for-Test means just that, taking test into consideration when creating the design. From a higher perspective, DFT is thought of as the methods, techniques, and tools that enable structural test. It is DFT techniques such as internal scan that make it possible to verify the topology of the manufactured chip.

Today, most design groups are at an intermediate level when it comes to the adoption of DFT techniques, known as "DFT maturity" [8]. For the average design group:

- Almost all designs use Mux D-flip-flop based full scan.
- Scan delay test is at an experimentation stage.
- About 50% designs utilize I_{DDQ} test.
- Almost 100% of memories are BIST'ed. About half-and-half internally developed vs. commercially available solutions.
- Very low utilization of logic BIST.

What makes structural test cost efficient and what is the right DFT for a LCST system? The answers lie in the practical implementation and application of DFT techniques. As we mentioned before, even though a LCST system is of lower cost than a traditional tester, overall test cost is not reduced if the LCST test program has significantly longer test time. Let's examine some common test techniques to elaborate on this claim.

Internal Scan (all fault models)

For scan-based test, test cost and test time depend on the amount of test data. This can be simply expressed in the following way:

$$\text{Test time} = (\text{scan chain length}) \times (\#\text{ test patterns}) \times (\text{test cycle period})$$

Scan chain length refers to the length of the longest scan chain in the design[12]. Therefore, balancing the chains so that they have as close as possible to the same length minimizes test time. In many design flows, one tries to avoid multiple clock domains in the same chain, which may worsen the balancing. Some testers have "scan options" that allow additional tester memory behind a certain number of pins (not all pins). If a design has too few scan chains, test time ends up being unnecessary long because the chains are very long. If a design has too many scan chains, one might not be able to utilize a tester's scan memory, the entire test program might not fit in tester memory, which in turn causes costly reloads, re-tests, or re-insertions.

The number of patterns typically depends on design structure rather than design size. Different ATPG tools offer different compression and pattern ordering techniques to help reduce pattern count. Notice that fault models beyond stuck-at typically require pattern counts that are much larger than those for stuck-at only.

The test cycle period (1/frequency) depends on both the tester and the design. In most cases the limiting factor is in the device. When data is shifted through the device, there is more switching activity than for functional operation. That, combined with the layout of the scan chain path (clock vs. data) limits the scan shift frequency.

Scan test in general is "low cost friendly", since scan patterns do not require testers with high timing accuracy or other factors that increase the tester cost (these factors are discussed later in this chapter). However, for scan test to be a cost effective solution, the scan chains must be configured to match the tester resources. DFT oriented systems do normally accommodate a large number of scan chains, but without knowing what the available resources are, it is possible to end up with a configuration that exceeds the capacity of the target ATE.

JTAG TAP and Boundary Scan

While boundary scan as a DFT technique was originally intended for board test, it is often used for IC test as well. The boundary scan Test Access Port (JTAG TAP controller) has been proven a useful control mechanism for internal test circuitry such as setting up test modes or invoking BIST. Commercial boundary scan insertion tools allow the user to add custom instructions to the TAP controller for this particular purpose. Therefore, while boundary scan patterns themselves can be applied on any test system, it is the test mechanisms controlled by the TAP controller (such as BIST) that are especially suited on a DFT-oriented tester.

Boundary scan can also be used to help facilitate *reduced pin count test* (*RPCT*), which is discussed later in this chapter.

Scan Delay Test (AC scan, scan-based at-speed test)

Scan-based delay test can replace costly functional patterns and enable DFT based characterization. However, when not applied correctly, scan delay patterns require the same expensive tester as functional patterns. It is important to understand what

---

[12] Each scan chain has independent scan in and scan out pins.

the clocking requirements are for the particular scan delay methodology. Typically, scan delay patterns shift data in at a low frequency, then a launch and a capture pulse are applied at higher frequency. In many cases, it makes sense to use an on-chip Phase Locked Loop (PLL) to generate the launch and capture clocks for scan delay test, rather than generating these clocks on the tester [7]. One thing that has great impact on the test equipment is how patterns targeting different portions of the circuitry have different timing requirements:

- From register (scan cell) to register within the same clock domain (CLK1 – CLK1). In this case, two clock pulses of one particular clock have to be delivered at a "high" frequency.
- From register in one clock domain to register in a different clock domain (CLK1 – CLK2). In this case, the two clock pulses must be applied on two different clock pins and the distance between their edges is the at-speed clock.
- From primary input pins to registers, also known as input side shadow circuitry (PI – CLK1). In this case, the timing of the input pins compared to the clock pulse is critical—a similar case is true from memory to logic. This is difficult to achieve when the high-speed clock is generated using a PLL.
- From registers to output pins, also known as output side shadow circuitry (CLK1 – PO). The location of the output strobe compared to the clock pulse is critical—again a similar case occurs from logic to memory. This is difficult to achieve when the high-speed clock is generated using a PLL.

If input and output pins are used to launch and capture transitions, it adds considerable requirements to the tester in terms of timing precision on input and output pins. Another key issue is to selectively generate the optimum pattern set. The most common fault models used for scan delay test are transition delay and path delay. While the generation of transition fault patterns is completely automatic, this fault model typically only detects gross timing defects. For path delay patterns, the art of selecting the best paths and the right amount of paths is not yet well understood by industry. If done incorrectly, and since most commercial ATPG tools test only one transition (slow-to-rise or slow-to-fall) on one path per pattern, the result can be high pattern count, insufficient coverage, and may therefore not be economically feasible.

When configured correctly, for instance by using a programmable on-chip PLL to generate launch and capture clocks, scan delay test methods makes it possible to perform at-speed test on slower, less expensive DFT oriented test systems. This makes scan delay test a key element in a DFT oriented test flow.

Memory BIST

For performance reasons, many designs contain many small memories rather than a few large memories. While choosing memory BIST as a test methodology for memory test is a no-brainer in most cases, this myriad of small memories can prove to be a test headache. Test time depends on the type of algorithm and memory size[13].

---

[13] A typical algorithm is "11n", which means that each address location is read from or written to 11 times. Test time can then be derived from the cycle time, memory size, and number of cycles required to perform a read or write operation.

Executing a March-based memory BIST test algorithm on several hundred memories in parallel can cause some very interesting power supply behavior. To optimize test time, memories should be grouped together such that as many as possible memories of similar size are tested in parallel. Depending on the design structure and power requirements, it may or may not be possible to run other tests in parallel to memory BIST.

While memory BIST could have significant impact on test time for a design with large and/or many memories, it is important to note that production memory BIST requires virtually zero tester memory (engineering memory BIST does produce fail data). Memory BIST is therefore a good fit for a system with low per-second cost.

Logic BIST

While logic BIST can help reduce test data volume, it does not necessarily reduce test time compared to scan [9]. Typically, logic BIST utilizes phase shifters so that a relatively small pattern generator can feed a very large number of short scan chains. However, due to the nature of the pseudo-random patterns, the increase in pattern count (compared to ATPG) is often such that overall, test time might be higher than for conventional scan. Just like memory BIST, logic BIST requires virtually zero tester memory. Therefore, logic BIST will in many cases reduce test cost only if applied on a lower cost tester, or if the reduction in tester applied pattern data can prevent a pattern reload.

SOC / core Test

While test of embedded cores and SOCs is discussed in detail in a separate chapter of this book, it is worth mentioning here that most, if not all, SOC test methods, where test structures and patterns are developed on a core basis rather than a top level, are DFT based and suitable for LCST systems. Memory BIST is very popular in core/SOC test, and many cores are available with logic BIST. Limited access to an embedded core can make it costly or impossible to test it functionally. The IEEE 1500 project is developing such a standard [10]. The Core Test Language (CTL) started out as a part of IEEE 1500, but is now part of the Standard Test Interface Language (STIL), as IEEE 1450.6 [11]. CTL is not tied to a particular test structure such as IEEE 1500, which makes CTL more flexible and may increase the adoption rate of CTL. The CTL language also allows test systems to "understand" the structure of an SOC without having access to a detailed (and often proprietary) netlist. It is expected that CTL will be a facilitator for core/SOC on LCST systems as it matures.

6.4.3 Test Development Automation

The test development process is more than ATPG. After generating patterns and achieving acceptable coverage for the desired fault model(s), the patterns must be verified, preferably using timing based simulation. The more thorough this verification process is, the less likely it is that problems will occur later in the flow.

The next step is to translate patterns to the format for the target tester. While commercial ATPG and BIST tools generate patterns in a variety of formats, most of these are not tester specific formats, but intermediate formats such as Waveform

Graphics Language (WGL) and Verilog. These patterns must be changed from an event order format[14] to a tester specific cycle-based format[15]. During this cyclization and translation process, any information describing the DFT structures is typically stripped out. Without this information, it is difficult to interpret tester results with respect to the DFT structures because the fail report is based on cycle number and mismatched pin. Such interpretation is not required for pass/fail manufacturing test, but is of great importance for bring-up, characterization, and failure analysis. Finally, multiple test pattern sets are merged together in a test flow together with additional required information, such as power and timing parameters.

Most LCST systems can read test patterns directly from the pattern generation tools in a format such as STIL[16] as shown in Figure 6-5, eliminating the need for translation. STIL is an open standard shaped by industry inputs from EDA, ATE, and customers [12]. STIL has explicit scan support allowing more sophisticated scan structures than any other pattern format. The scan structures include not only the scan pins and scan chain length, but a complete description of all the scan cells in the chain, and the order in which they are connected. That enables the mapping of a particular failing bit onto a specific scan cell. While the *ScanStructure* statement (which describes the order of connection of scan cells) is not mandatory in STIL, it is utilized by virtually all commercial ATPG tools.

Pattern data and pattern protocols can easily be separated in STIL. The protocols, such as *signal group* and *waveform table* are more powerful and flexible than other formats. The *macro* and *procedure* definitions allow data to be passed into them, something which is very convenient for scan patterns. For instance, the scan load operation can be defined as a procedure, and called with different data for each scan pattern in a test set. Complex test programs are generally made up of multiple selectable pattern blocks. ATPG tools that create patterns in a tester-ready format reduce the schedule time of the overall process, especially during debug when new patterns must be iteratively generated.

The recent 1450.1 extension makes it an even more efficient language in the DFT insertion process as a language that can carry DFT information from the DFT insertion tool to the pattern generation tool [13]. One detail lacking in the original version of STIL, but which exists in 1450.1, is a recommended strategy for how to use labels to mark the functions of and boundaries between test units in a pattern set. This is a feature that makes it easier for LCST systems to provide feedback to the user and to the EDA environment for debug and diagnostics purposes. This extension includes features to support "fail feedback", allowing STIL to be used not just as a pattern format (as shown in Figure 6-5), but also as a format for test results

---

[14] In an event order format, the times for which signals change state (events) are explicitly noted. For instance signal A goes high at 100ns, low at 212ns, high at 298ns, etc.

[15] In a cycle based format, all events are described as a sequence of cycles (vectors). The timing information is described in timeplates, and each cycle references a timeplate. There may be as little as one timeplate in the vector set. For instance, all cycles have period 100ns, all inputs change at the beginning of the cycle all clocks are pulsed at 30ns with a 20ns wide pulse.

[16] Standard Test Interface Language for Digital Test Vector Data, IEEE 1450-1999.

from the test system to the EDA tool. Such failure information is currently provided in simple, but tool-specific, formats.

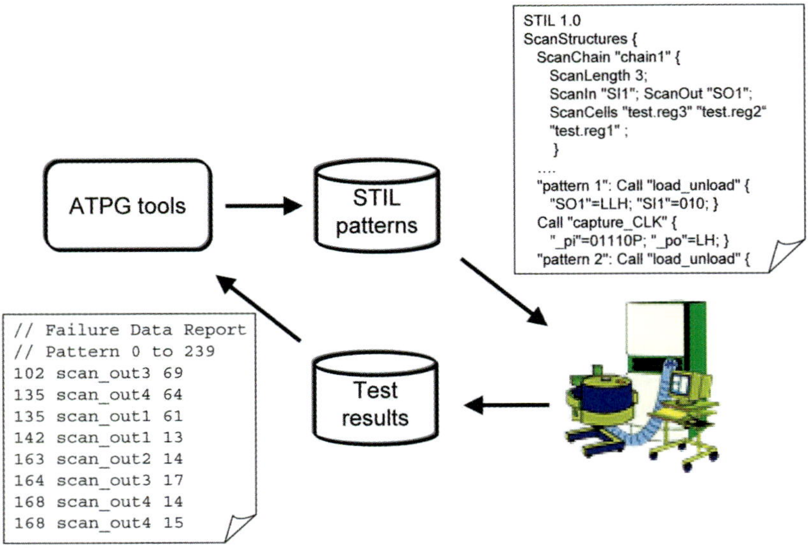

Figure 6-5: Test flow with LCST: direct paths for patterns and results.

Since STIL is developed with DFT and DFT based test flows in mind, it is an ideal pattern format in a test environment including LCST systems.

6.4.4 Defect Coverage and Fault Models

The value of structural test methodologies depends on how accurate the fault models are. In the era of nanometer design, traditional fault models may no longer ensure sufficient defect coverage. Fault models represent a target for automatic pattern generation[17]. For some fault models, generating the fault list is a non-trivial task. While fault models like stuck-at simply adds fault sites on all nodes in a design, other models, like path delay, depend on data from tools other than the ATPG tool[18] to identify fault sites. The latter model is more complicated to use in an automated flow. Another important characteristic of a fault model is how the metric used to measure detection of that particular fault model can predict actual test quality – the defect coverage.

---

[17] Fault models are generally applied at the level of design abstraction known as the gate-level. This is two levels up from defects which are at the physical level and may be modeled at either the physical level or at the transistor level. Functional vectors are generally created at an even higher level of abstraction, against the HDL or RTL model at the behavioral level.

[18] The path delay fault model depends on timing information from a static timing analysis (STA) tool. Some bridging fault models require layout information that is otherwise not used during ATPG.

The stuck-at fault model has been used in DFT since the very beginning. Even though the stuck-at model does not always model the behavior of a faulty circuit, it has been shown that a test set developed to test stuck-at faults will also cover many other defects that do not behave as stuck-at faults (serendipitous fault coverage). While higher stuck-at coverage typically means higher defect coverage, it is virtually impossible to predict or characterize how and when these "other defects" are covered. Given two pattern sets with the same stuck-at coverage, which one is the best?

To ensure that most types of defects are accounted for, a good structural test program targets several different fault models. Using today's tools and technology, this typically means stuck-at, transition, path-delay, and I_{DDQ}. This is illustrated in Figure 6-6. In this experiment, a study examined how many defective parts were "caught" by four different pattern sets: one functional pattern set and three structural pattern sets representing three different fault models. Many defects were commonly detected by all four pattern sets, but some defects could only be detected uniquely by one set (the outer 4 corners of the left diagram). With only structural patterns, using the three fault models, most of the defects would be detected. The exact effectiveness of each fault model depends on many factors including fabrication technology, design structure, and the effectiveness of the pattern generation tool.

Figure 6-6 shows the Venn results of two well-known industry experiments: both diagrams show that the majority of detected defects are caught by structural patterns – the upper right corner of the left diagram shows out of 3140 total fails, only 14 were uniquely detected by functional vectors. The diagram on the right more clearly shows only 100 defects out of 5417 were uniquely detected by functional patterns.

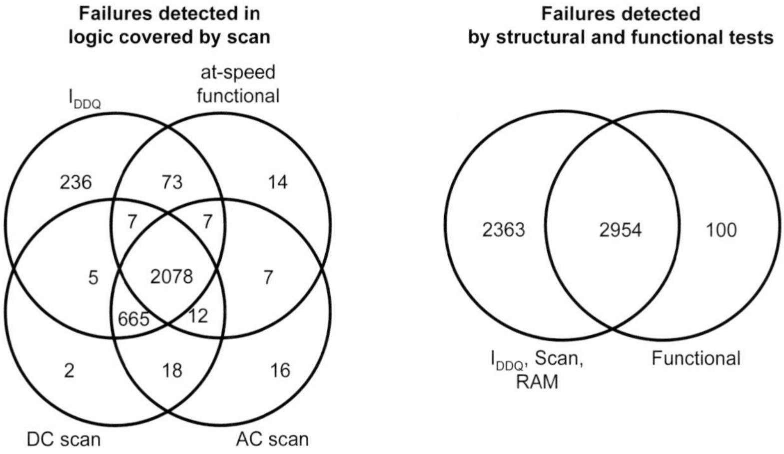

Figure 6-6: Effectiveness of different fault models [14].

As at-speed test becomes increasingly important, scan-based ATPG solutions for at-speed test are becoming more mature and capable of ensuring high test coverage and without the high development effort required to develop functional at-speed patterns. Today, even *speed binning*[19] can be performed based on scan-based test [15].

Current based testing such as I_{DDQ} is becoming more and more difficult to do, as the difference between nominal background current and the additional current caused by a short is shrinking due to increased transistor count and reduced feature sizes. To make I_{DDQ} successful in this environment, great care has to be made in designing I_{DDQ} friendly test modes (where all buses have one driver, memories are turned off, and there is no contention) and in the selection of I_{DDQ} patterns. Experiments have shown that I_{DDQ} is still an effective test technique, but that larger numbers of I_{DDQ} measurements might be required. As classic I_{DDQ} threshold measurements become ineffective, new techniques such as delta I_{DDQ} [16], current ratios [17], and nearest-neighbor residuals [18] are becoming more popular. Recent experiments [19] show that in combination with the proper measurement strategy, there is a future for production I_{DDQ} testing. The quality of the I_{DDQ} measurement equipment is an important factor affecting the screening efficiency.

Devices manufactured using newer processes with smaller feature sizes and higher densities behave slightly different than devices manufactured using older processes. It is therefore likely that new processes are susceptible to different types of defects. One can therefore question whether the fault models available in today's pattern generation tools are representative in technologies below 100nm. In general, there are two ways to improve test quality. One solution is to enhance the fault model and/or add new fault models that better describe the defect behavior in a way that is suitable for the ATPG tool. The other approach is to utilize the existing fault models and apply the same ATPG algorithm to generate more patterns that increase the probability of detecting non-modeled defects. There have been several attempts in modeling bridging defects. Bridge defects can occur between signals and power or ground, or between signals. When there is a bridge between two signals, this defect might behave as an OR, AND, or one net might dominate another. One particular challenge is to come up with a method that is practical to implement with an ATPG tool. One might expect the best models to be based on layout information, but this adds restrictions to 'when' in the process patterns can be generated and also leads to an explosion in the problem's complexity. Another approach, which has shown to be effective, and is significantly easier to implement in a tool flow, is to use a "multiple detect (or N-detect) stuck-at model", where each stuck-at fault is detected multiple times in several different ways (with different control and observation points) [21].

No matter what exact fault models are currently in use, the future of LCST relies on adequate fault models for defects in today's and tomorrow's devices because fault modeling is one of the most important parts of structural testing.

For a more detailed discussion about fault models and defect-oriented testing techniques the reader can refer to other chapters of this book.

---

[19] Speed binning is a test process commonly used by microprocessor manufacturers to determine the speed grade (frequency) of each individual unit. For instance, the binning will determine whether a unit can be sold as a 3.0GHz or a 2.6GHz unit.

6.4.5 DFT and First Silicon Validation

Earlier in this chapter, we introduced three tester use models. One of these, the design-only space use model, includes applications for first-silicon bring-up. In this situation, it is important to quickly validate that the device functions as expected and that the test program works as expected. This validation must take place if the test program is functional, structural, or a combination of both. The next step is to deliver qualified samples to internal and/or external customers. This work is done under tremendous pressure, because the sooner this process can be completed, the sooner manufacturing can start.

When problems occur during bring-up, it is important to get feedback indicating what the problem is. The advantage of DFT is that such debug can be done more quickly than debug of functional tests – provided that the tools and methodologies are there to support the effort. It is also important to execute the particular test components in the right order, so that the problem is first observed when it is easier to debug. For instance, if the scan shift path is broken, mismatches will occur for all types of scan-based tests. However, it is easier to isolate this particular problem during a scan chain integrity test (chain test). Similarly, if stuck-at scan patterns pass and at-speed scan patterns fail, one knows that the problem is speed related, without having to modify or change the original patterns. An example test flow for first silicon validation is shown in Figure 6-7. Here, the functional patterns (assuming they still exist) are included at the end, because it is perceived that the other tests are easier to debug[20].

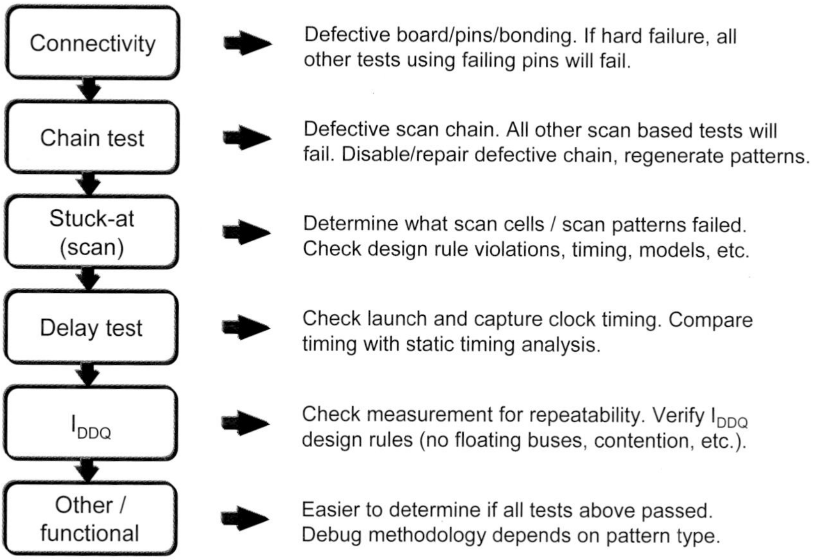

Figure 6-7: Test flow for DFT based first silicon bring-up and debug.

[20] Building a test program for production is driven by other goals, such as the reduction of test time. In this case, the order of tests is based on the "first most likely failure" so the test program can be "stopped on first fail" to shorten the "in-socket time."

In addition to being able to differentiate between components in a test flow, the ideal test system for a first silicon validation situation "understands" the DFT structures and has capabilities to help the user debug various problems. Figure 6-8 illustrates the problem. On a traditional ATE, test patterns for internal scan (such as stuck-at) are treated as any other test patterns, i.e., everything is represented in terms of pins and cycles. The test system has no understanding of the device's internal structures. When mismatches occur (the underlined bits coming out of the scan output pin in Figure 6-8), these are presented as failing pins and cycles. It is then left as an exercise to the user to interpret this data. Is it the same scan cell that fails multiple times, or are mismatches observed in multiple cells? Are the failing scan cells in the same part of the design? Traditional ATE can't answer these questions.

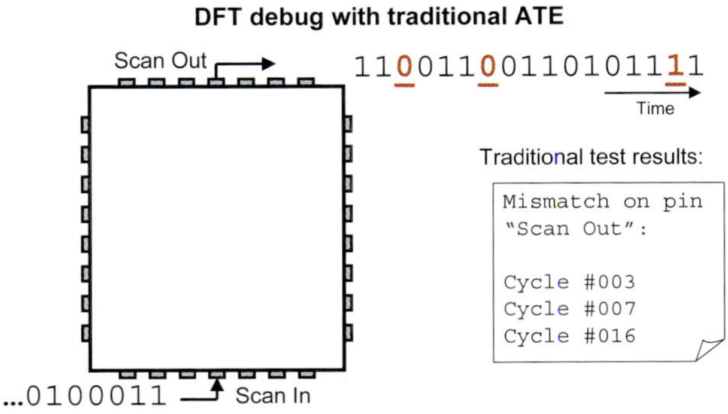

Figure 6-8: DFT debug with traditional ATE.

Figure 6-9 shows how a LCST system would handle the situation differently. It is much more powerful than a traditional ATE because it understands "what scan is" and can map a failure directly to a scan bit. This capability is enabled by the structural information in the ATPG patterns. The two key items to keep track of are the scan chain structures and how the test set is divided into individual scan patterns. When this information is utilized, the test system can tell the user which scan patterns and which scan cells failed. In this example, the same scan cell failed 3 times.

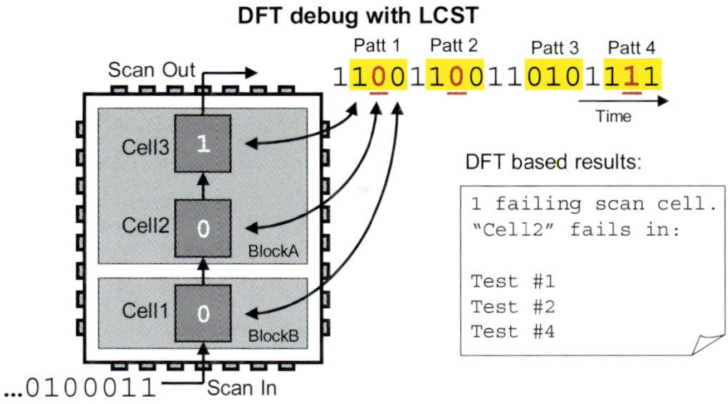

Figure 6-9: DFT debug with LCST that recognizes scan structures.

In some cases, ATPG based diagnostics can be used to help determine the cause of the mismatch. In this case, a direct feedback path to the ATPG tool as illustrated in Figure 6-5 is required.

6.4.6 DFT and Device Characterization

Also related to the design-only tester use model is device characterization. Various experiments are executed to determine the performance of a device in terms of voltage, speed, and temperature.

To be able to do such experiments, one must be able to output and understand failures when they occur, similarly, but to a greater extent than what is done during first silicon validation (the goal is not so much to find failures but to use the failure to identify the limits of operation). For instance, when the voltage is decreased, the device will perform slower and slower until a certain point, where it will no longer work. During characterization, it is important to understand the exact behavior under this scenario, and determine if the device behaves as expected. The LCST characterization system needs to have the means for supporting such experiments, such as Shmoo plots and the ability to extracting logic cones. A Shmoo plot is used to graphically represent test results when parameters such as voltage and timing are varied. This is used during characterization to determine, for instance, the maximum frequency or voltage/frequency relationship. Logic cone extraction is used to better understand origin of failures. For instance, when a scan cell captures incorrect results, cone extraction is used to determine which logic gates fan in to this cell. On the design side, the following DFT structures and methodologies need to be in place to allow DFT based characterization:

Scan

Conventional scan lends itself to both IC debug and characterization. On the pattern generation side, one should make sure that the pattern set includes a *scan chain integrity test* (sometimes referred to as a *chain test* or *flush test*) to better allow differentiation between scan chain failures and failures in the regular scan tests. For designs using multiple capture clocks and/or multiple pattern types (combinational, sequential, etc.), these different pattern types should be grouped so that it is easy to determine if failures relate to a particular type of pattern or a particular group of patterns.

Scan Delay Test

Especially for path delay test, it is of interest to understand which specific paths and clock domains are part of a failure. This information is usually not included in pattern formats such as STIL. The user must therefore set up an environment where this information is recorded as part of the pattern generation process, so that when the tester reports that pattern 3 failed, the user can easily look up which path was involved, the control and observe points, and the launch and capture clocks[21]. In a

---

[21] A path is a described collection of specific gates and nets that starts on an input pin or internal register, describes the propagation route of a signal transition, and ends on an internal register or output pin. This is exactly the output of a static timing analysis tool.

path delay test, there is one particular path that is exercised for a pattern. Additional scan cells are loaded with particular values (care bits) to sensitize the path. The remaining scan cells are typically filled with random values (random fill), and a fault simulation is performed to utilized the patterns for increased test coverage. This is all well and good, but typically makes the debug process more complicated. Due to the random fill, there is nothing in the patterns that indicate which bits are care bits, and which bits represent the random fill. With most commercial tools, the user must therefore keep track of this information separately (through tool log and report files) and have it available for the characterization process. The tester needs to provide a clock with the ability to create an at-speed test interval and to vary that interval (clock margining).

Logic BIST

In a traditional logic BIST design, the user observes a relatively short signature from a Multiple Input Signature Register (MISR). From this signature, it is impossible to interpret anything at all about the particular failures, such as which scan cells were involved, whether it was an at-speed problem, a chain integrity problem, etc., unless the MISR equation is known and much pre-work has been done. Therefore, different logic BIST circuitries typically include special modes and methodologies for debug. It is important that this flow is completely understood up front. One solution is to use a bypass mode where the circuit is operated as a regular scan device. Other schemes include a binary search technique to determine the pattern that fails, followed by dumping the scan chains in a conventional scan diagnostics scenario; or the inclusion of logic to mask scan chains feeding into the MISR to narrow the search space during operation. This implies that the test system (the LCST) used for diagnostics and characterization is capable of handling a different (larger) set of patterns than what's used for production. The tester needs to support capture memory for the incremental signatures or the scan chain dumping required for diagnosis. This is a data dump rather than a comparison.

Scan Compression

Different scan compression techniques have different debug capabilities, but similarly to logic BIST, the case is that it is often not a 1:1 correspondence between a specific bit on an output pin and an internal scan cell. Therefore, one often has to either have a separate pattern set (which, similar to logic BIST bypass patterns is much larger than the original pattern set), or one has to do more of the diagnostics task in the compression tool (rather than on the tester).

Memory BIST

To be able to characterize memory BIST failures, the BIST circuitry needs to have additional diagnostics capabilities beyond the traditional pass/fail flag. The goal is to build a memory bitmap to enable rapid visualization. Several memory BIST engines have the capability to scan out data such as failing address, data, and the step in the algorithm for all failures. To be able to benefit from this capability, it is important that the top level circuitry controlling the various memory BIST controllers can invoke and differentiate the individual controllers, and capture the data. On the test system, software is required to collect and interpret this information in communication with the top level memory BIST circuitry. Some memory BIST controllers have the capability to run different memory test and diagnosis algorithms, something that can be very helpful for characterization purposes.

6.4.7 DFT and Yield Learning

The correlation tester use model relates to yield analysis, conducting debug and diagnosis off line with results correlating to manufacturing results. Yield has great impact on profit, and is important especially to cost sensitive devices (see Figure 6-1). Yield improvement is important both for a new process, as well as for a new device manufactured with a mature process. Initially, when ramping a new device, the failures aren't necessarily the device as the environment such as loadboard, socket, or the vectors. These are systemic issues that affect yield and can be fixed without a re-spin of the device. There are many steps in the design process that impact yield. DFT can be used for yield improvement by utilizing automated failure analysis techniques. Failure analysis can utilize many of the same tools used for debug during bring-up and characterization. A scan diagnosis tool can be used to determine the gate level candidate(s) causing a particular failure for a particular device.

For IC designers, it is important to know what to adjust in the design for better performance, yield, or simply fix design bugs. Failure analysis is a complex flow that includes data from multiple sources[22]. DFT can greatly simplify this flow. ATPG tools have for a long time had the capability to determine what the most likely defects are based on the gate level netlist, the test patterns, and failure data from the tester. To make this process work seamlessly, it is important that the failure data can be read back into the ATPG tool. In traditional test flows, since ATPG tools usually think in terms of scan test numbers and scan cell numbers rather than pins and vector cycles, this is difficult. For this diagnosis process to work, a large amount of failure data is required, often more data than what traditional ATE capture memory can handle. Even with a seamless flow, there are still issues to resolve. ATPG–based diagnostics have proven to be effective for single defects, but often run out of steam when handling multiple defects and defects that do not match the stuck-at fault model.

The scan-based failure analysis flow shown in Figure 6-10 is based on the test flow introduced in Figure 6-5. The foundation for the method is a seamless flow of pattern data and failure results between the ATPG tool and the tester. Once the defect candidates are properly identified at the gate level by the ATPG tool, the layout information can then be included to find the physical location of the defect candidates in the chip on the silicon die. While the defect location described by the diagnostic tool is a "node", in layout this location can be represented by one or more traces. To further improve the resolution, it is common to overlay this information with the in-line defect inspection data from the associated process layers in a process referred to as logic mapping. The identified location can be used to perform physical failure analysis of the part [22].

[22] Failure analysis is typically based on data from ATE and in-line defect inspection data. In addition, the process involves design data at logical and layout level.

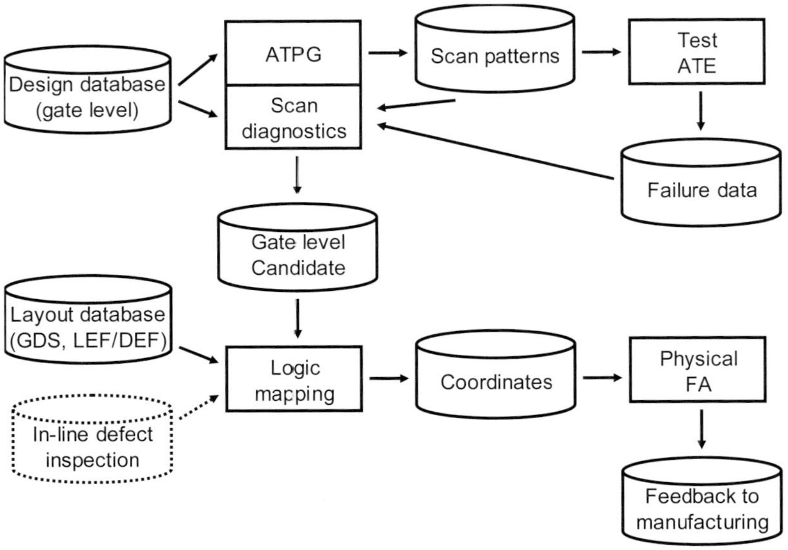

Figure 6-10: Scan-based failure analysis flow.

To benefit from the full potential of LCST, the design and test flow must provide these capabilities discussed above.

6.5 WHAT DOES LOW COST HAVE TO DO WITH THE TESTER?

We have previously investigated how DFT can reduce test–related costs. DFT–side or design-side costs and expenses are not as implicit and direct as the cost of an asset and cannot be modeled or tracked easily—these test costs are very complex and cannot be described in simple terms. Tester hardware, however, can have one of the most significant impacts on the cost of test; it is, at the very least, one of the most visible components of cost of test. In reality, the test equipment and the operation and overhead costs of a test floor make up one of the most well-known costs involved with testing, the "per second" test cost. In this section, the tester will be evaluated for its contribution to test-side test costs.

6.5.1 What Makes a Tester Expensive?

One of the things not well understood by those not in the ATE industry is "why is an ATE or 'Big Iron' so expensive – millions of dollars expensive?" When asked, the answer from most ATE providers may seem vague and evasive, but the truth is, what drives the cost is a whole range of effects – it really isn't easy to point to one thing or another, but we will endeavor to identify the most important factors in this section.

The main problem that has historically plagued ATE is what is commonly called the 'treadmill' effect. If the ATE were viewed as a police car, and the part to be tested as a speeder, then the police would claim that they had to have faster cars than the speeder in order to chase them down and catch them. The problem is that some of the parts to be tested are on the leading edge of the 'Gordon Moore' curve and are supposedly the fastest, most high-performance devices made. This means that most ATE offerings go outside of normal CMOS silicon development to build machines that are capable of testing the latest semiconductor offerings by using exotic technologies such as gallium arsenide driver hardware.

Another cost problem that is often stated is the '*low volume*' problem. The ATE providers are building testers in the hundreds, where semiconductor providers want to sell semiconductors in the tens of thousands or millions. So, getting leading-edge custom chips, and sometimes exotic chips, made for the ATE, and in low volumes, is a very expensive proposition.

The main cost factor, though, is because the ATE providers are still making machinery that is attempting to test the part functionally. Supporting functional capability requires the ATE architecture to be able to react to any sort of 'end user' environment – from a microprocessor to a DSP engine to a graphics processor to a wireless communication chip. This means that the tester must support complex sequencing, must have an amazing range of capabilities per channel and must have a way to coordinate and synchronize the different pins to a high degree. To meet these conditions, many ATE providers[23] have moved to the "per pin" architecture where each pin channel is a tester unto itself with complex sequencers, waveform generators, associated high speed memory, and high-precision channel drivers – but the overall grouping of channels has another layer of control to coordinate each channel to the timing and sequencing of the other channels, also with high accuracy and precision.

An example of the requirements that this type of architecture must support would be that the voltage on any given pin channel could range from minus 1 V up to plus 12 V, and must change in increments of a few mV; that the period window be able to vary from KHz to GHz; that the signal edge driver must be able to place a rising or falling edge within the period window with an accuracy of 25 ps and with a step resolution of 2.5 ps; and that the relationship between edge placement on any two pin channels in the tester be within 250 ps. In order to meet these specifications, the tester must maintain a high level of thermal equilibrium (because these specifications are very sensitive to temperature variations) – which is why most of them have complex cooling architectures.

The terms used to describe the example ATE requirements were 'voltage resolution', 'signal placement accuracy and precision', 'frequency range', and 'temperature stability'. However, another way to say this is that the tester must have a high degree of *precision*, *accuracy*, and *flexibility* – and these three items are what drives the high cost of multi-functional ATE.

---

[23] Some high-end ATE providers as of this writing are Advantest, Agilent, Credence, LTX, and Teradyne. The offerings from these providers range from the multi-million dollar range (e.g. $5M) down to the $500K range.

The direct way to reduce the cost of the tester is to no longer require the high degree of precision, accuracy, and flexibility. Structural test accomplishes this by applying a known, non-changing set of test sequences that have much looser voltage and timing requirements. For example, scan is a well known protocol and is the application of vectors onto scan chain input pins at a slow data rate (such as 50 MHz) that doesn't overly stress the chip's power structure; and with loose setup-and-hold timing (for example, 2.5ns) because the goal is to deliver data into the part, not to evaluate the timing of the scan input and output pins. All pins that are not scan inputs, outputs, or clocks, change once per scan. In addition, since the shift rate (which represents the data rate) is slow, then the memory used as scan tester memory can be low performance memory such as plug-in computer DRAM memory (PC DIMMs[24]), and much more of it can be supported. Figure 6-11 shows an overview of a typical DFT tester architecture.

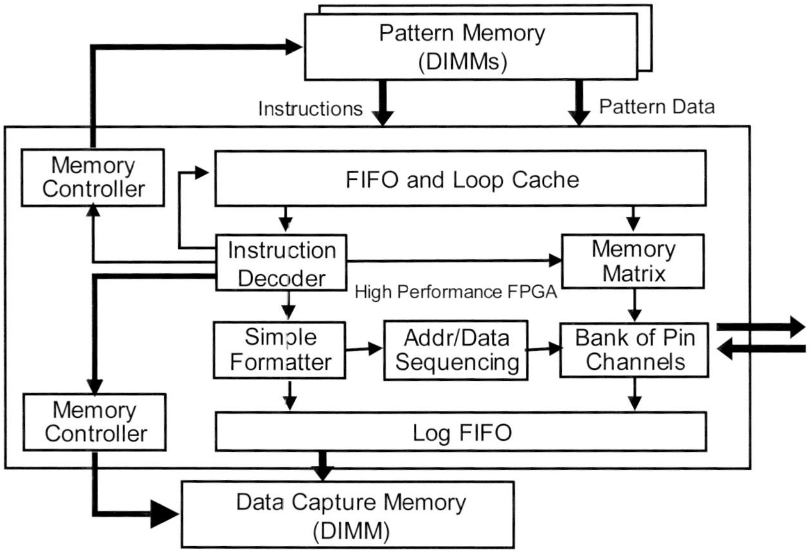

Figure 6-11: A generic DFT tester architecture.

To more fully understand the differences between a high-end tester and a DFT tester, it might be useful to describe the contrast between Figure 6-11 and a typical high-end tester architecture. Most high-end testers are "per-pin" architectures that are built as high-performance channel cards – a channel represents one driver and receiver pair. High-performance means that the drive data must be delivered at the same frequency as the pin data-rate, so generally, it is high-performance exotic memory that is implemented with some form of multiplexed-access in order to present drive data at-speed – and high-performance exotic memory is usually not very large (for example, only 1 MB to 4 MB is common). The same requirement is true for capture memory. In addition, each channel in a "per-pin" architecture must allow flexible waveform formatting, so there is usually a complex waveform driver

[24] Dual In-line Memory Modules.

made as a custom Application Specific Integrated Circuit (ASIC). These drivers allow very precise edge-placements, pulse-width sizing, frequency ranges, data rates, waveform types (NR, SBC, NRZ, R0, R1, etc.), and incremental voltage adjustments.

In contrast, a DFT tester architecture is not a "per-pin" architecture, but a bank of channels (for example, 64 pins) that shares a waveform formatting unit and a bank of memory. The memory architecture is flexible so that it can be shared equally among all of the supported pins, or it can be allocated to just a few pins. Since the data rate in a DFT tester is defined to be much slower – it only has to achieve a reasonable scan shift rate – then standard off-the-shelf personal computer DIMM memory can be used. Standard PC DIMM memory actually comes in large sizes (for example, 256 MB to 1 GB is common), so the drive and the capture memory can be quite large. Since waveform generation is less complicated, and the timing and data rates are less rigorous, then the majority of the channel complication is in the assignment of large amounts of drive and capture memory to a particular channel. All of this can be programmed into a high-performance Field-Programmable Gate-Array (FPGA) as opposed to being implemented in a custom IC. However, the voltage adjustment capability is limited to the output drive capability of the FPGA.

Basing a tester on the requirements needed to implement scan and BIST reduces the voltage resolution requirements (for example, only supporting fixed voltages of 1.8V, 2.2V, and 3.5V as opposed to a broad range of 0V to 5V with mV adjustment capability), the frequency range requirements, the data rate, and the accuracy and precision requirements – and having simpler electrical requirements reduces the sensitivity to thermal variation. All of this can significantly reduce the cost of a tester. Having large quantities of inexpensive memory or flexible and configurable memory (such as a sea of memory concept that shares memory across pins) can significantly reduce the application cost of a tester by reducing the number of re-insertions or test vector memory reloads.

It must be noted that it is possible to make a very high cost structural tester that supports gigahertz clocks and data rates; supports a rich feature set such as analog, mixed-signal and complex digital sequencing; and has a high degree of precision and accuracy for applied data and clocks. However, this case is not explored in this chapter [19].

6.5.2 Achieving Test Goals Without Precision, Accuracy, Flexibility

If the way to reduce the cost of a tester is to remove the precision, accuracy, and flexibility, how then can test be accomplished to the same level of quality? The short answer is that removing capability from the tester requires making up for it by including DFT in the part. A little logic in the chip can alleviate a lot of bandwidth and capability requirements placed on the tester. There are two tradeoff drivers that must be explored when using an LCST; one is to understand what is problematic or costly to test using high performance ATE; and the other is to understand what the target goals of testing really are.

For example, is the best way to test an embedded core that runs hundreds of megahertz, to apply a full-speed clock from the tester into a pin and to apply at-speed vectors though a bus interface meant to operate at sub-100 MHz data rates? To conduct this test would require an expensive clock pad and over-engineering the bus interface to handle pass-through signals for test at high frequencies. In this case, it is definitely better to use a slow test clock and to deliver scan vectors to the embedded core by shifting the vectors at the natural designed data rate for the bus interface – an AC test can then be accomplished by using the chip's PLL with AC delay fault models. This is a case where DFT and structural test eases the design requirements for test on the chip.

The other tradeoff driver begins with understanding the goals of test and to then map those goals onto tests that can meet them. What type of test is to be replaced in the functional space? For basic testing and simple stuck-at fault coverage, the choice is either fault-graded functional vectors, or the more efficient and automated DC scan in conjunction with automated vector generation. For frequency assessment, speed binning and delay defects, AC scan can replace at-speed functional vectors[25]. For reliability screening, I_{DDQ} vectors and leakage measurements, in conjunction with BIST for burn-in have proven to be much more effective than functional operation at burn-in. And one last item that is common in functional test programs, pin parametric testing. Parametric testing is generally conducted using a parametric measurement unit (PMU) per pin, but for cost containment, structural testers that do support parametric testing do so by providing a PMU per domain – were a domain is some grouping of pins such as 32 or 64 – which may lengthen the test time.

Not all cost savings or tradeoffs are the simple tool-driven DFT techniques that are available today. There are test challenges today that can be viewed as second generation structural test techniques; or as solutions that can only be enabled by structural test. For example, a major test issue in the embedded IP (intellectual property) core space is the preservation of proprietary interest. What this means is that the IP core provider does not wish for any customer or foundry to be able to use test to reverse engineer the proprietary and trade secret content of the delivered hard core using test techniques. In this case, the netlist can be delivered as a binary ATPG image (which represents an encrypted model) that can be used by diagnostic tools to report an error on a logic gate that can be sent back to the core provider, but does not allow any further investigation.

Since scan vectors also carry a large amount of inherent design information, the vectors must also support some form of encryption. This has been done quite effectively by merging the best aspects of both scan and logic BIST to create

---

[25] Frequency assessment is the verification that a device or some portion of the device operates at the specified frequency – this is important because many logics are defined by IEEE or ANSI specifications to have a defined bit rate or a frequency of operation. An example would be the IEEE 1394 Firewire specification which states that the logic must operate at 78 MHz. Either a device operates at the required frequency, or it is discarded (pass/fail). In contrast, speed-binning is the separation of devices that operate at some high-range of frequency from devices parts that operate at a lower range or a series of lower ranges of frequency. This is most commonly applied in the microprocessor business where process variations may produce a range of performance, and since the highest performance parts can be sold for a higher price, identifying them is a major business requirement.

deterministic embedded structural tests. In essence, the vectors that are delivered to the tester generally represent the seeds to a linear-feedback shift-register (LFSR) being used as a pseudo-random pattern-generator (PRPG) and these seeds were calculated by an ATPG engine that made deterministic fault-model based, scan-based tests – and then calculated what seed in an LFSR could place these bits in the scan chain. The output data can also be encrypted by using a multiple-input signature register (MISR) or other compactor structure to compress the output vectors. These 'deterministic compression' techniques reduce the bandwidth requirements of the tester by significantly reducing the vector image, and also have the effect of encrypting the vectors [23].

There are other test goals that can only be cost-effectively supported by structural test techniques and can be considered as complex or compound second generation applications. For example, in order to successfully implement multi-site testing of digital chips, without having an expensive high-pin-count probe head, the chip must support some sort of reduced pin/pad signal interface. This is commonly known as *reduced pin-count testing* (*RPCT*). The easiest method is to support just a few scan chains, or to embed a BIST in the chip. In the case of just a few scan chains, the scan data could then become large compared to the tester, either in vector depth, or in vector load or application time – so some sort of vector compression or compaction may be required. The combination of vector compression and RPCT enables low cost multi-site probe for digital logic chips better than any other technique.

6.5.3 The Next Step in Test Cost Reduction – the Test Interface

The one step that goes beyond today's DFT and the current methods of use of structural testers, to achieve lower test cost, is the adoption of an industry standard tester interface. Currently, there are still a wealth of vector data formats, even though STIL (IEEE 1450) is quickly becoming a supported industry standard in this area. This is one of the first steps to reducing the vast number of variations in tester requirements. The next step after that is to support an industry standard on-chip test interface that would allow the protocols involved with test control and vector application to be standardized. Even though this might be the right direction to take, the detractors claim that this will minimize the uniqueness of each of the different tester company's offerings. Different testers have different strengths and weaknesses, and the way to highlight these is to have proprietary languages and operating systems that are closely coupled to the hardware tester architecture.

However, supporting one test language (STIL) and adopting a standard on-chip test interface or test interface protocol allows the structural tester to limit its flexibility to known bounds (and the reduction of precision, accuracy, and flexibility is the key to cost reduction). There are multiple protocols that are available or in work at this time, such as the IEEE 1149.1 and the IEEE 1500 standards, but no effort has been made to link low cost structural tests to a particular standard or to define a structural tester standard.

Creating a standardized structural test interface could lower tester costs even more, but there is still enough variation in the different structural techniques to stall any consideration of standardization. For example, Mux-D flip-flop scan versus level-sensitive scan-design (LSSD) requires the same scan data, but the clocking may be completely different; there are many different memory BIST fail data output definitions and protocols used to build memory bitmaps; and there is no definition at all about test scheduling priority – does scan run simultaneously with logic BISTs and memory BISTs? How many BISTs operate in parallel? If multiple failures occur at the same time in different BISTed blocks, which block reports its failure first?

There is also not any agreement on what such a chip-level interface could be. In many places the IEEE 1149.1 interface is being used as a chip test controller, even though it was proposed as a board test and board integration interface. However, since the JTAG interface was not initially designed with modern chip-level structural test in mind, there are detractors that believe it cannot handle modern structural applications such as AC scan-based delay tests, and that it cannot easily grow into the future. Alternately, there is the IEEE 1500 standard, which has considered multi-core, multi-memory, partitioned and embedded test, and modern structural test types – but this assumes that the chip making world will all be making SOC-like devices in the future[26].

One of the other issues involved with the consideration of a standard chip-level test interface is its sizing – does the interface define just a subset of pins that need to be touched during test, or does the interface define the role of every pin on the part? The answer depends largely on the concept of there still being a functional or all-pins structural insertion during the test process (some test somewhere must verify connectivity, continuity, and to some extent, pin parametrics). However, for some insertions, such as production probe, just defining a few pins to touch and allowing the other inputs to float and the other outputs to be ignored, enables reduced pin-count testing, which further enables multi-site and massive multi-site testing.

Most internal scan based designs do require that the non-scan pins must be used during a scan test. This is done to allow testing of the input side shadow circuitry (the logic between non-scan-interface input pins and internal scan cells) and output side shadow circuitry (the logic between internal scan cells and non-scan-interface output pins). To facilitate reduced pin-count test (RPCT), the need for touching and accessing non-scan-interface pins can be obviated with "wrap-I/O" techniques or the correct utilization of JTAG boundary scan cells. By using one of these two techniques, a 1000 pin device can be tested with just a few tester pins (reducing the resource support requirements of the tester)[27].

Figure 6-12 shows an example where, in a dedicated test mode internal scan pins are accessed directly, while non-scan-interface pins are accessed through boundary

---

[26] A reuse-based design flow has been a requirement for several editions of the International Technology Roadmap for Semiconductors (ITRS) and it has not yet permeated the system design process. At the 90 nm to 65 nm transition, this requirement is expected to be critical [4].

[27] Currently, the practical limitation of reduced pin count testing is the planarity of cantilever probe needles across multiple die on a wafer, so in most cases testing is limited from 256 to 512 pins. Common multisite configurations with current technology are by-2 and by-4, so the range of pins targeted to be touched are generally driven by either the size of the scan interface or the limitations placed by multisite probe. Common RPCT interfaces are 16, 32, and 64 scan chains (35 to 130 signal pins touched for each chip).

scan during test. This enables the use of more than one scan chain during test. In this example, boundary scan is used, but the principle is the same for wrap-I/O, where dedicated circuitry is used to control the input pins and observe the output pins. One of the most common non-JTAG wrap-I/O techniques is to make every input pin a bidirectional pin by using a three-state or high-impedance driver – during test, all input pins become output pins and their values are sampled in internal scan registers.

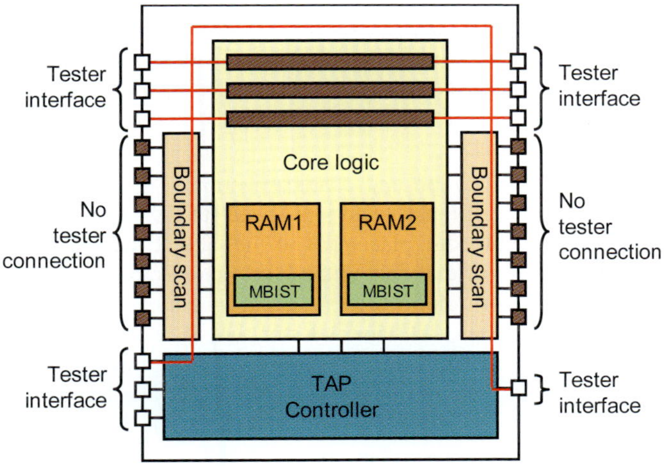

Figure 6-12: Reduced pin count test using boundary scan.

When reduced pin count test is facilitated using boundary scan, special boundary scan cells must be used, such as the one illustrated in Figure 6-13. The cell below disables the usual Parallel-in to Parallel-out pathway that is normally used during functional operations or full-pin count testing and allows the Serial-in to Parallel-out path for reduced pin count testing. The serial-in pathway requires that the cell be operated by the sequencing of the 1149.1 logic to serially fill the boundary scan ring with pin drive data – the data in the second flip-flop (the Update Cell) represents the data that would normally be on the parallel-in pin.

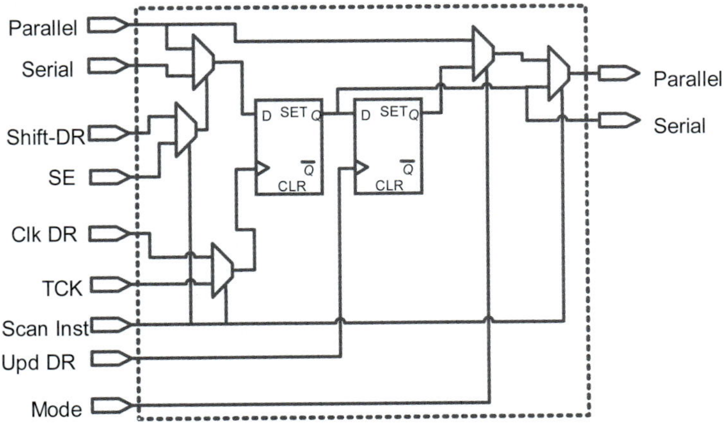

Figure 6-13: Special purpose boundary scan cell for RPCT [24].

6.5.4 The LCST is Not the Silver Bullet

The LCST achieves its cost savings by reducing the functionality of the tester and makes up for it by relying on DFT included within the chip. However, DFT cannot make up for all of the lost functionality; and the structural test methodology is not needed by, or allowed in, all designs.

The LCST is not meant to completely replace a high-performance functional tester – the high-performance functional tester still has its place. For example, there is currently no standard or widely adopted mixed-signal, analog, and RF DFT strategy that can be used to alleviate the need for analog instrumentation in the tester. Some of the items that *cannot* be accomplished by the low cost structural testers of today are:

- High-speed I/O verification and characterization – usually accomplished by the application of functional vectors at-speed (at the rated bus interface data rate) with a very fine edge placement of the data at the specified setup time and the complement of that data at the hold time. Traditionally, this has been done using a waveform defined as *surround-by-complement*. This testing can be accomplished using a structural tester, but it requires having a clocked launch register on the loadboard, and then conducting a path delay test through a described critical path and into a scan capture cell inside the device that is based on a different clock (but provided by the tester).
- Any non-digital type of test that can't be provided through the shared PMU (PMUs often have low resolution sampling circuits) or a low speed frequency counter.
- Any test that requires multiple edge sets, or multiple time-sets (aside from the AC scan launch-capture interval), or highly precise signal or clock edge placement. For example, testing a high-performance bus interface for different modes of use that have different timing requirements – interfacing to a memory in one case and interfacing to another processor in another case.
- Any test that requires a wide voltage range and/or fine grained voltage adjustments. For example, providing analog waveforms with continuous voltage modulation; or providing adjustable voltage ranges for any given pin on a single part that supports different voltages for power, signals, and even programming of internal Flash or EEPROM memory.
- Ultra-high-speed low voltage differential signals (LVDS) – some of these tests are outside of the purview of both structural and functional testers. These types of logics are commonly tested by using a loopback either inside the chip, or on the loadboard with some sort of phase or jitter control placed on the loadboard.

The structural test methodology is not needed if the business space requirements of the chip are met. For example, if the chip is a microcontroller based on a fixed instruction set architecture (ISA) and completed new chips must be compatible with older members of the family. The vectors can then be the legacy code used on the previous family member, the vectors can be delivered before the chip returns from the fab, the fault coverage can be in the high 90s percentile, and the test program may be fully applied within the handler sequencing time. In this case, all the needs are met by the existing or legacy functional test methodology. However, as soon as this same part is converted into a core to be embedded within other chips with limited access to its pin map, then structural test becomes the only portable solution.

The LCST is not meant to replace the high performance functional tester, but to augment it and to offload some of the capacity onto the more efficient and more cost effective structural testers. The decision to adopt an LCST for a portion of the testing insertions, or to replace a functional tester, strongly depends on the business model and markets of the chip designing company [1].

6.6 LIFE, THE UNIVERSE, AND EVERYTHING

No matter what it is called, LCST, LCDT, DOST, or VLCT, the common theme is that a structural tester is optimized for DFT techniques and the application of structural tests – as opposed to the application of functional, behavioral, or design verification tests. The structural tester is cost managed by making intelligent bandwidth decisions and placing certain capabilities within the chip or on the loadboard as opposed to within the tester.

Structural tests have been aligned with DFT through EDA tools and other automated methodologies and are based on the use of deterministic fault models. Simple structural tests are based on scan (DC and AC), logic BIST, memory BIST, and current leakage measurements. Compound or second generation structural test methodologies are items like multi-site testing using a reduced pin-count (test port) methodology in conjunction with vector compression techniques; encrypted netlists and encrypted vectors for IP core diagnostics that retains proprietary interest; and embedded deterministic vector generation (a mixture of the best optimizations of scan and BIST) to reduce the overall vector image in the tester.

Figure 6-14: Structural test systems Teseda V500 and Inovys Personal Ocelot.

Organizations are moving to structural test for other reasons than to use structural testers (two examples of desktop structural testers are shown in Figure 6-14). The ability to use the structural tester is not the driving force for the change – but now that they are available, this closes the loop on the long said quote, "structural test can reduce the cost of test." The use of structural test does lower cost, and in many different ways, subtle, explicit, implicit, and overt:

- Even though there is schedule and effort added in installing DFT, the DFT reduces the vector generation and qualification time, reducing overall schedule time – the tradeoff is that a task once done exclusively post-silicon has been significantly reduced by moving a portion to the pre-silicon side of the equation.
- The DFT allows more observability and controllability and enables automatic vector generation, providing more and more varied fault coverage and therefore a higher measured quality level.
- The deterministic nature of structural vectors carries more targeted and specific design and fault information, providing more rapid and accurate detection, debug, diagnosis, and isolation.
- The lesser electrical, sequencing, precision and accuracy requirements of structural test minimizes the requirements placed on the tester, reducing the cost of the test equipment and all of the tangible and intangible maintenance factors (electricity consumption, floor space, cooling, operator expertise, etc.).

The use of structural test has evolved to second generation effects – capabilities that are only possible because of structural test. For example, portable embedded IP vectors; multi-site logic testing; and cost effective comprehensive DC and AC testing at probe.

There is a wide variety of use spaces for structural test – design validation, first silicon bring-up, device characterization, failure analysis, burn-in, yield learning, production probe, and final packaged production testing – and the goals, application and methodologies of structural test are different in each space. This is also true for structural testers, they can be used by design and development organizations, test development organizations, and production test organizations – and the goals, drivers, and applied methodologies of use are different in each of these spaces. The design-side space is driven by 'needs and capabilities', the test development and correlation space is driven by 'time-to-result', and the production-side space is driven by 'throughput and capacity'.

Structural test has come a long way, but it is still not the complete solution – it does not magically solve all problems for everyone. It is not the 'magic pill', 'silver bullet', or 'eternal panacea' of legend and lore – it is a practical and cost effective solution to many of the current problems associated with testing, assessing, diagnosing and measuring the quality of modern silicon devices. And, despite all of the pro and con and tradeoff analyses done, there is not an effort underway for structural testers to completely replace functional testers. Even though there is a rich set of applications, structural test and structural testers cannot fully replace functional test and functional testers – there are still high-speed interface tests and analog and mixed-signal tests that have not yet been addressed by any standardized DFT or structural test and tester methods.

As time goes on, and silicon designs continue their relentless march up the Moore curve, structural test will be come a comfortable and familiar part of the test methodology applied to devices of many different types and in many different market spaces. As the second generation compound structural techniques become common and well-adopted and mature technologies, then third generation techniques will be invented to handle the massive volume and miniscule, but massively integrated, complex dies of the future (such as the new class of devices known as *Networks on a Chip – NoC*). The hope, however, is that as the future gets more complex, the structural tester doesn't fall into the same 'treadmill' pattern that functional testers suffer from today.

REFERENCES

[1] K. Tumin et al., "Scan vs. Functional Testing – A Comparative Effectiveness Study on Motorola's MMC2107™", *Proc 2001 IEEE International Test Conference*, pp. 443-450.

[2] J. Bedsole, R. Raina, A. Crouch, M.S. Abadir, "Very Low Cost Testers: Opportunities and Challenges", *IEEE Design & Test of Computers*, vol. 18, no. 5, pp. 60-69, Sept.-Oct. 2001.

[3] J. Johnson, "Is DFT right for you?", *Proc 1999 IEEE International Test Conference,* pp. 1090-1097.

[4] International Technology for Semiconductors Roadmap 2004 Update. Available http://public.itrs.net/

[5] G. Eide, K. Posse, P. Decher, "Is DFT Ready for Prime Time", Test & Measurement World, 3/26/2004. http://www.reed-electronics.com/tmworld/article/CA406081.

[6] R. Garcia, Wayne Needham, "How to Succeed With Structural Test", *EE-Evaluation Engineering,* November 2001.

[7] T. McLaurin, F. Frederick, "The testability features of the MCF5407 containing the 4th generation ColdFire(R) microprocessor core", *Proc 2000 IEEE International Test Conference,* pp. 151-159.

[8] K. Posse, G. Eide, "Key Impediments to DFT-focused Test and How to Overcome Them", *Proc IEEE International Test Conf.,* pp. , September 2003.

[9] G. Hetherington, T. Fryars, N. Tamarapalli, M. Kassab, A. Hassan, and J. Rajski, "Logic BIST for large industrial designs: real issues and case studies," *Proc. IEEE International Test Conf.* pp. 358-367, 1999.

[10] IEEE 1500 Web Site, http://grouper.ieee.org/groups/1500.

[11] IEEE 1450.6 Core Test Language (CTL) Web Site, http://grouper.ieee.org/groups/ctl/.

[12] IEEE Standard Test Interface Language (STIL) for Digital Test Vector Data, IEEE 1450-1999.

[13] IEEE 1450.1 Web Site, http://grouper.ieee.org/groups/1450/dot1/.

[14] P. Maxwell, I. Hartanto, and L. Bentz. "Comparing Functional and Structural Test" *Proc. 2000 International Test Conference,* pages 400– 407.

[15] B.D. Cory, R. Kapur, B. Underwood, "Speed Binning with Path Delay Test in 150nm Technology", *IEEE Design & Test of Computers*, vol. 20, no. 5, pp. 41-45, September-October 2003.

[16] T.J. Powell et al., "Delta I_{DDQ} for Testing Reliability" *Proc 2000 IEEE VLSI Test Symposium*, pp. 439-443.

[17] S.S. Sabade, D.M.H. Walker,"Comparison of effectiveness of Current Ratio and Delta-I/sub DDQ/ tests". *Proc 2004 International Conference on VLSI Design,* pp. 889-894.

[18] R. Daasch et al., "Neighbor Selection for Variance Reduction in I_{DDQ} and Other Parametric Data," *Proc 2001 International Test Conference*, pp. 92-100.

[19] M. Mayberry, et al., "Realizing the Benefits of Structural Test for Intel Microprocessors", *Proc 2002 International Test Conference,* pp. 456-463.

[20] A. Fudoli, A. Ascagni, D. Appello, H. Manhaeve, "A practical evaluation of I_{DDQ} test strategies for deep submicron production test application. Experiences and targets from the field," *Proc. IEEE European Test Workshop*, pp. 25-28, 2003.

[21] B. Benware, C. Schuermyer, N. Tamarapalli, K.H. Tsai, S. Ranganathan, R. Madge, J. Rajski, P. Krishnamurthy, "Impact of multiple-detect test patterns on product quality" *Proc. IEEE International Test Conf.* , pp. 1031-1040, 2003.

[22] A. Kinra, "Scan Diagnosis and Logic Mapping: A Practical Guide to Performing Electrical Fail Analysis", TTTC Test Technology Educational Program (TTEP) Tutorial, *2002 International Test Conference.*

[23] J. Rajski, J. Tyszer, M. Kassab, N. Mukherjee, R. Thompson, K.H. Tsai, A. Hertwig, N. Tamarapalli, G. Mrugalski, G. Eide, J. Qian, "Embedded deterministic test for low cost manufacturing test", *Proc. IEEE International Test Conf.* pp. 301-310, 2002.

[24] Mentor Graphics. Reduced Pin Count Testing using BSDArchitect. Mentor Graphics User Group Conference, March 2003.

RECOMMENDED READING

- A. Crouch, Design for Test for Digital ICs and Embedded Core Systems, Prentice Hall PTR, Indianapolis, 1999, ISBN: 0130848271.
- A. Crouch, "Structural Test Simplified", Whitepaper, available http://www.inovys.com
- IEEE Design & Test of Computers: Speed Test and Speed Binning for DSM Designs. vol. 20, no. 5, September-October 2003. This entire issue is devoted to structural at-speed test.
- R. Kapur, CTL for Test Information of Digital ICs, Kluwer Academic Publishers, Boston, 2002, ISBN: 1402072937.
- G. Maston, T.R. Taylor, J.N. Villar, Elements of STIL: Principles and Applications of IEEE Std. 1450, Kluwer Academic Publishers, Boston, 2003, ISBN: 1402076371
- S. Morris, "The DFT-Focused Tester", Whitepaper, available http://www.teseda.com
- J. Saxena, et al., "Scan-Based Transition Fault Testing – Implementation and Low Cost Test Challenges", Proc IEEE International Test Conf., pp. 1120-1129, 2002.

Chapter 7

Embedded Cores and System-on-Chip Testing

Rubin Parekhji

As design paradigms evolve to construct complex *Systems on Chip* (SOC) using a diverse collection of pre-designed cores, so must the test paradigms evolve too. This chapter describes the various Design-for-Testability (DFT), Automatic Test Pattern Generation (ATPG) and Built-In Self-Test (BIST) techniques for these new paradigms for the test of embedded cores and SOC architectures.

The chapter is organized into nine sections. Section 7.1 gives an introduction to embedded cores and SOCs. Section 7.2 outlines the design paradigm and test considerations for the creation and use of embedded cores and SOCs. Section 7.3 elaborates on how conventional DFT techniques have to be augmented for application in these newer paradigms. These include choice of appropriate scan and clock control architectures, and design partitioning. Section 7.4 describes various test access mechanisms using standard test interfaces. The IEEE 1500 test interface for embedded cores is described, together with its usage with different test architectures. Section 7.5 introduces the new problems in ATPG using embedded cores and their impact on ATPG of the overall SOC. It also proposes some solutions for them built on well-known techniques. Section 7.6 discusses the various SOC test modes, and shows how these test modes contribute to the test efficiency and test

D. Gizopoulos (ed.), Advances in Electronic Testing: Challenges and Methodologies, 217-261.

quality in an SOC. Section 7.7 is devoted to at-speed testing using ATPG techniques. Various design and pattern generation techniques are described and compared. Section 7.8 discusses the design and implementation issues in BIST of memories and logic in embedded cores and SOCs. Conventional BIST, BIST enhancements and processor based BIST approaches are explained. Section 7.9 concludes this chapter with a discussion on newer design and test challenges, tradeoffs and optimizations, and their influence on SOC testing.

7.1 EMBEDDED CORES AND SOCS

As the design complexity increases, the construction of individual design modules takes longer time. This increase in the design time also correspondingly increases the time required for the construction of the overall system built using a collection of such modules. Technology advances and end application markets have, however, rendered traditional design techniques unviable. It is no longer feasible to rapidly build systems using modules that are themselves being designed from scratch. On the other hand, there is an increasing need to build a repertoire of such modules which can then be integrated together to realize a system.

Such a module must be designed with a set of comprehensive specifications making its usage amenable across various designs. When used inside a larger design, such a module is embedded, and hence the term *embedded core*. The larger design is now a system consisting of such modules or cores. With present levels of integration, such a system can be built inside a single chip, and hence the term *System-on-Chip* (*SOC*). An embedded core is also referred to as an *Intellectual Property* (*IP*) *core,* in case the core's implementation and construction are proprietary and are available only with the designers of the core, and not to its users. The reader can find a more detailed description of the characteristics of embedded cores and SOCs in [1].

In this chapter, the SOC construction process refers to the design, integration and test of such a system using a collection of one or more such cores[1]. The focus is on manufacturing test techniques for such cores and systems. Unlike discrete standalone modules which benefit from unrestricted controllability and observability, embedded cores and SOCs have several other test constraints. These include those imposed due to design characteristics of the modules, their interfaces and that of the SOC. Though these characteristics, in an isolated context, may cause no additional problems for test, in the SOC context, they impose problems that disrupt the use of well-established test practices. This chapter discusses the problems and proposed solutions in the test of such embedded cores and SOCs[2].

---

[1] The terms *IP cores* and *embedded cores* are used interchangeably. The term *modules* is used to refer to various blocks inside an SOC. The terms *SOC, chip* and *device* are also used interchangeably.

[2] The term *test* is used as distinct from *verification*, and refers to manufacturing time testing.

7.2 DESIGN AND TEST PARADIGM WITH CORES AND SOCS

This section explains the design paradigms and test considerations for the creation and use of embedded cores and SOCs.

7.2.1 Classification and Use of Embedded Cores

Embedded cores can offer several alternative views to their users. The first view is a physical view, e.g. in the form of a geometric layout, with the placement of core inputs and outputs (I/Os) having been fixed and fully characterized electrically. No modifications to such a core structure are possible. This is a *hard* core. The second view is a structural view, e.g. in the form of a netlist, with the core logic having been fully specified. The core functional implementation cannot be altered; however, other parameters like performance, power, physical routing, I/O locations, etc., can be set to fit in a particular design in which it is instantiated. This is a *firm* core. The third view is a behavioral view, e.g. in the form of Register Transfer Level (RTL) code, with the core functionality alone having been fully specified. Its realizations can be many, and an appropriate one can be created to fit in a given design. Being a behavioral specification, the core functionality can also be altered if required. This is a *soft* core. The test requirements and considerations of these views/types of embedded cores, that of logic and modules around this core, and that of the system in which they are instantiated, vary with the type of the core.

Existing design and tes methodologies use one of these core views for different steps in the design flow. For example, ATPG uses the structural view, floor-planning uses the physical view, and high level simulation uses the behavioral view. While functionality is often available through all the three types of cores, the use of firm cores speeds up the SOC creation process, since these cores are ready for direct integration. Firm cores are, therefore, in widespread use. The main focus of this chapter is on firm cores which present a structural view, since most of the techniques and issues described here are based on structural test methods. A hard core will also have a structural view, and a soft core has to be synthesized to create such a view. Functional tests can also be created using descriptions of firm cores.

Design Considerations and Design-in Methodology

Embedded cores have to be designed keeping in mind the various chips and systems that they will be designed into, and their end applications. For example, consider four contrasting requirements for: (a) highly reliable systems, requiring very high fault coverage[3], (b) high performance systems, requiring very high operating frequency, (c) portable systems, requiring very low power consumption, and (d) low-end consumer systems, requiring very low cost. A hard core designed for application (b) above may not be cost effective for application (d). However, if

[3] The terms *fault coverage* and *test coverage* are used interchangeably. More specifically, fault (test) coverage is the percentage of faults detected from the set of all modeled (all detectable) faults.

designed suitably, such a core can still be used in various applications, since the additional cost of any form of over-design, either for higher fault coverage or for higher performance or for low power, can be lower than the cost of re-design from scratch.

Embedded cores must, therefore, incorporate features permitting re-use in different applications, thereby enabling a *design-in methodology*, whereby a core is designed smoothly into a system, as against having to be designed anew for every system. Other design considerations include the choice of the cell library for hard cores, I/O characteristics, interfaces to other system modules, ease of physical and electrical integration, clocking mechanisms internal and external to the core, etc.

Test Considerations for Embedded Cores

The important test requirements for embedded cores include (i) standard test access mechanism (TAM) for ease of integration and test, (ii) high fault coverage and high test quality for better SOC test quality, (iii) restricted number of test control pins since in an embedded context all the core I/O pins may not be available in the test modes, (iv) test mode isolation of the core for ease of test application independent of the state of the logic around the core, (v) control of logic around the core to facilitate its testing, (vi) test pattern re-use for ease of system test, and (vii) test optimizations in terms of test data volume, test cycles, test power, etc., for ease of system test [2].

Each of these requirements vary depending upon whether the system is core dominated or not[4]. In the former case, the core must lend itself for test of the entire system. In the latter case, however, the test architecture of the system can be separated from the core. Requirements (i) to (iv) are often mandatory. Those of (v) to (vii) may be optional, depending upon the type of the core. These requirements are further described in different sections of this chapter.

7.2.2 Components of an SOC

A block diagram of an example SOC is shown in Figure 7-1(a). Instances of hard cores therein include processors and memories, and of firm (soft) cores include peripherals with fixed (programmable) configurations[5]. Figure 7-1(b) is a photograph of an implemented SOC[6].

---

[4] Not the entire IC may be core-based. Also, there may be multiple cores of varying sizes.

[5] Additionally, some analog modules are also shown. While they are also being built as cores, the emphasis and coverage in this chapter is mainly on digital modules, which are in widespread use and which lend themselves to modular design, integration and test.

[6] The SOC is a Texas Instruments' low-cost single chip ADSL modem device, consisting of about 15 million transistors, in 0.13 um five metal layer CMOS process. The chip includes a digital sub-system consisting of DSP and other IP cores, peripherals, and memories, and an analog sub-system consisting of transmitter/receiver channels, ADC and DAC circuits, and power management unit.

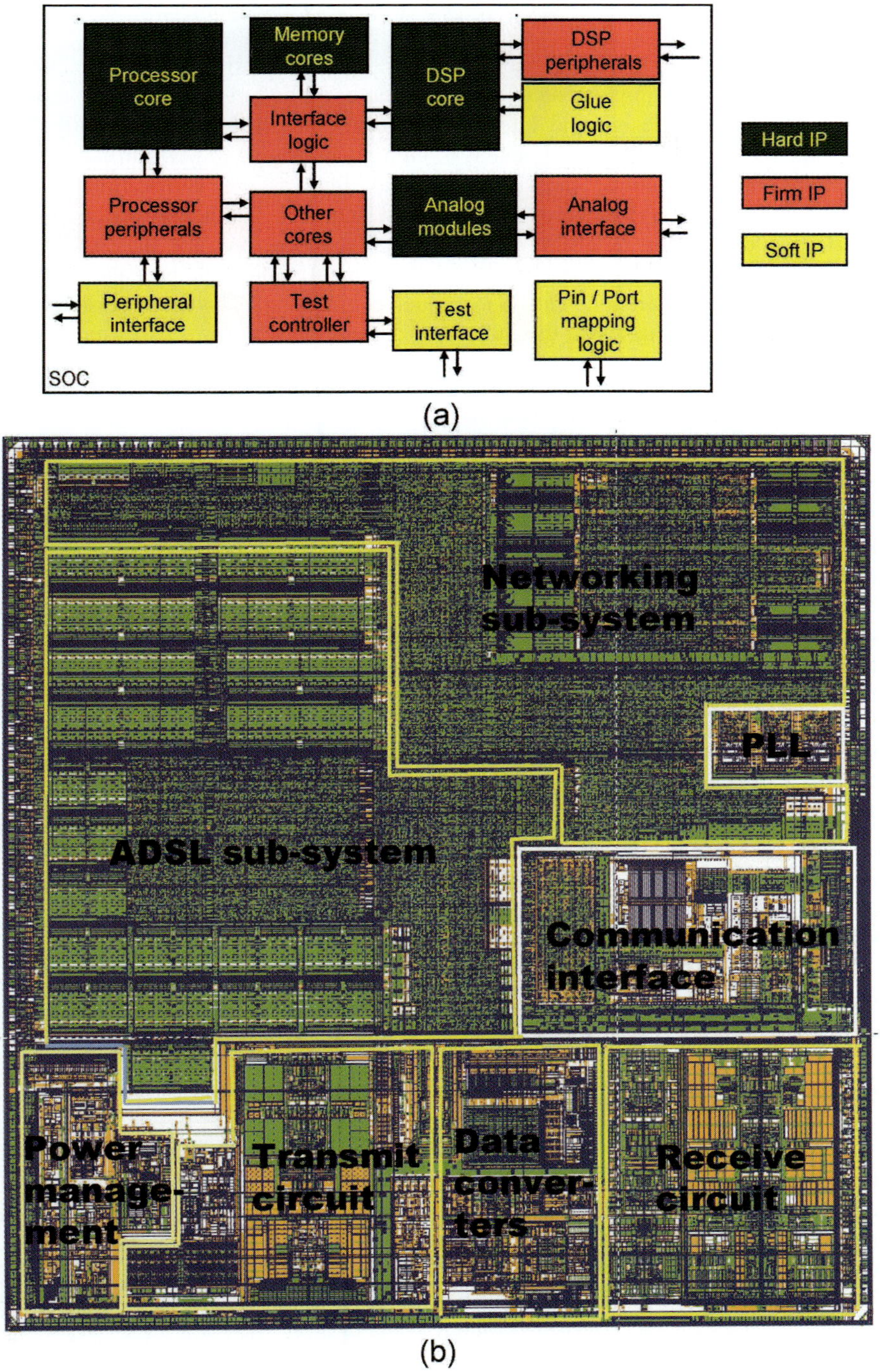

Figure 7-1: (a) Example SOC block diagram, (b) Photograph of SOC.

SOC Integration

The high level SOC design process includes: (i) the identification of functional blocks which can be implemented using existing cores and others which have to be designed, (ii) their design and integration into an acceptable floorplan, (iii) their interface requirements, (iv) assignment of chip package pins to perform the system functions and to control individual modules therein, (v) verification of the functionality of individual cores and modules, and their interaction at the system level, (vi) design for testability for individual modules and for the SOC, (vii) physical design of the entire system, (viii) all electrical, timing and package analysis, (ix) test pattern generation, and (x) product engineering. This is an example of concurrent engineering where the individual designs steps and processes must often be executed concurrently.

Test Considerations for SOCs

The test architecture of an SOC is greatly influenced by the corresponding test architectures of its constituent cores. An SOC may be easy or difficult to test depending upon the extent of DFT techniques adopted in the cores and the sophistication of their test interfaces. Important SOC test considerations include: (a) access to the test interface for individual cores through the SOC test interface, considering that not all core interface pins are directly accessible through the I/Os of the SOC, (b) composite coverage of the individual cores, the logic in their interfaces, and the overall SOC, (c) scan control and clock control mechanisms for serial and parallel test of the individual cores and the logic around them, for test time reduction, and (d) various test modes for test generation and test application for different cores, and for the device I/Os [3].

These requirements significantly influence the design and test of the SOC. In fact, test is often the stumbling block in an SOC design. Hence good DFT and good test are also differentiators. This is highlighted through various techniques and optimizations which are discussed in the subsequent sections.

7.3 DFT FOR EMBEDDED CORES AND SOCS

This section explains how conventional DFT techniques have to be augmented for application in the new design paradigm using embedded cores and SOCs.

7.3.1 Conventional DFT Techniques

Several DFT techniques have been integrated into the design process over the last two to three decades, driven by the increasing demands on quality through cost effective test mechanisms, and through improving support in design automation tools for DFT insertion, DFT analysis, ATPG and BIST. Well-known DFT techniques include those for improving the controllability and observability for all logic structures, such as scan design for sequential circuits, parallel module test for embedded blocks, (e.g. memories), control and observe logic for I/Os, test mode control for hard-to-test logic, (e.g. tristate buffers, clock control network, latch based designs, etc.), and timing synchronization for asynchronous portions of designs. The three main benefits of good DFT, namely, (i) detection of hard (permanent) faults, (ii) easier detection of other faults, and (iii) cleaner design rules for faster and efficient test generation, have to be further emphasized when incorporating DFT techniques in an SOC.

7.3.2 DFT for Embedded Cores

While DFT techniques like scan design are well-known, specific considerations for DFT in embedded cores need some additional analysis.

Scan Control

The logic inside the core may be partitioned into groups, with each group having one or more scan chains[7]. Two generic situations based on this partitioning emerge.

(i) Each scan group has its dedicated test interface, in terms of the scan chain inputs, outputs and control. Here all the scan chains can be operated in parallel.

(ii) Multiple scan groups share a common test interface. Here the individual groups and the scan chains therein must be operated sequentially.

The type of test access mechanism, (dedicated or shared), impacts the test time. However, based on the clocking mechanism for the individual scan groups, it may also impact the test coverage. It is important to partition the logic based on the *clock domains*[8], since mixing scan flip-flops from different clock domains into one scan chain may otherwise create timing conflicts in the scan shift and capture modes, resulting in either erroneous scan chain operation or loss of coverage. This can be corrected through more complex scan clocking or *over-design.*[9]

Clock Control

The effectiveness of scan design is closely coupled to the clock control mechanism. A dedicated test clock is recommended for each scan group. Consider the following two cases[10]:

(a) The clocks for individual groups can be simultaneously applied. Scan shifts and scan captures can be applied in parallel. However, inter-clock domain paths have to be suitably handled for timing violations, if any, through appropriate masking in the capture flip-flops or through over-design.

---

[7] Standard scan design guidelines covering asynchronous preset and reset controls, positive and negative edge triggered scan flip-flops, non-scan logic and their initialization and control, non-inversions in the scan chain, non-sharing of flip-flops across multiple scan chains, etc., must be followed.

[8] Clocks may be distributed across different groups, called *domains*, differing in their frequencies, duty-cycle, operating edges and skews.

[9] *Over-design* in this context refers to stretching the design requirements beyond the goals for ease of integration. For example, logic in a lower frequency domain may be synthesized to operate at a higher frequency domain to facilitate homogeneous test operation.

[10] There are several combinations that may have to be considered in large cores. However, they can be further analyzed in terms of these two cases.

(b) The clocks for individual groups can only be applied in sequence. Scan shifts and scan captures must, therefore, be done group-wise. It is important to build in clock isolation across groups such that a shift or capture in one group does not invalidate the bit stream shifted or captured into the previous group.

Such control mechanisms are illustrated in Figure 7-2 for two scan chains. The external interface can support serial or parallel access to the two scan chains. Separate captures must be performed for each scan chain. Shift in one scan chain must not disturb the contents of the other. Simultaneous captures cannot be performed for different clock domains with a single clock.

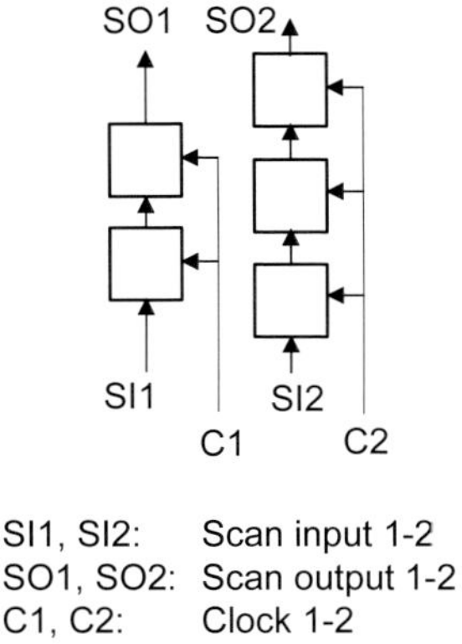

Scan Access	Clock Control	
	Combined	**Individual**
Serial	Separate shift. Separate capture.	Separate shift. Simultaneous capture.
Parallel	Simultaneous shift. Separate capture.	Simultaneous shift. Simultaneous capture.

Figure 7-2: Scan and clock control mechanisms for series / parallel operation.

Embedded Core Operating Modes

A functional module may have several operating modes. When used as an embedded core, the operating modes of this module must also be compatible with those of the neighboring peripherals and those of the overall SOC. As an example, consider the modes of operation for TMS320C28 DSP (digital signal processor) core (from Texas Instruments, Inc.) shown in Table 7-1. Within each of the four operating modes, there can be other design and test requirements, e.g. different scan configurations for test, emulation and debug, different clock operations for the four clock domains, etc.

Domains	Operating Modes					
	Normal	Debug	Idle	Test		
				Functional	ATPG	
					Core	Peripheral
Test clk	Off	On	Off	On	On	On
CPU clk	On	On	Off	On	On	Off
Emulation clk	On	On	On	On	On	Off
System clk	On	On	On	On	Off	On
System clock drives the peripherals. This illustration can be extended to multiple cores (CPUs).						

Table 7-1: Different operating modes for an embedded core.

Other DFT Requirements

Other important DFT requirements for embedded cores include the following and are illustrated in Figure 7-3.

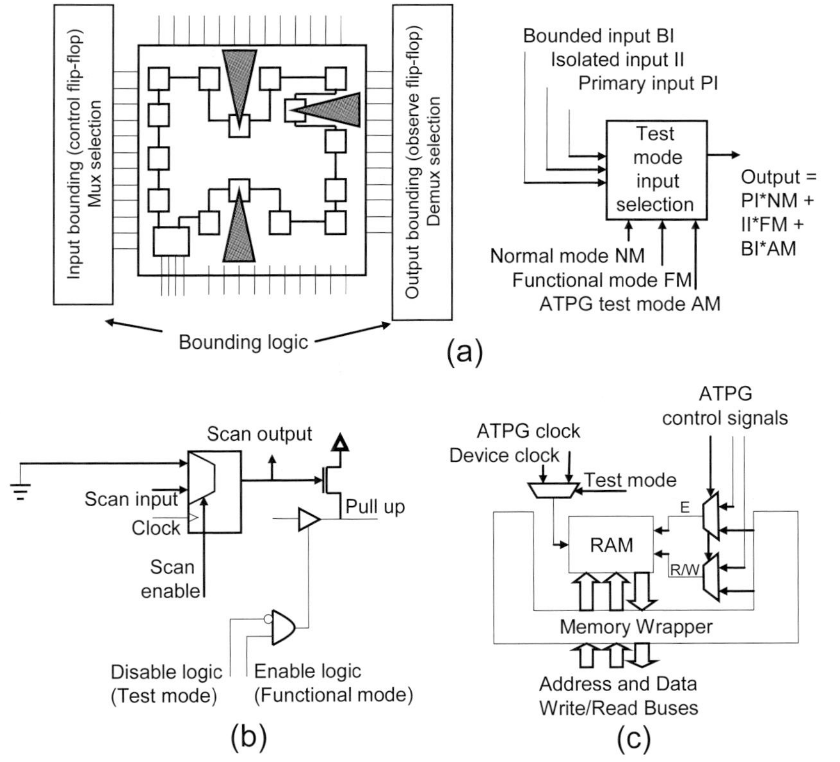

Figure 7-3: Examples of DFT in embedded cores (a) Input/Output bounding logic and input value selection, (b) Tristate logic control, (c) Control for embedded memories.

(a) *Bounding logic at the inputs and outputs for coverage improvement and isolation*: The bounding logic construction shown in Figure 7-3(a) is well known. Note the selection of the input values using the bounding logic for different operating modes. In the normal mode, test / ATPG mode, and functional verification mode, the primary input, scan input, and constrained input is selected, respectively, depending upon whether core is reading the inputs, is tested independent of the inputs, and is verified in isolation without being affected or interrupted by its inputs. This is an important requirement; correct standalone operation of the core for test and verification cannot be otherwise guaranteed.

(b) *Internal tristate logic control*: The tristate logic inside the core must be suitably controlled to disable contention during shift and capture operations, while still enabling fault detection for the logic at the two inputs and one output of a tristate buffer using ATPG techniques – Figure 7-3(b). Correspondingly, control is required during the intervals of shift, between shift and capture, upon capture, and between capture and the next shift. This form of control is also required when the logic around the core is being tested, which may result in the application of non-functional inputs to the core, causing spurious operation of the internal tristate logic.

(c) *Handling embedded memories*: Testing logic around memories requires some form of sequential ATPG, since the memory read bus can only be driven after a prior memory write operation. ATPG tools support this sequence through memory read/write control operations – Figure 7-3(c). Alternately, a bypass mechanism can be implemented to provide direct write control into the read bus. This technique can also be applied to other embedded logic modules, including analog blocks, whose internal test can then be decoupled from the test of the logic around it. Such DFT techniques, referred to as *parallel module test* (PMT) or *visibility module test* (VMT) are often mandatory for test application when only black-box views of embedded cores are available (for reasons of complexity or IP protection).

7.3.3 DFT for SOCs

The techniques described in the previous subsection can also be extended on a broader scale to SOCs, as described below.

Scan Partitioning and Control

Consider again the scan control requirements explained in Section 7.3.2. In an SOC, this can be further generalized. A given set of N scan groups can be operated in up to N^2 different ways, using one or more groups together. This is determined by the test access mechanisms and by the different clock domains in which these groups lie. Consider the configuration with three scan groups shown in Figure 7-4, with the corresponding clock control. The three scan groups can be operated in parallel through ports G1, G2 and G3 using test mode control TM_ALL. For a single scan group test interface, (e.g. through the ports for group G3), however, these groups can be operated only one after the other using test mode control TM_1, TM_2 and

TM_3[11]. However, the order and sets in which these groups are selected for a particular scan pattern set, is controllable. While a general formulation can be derived, it is sufficient to state that the test effectiveness varies with the formation and operation of these different groups. Coverage C1 obtained with N1 scan groups with P1 patterns can be higher than coverage C2 obtained with N2 scan groups with the same number of patterns P1, for N1 smaller than N2 [3]. A particular form of scan grouping can accordingly be selected.

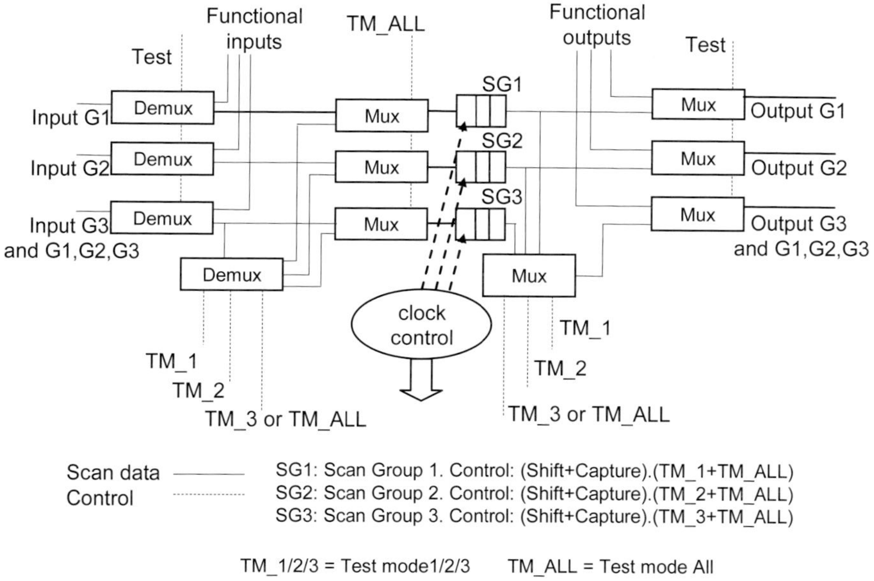

Figure 7-4: Scan and clock control for multiple scan groups in SOC.

Clock Partitioning and Control

The clock control mechanism must match the corresponding scan configuration. Apart from the scan integrity requirements, (scan chain must not be disturbed due to different clocking sequences, etc.), additional requirements at the SOC level include support for at-speed testing through application of different clocking sequences, selective gating of clock domains for power control during test, routing single or multiple tester clocks to individual embedded cores, clock separation for I/O modules for high frequency characterization, etc. These requirements are discussed separately in Section 7.7.5.

[11] To that extent, the inputs (outputs) for scan group G3 are shared with inputs (outputs) for scan groups G1 and G2, when there is a single scan group test interface. The corresponding routing between the external scan group inputs and outputs to and from the internal scan chains is done using the multiplexer and de-multiplexer logic shown adjacent to the clock control block in Figure 7-4.

7.4 TEST ACCESS MECHANISMS

Test access mechanisms (TAMs) through standard interfaces are described in this section.

7.4.1 Test Interface Control Requirements

Embedded cores have a large number of I/Os to support different peripherals with different bandwidths. These I/Os, however, cannot be mapped to the restricted number of I/O pins in the packaged chip. Also, a larger number of pins are required for diverse power requirements, logic running off different voltages, power isolation across analog and digital modules, dedicated power grids for reduced IR drop[12], etc. As a result, embedded cores inside an SOC render themselves increasingly difficult to test on account of diminishing controllability and observability at the chip level. Various solutions have been proposed to address this problem. These are illustrated in Figure 7-5.

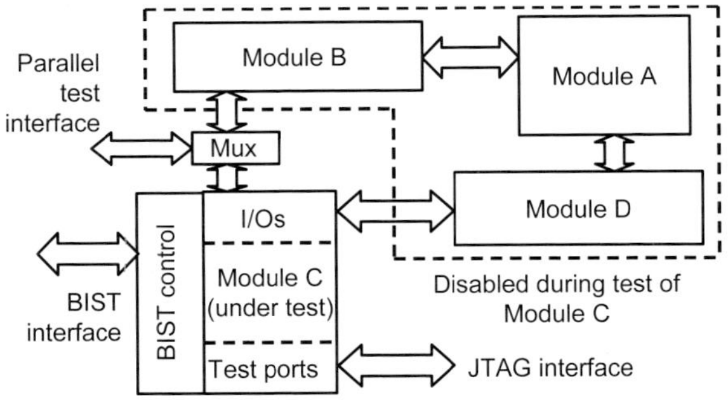

Figure 7-5: Test control for an embedded module.

(a) The core I/Os are brought out at the chip level in the form of a parallel test interface. (Refer to Section 7.3.2, *Other DFT Requirements*, point (c)). This approach is typically restricted to analog modules for which test application through the chip interface is not feasible, and is unviable for large chips with a large number of cores. Also, the individual cores can only be tested in series using shared pins and at the frequency at which the pins operate.

(b) The core I/Os, (all of them or a subset, depending upon which of them are bound or wrapped), are accessed through a standard test interface. Commonly used test interfaces are the IEEE 1149.1 JTAG (Joint Test Access Group) TAP (test access port) interface [4] and IEEE 1500 standard for embedded core test interface [5][13]. The latter is the preferred solution at the SOC level, wherein a

[12] IR drop is the drop in supply voltage caused by sudden increases in current consumption across a resistive power supply network.

[13] At the time of printing of this book, the IEEE P1500 proposed standard for embedded core test has just been accepted. The new standard, IEEE 1500, is being named "IEEE Standard Testability Method for Embedded Core-based Integrated Circuits". The official standard document is in the IEEE editorial and publishing process.

diverse set of test methodologies for heterogeneous cores can be supported. The SOC test interface, in turn, provides the test data and test control access to individual cores.

(c) Increasing adoption of BIST techniques where the test is initiated, applied and evaluated internally. Commonly used BIST techniques are described in Section 7.8. However, even these techniques still require a simplified test access mechanism of the form in (b) above across a restricted number of pins and restricted number of test cycles.

7.4.2 1149.1 JTAG TAP Interface

The IEEE 1149.1 JTAG TAP interface provides a five pins[14] TAP, the communication over which is controlled through the TAP state machine [6]. Data in user defined data registers can be scanned in and scanned out using instructions loaded into the instruction register. Examples of data registers include scan chains in the logic under test, boundary scan chain, status registers, emulation registers, debug registers, etc. The standard requires that a set of mandatory instructions are implemented, while providing support for additional instructions. The standard is scalable in the number of data and instruction registers, access to which is restricted over the five pin test interface[15]. The IEEE 1149.1 standard has also been extended to IEEE 1149.4 standard for mixed signal (analog) test and IEEE 1149.6 standard for AC test.

JTAG TAP Interface Control

Embedded cores have their dedicated test and emulation debug interface. The former is used for test application and the latter for self-debug or application debug during development. The JTAG TAP interface serves as this common interface[16]. An SOC with multiple such embedded cores must provide test and debug access to each core. As the number of pins available at the chip level is restricted, a similar JTAG TAP interface is also built at the chip level. This interface has three functions, namely: (i) it controls the overall chip boundary scan chain operation, (ii) it provides test access to the individual cores, through their dedicated JTAG TAP interfaces or otherwise, and (iii) it helps to test other distributed logic in the SOC outside the collection of these cores.

It is, therefore, evident that a need exists for controlling multiple such interfaces through one interface at the SOC level.

---

[14] Only four pins are mandatory. The function of the fifth pin can be derived using them.

[15] The standard was originally formulated to support boundary scan on chips in order to address the problem of digital testing of boards containing multiple such chips. Several extensions have been reported on how this standard has been used to test and debug embedded cores and SOCs.

[16] Test and emulation are just two of the core operating modes controlled through this interface. Other normal functional operating modes are independently set through I/O pins.

Integrating Multiple JTAG TAP Interfaces

Three different mechanisms exist:

(a) Provide additional control pins to select one amongst a larger set of such interfaces. Multiple interfaces can be connected in parallel to the same set of five pins. (The use of additional control pins is outside the recommendation of the standard).

(b) Connect the interfaces in *tandem*. This mode of operation is facilitated by the standard in the form of the *bypass* operation of individual interfaces. Automation aids exist to compose the test sequences at the chip level, given the test sequences and test interface details (through BSDL – *boundary scan description language*, or equivalent) for individual cores.

(c) Operate multiple interfaces in tandem. In this case, more than one TAP can drive the interface; however, only one TAP can be active at a time. Two approaches have been suggested: (i) The approach in [7] uses a *TAP linking module* (*TLM*) which provides connectivity between the TAPs. (ii) The approach in [8] uses an *hierarchical TAP* (*HTAP*) wherein the *master TAP* FSM controls the operation of a *snoopy TAP*, and, through it, the individual core TAPs. (In both the approaches, no change is required to the five pin standard external interface).

The above approaches focus on the re-use of the same five pins for multiple interfaces. The operation of individual interfaces sequentially or simultaneously depends upon the choice of data and clock control. The implementation for BIST methods is easier than for scan based methods which are heavily data dependent.

7.4.3 IEEE 1500 Standard Test Interface

The JTAG TAP interface is restricted to five pins, and all test data transfer to actually only two. All other chip I/Os are bound using boundary scan flip-flops. This poses significant test access restrictions when testing several embedded cores in an SOC, as explained in 7.4.2. Either all test access is serial, or any form of parallel test access is specific to the actual assembly of cores. The IEEE 1500 standard test interface proposes a mechanism to test such embedded cores through the definition of core test access mechanisms using a standard test interface in the SOC, (e.g. IEEE 1149.1 or IEEE 1500).

The individual cores can have their own TAM ports. Figure 7-6 shows how the IEEE 1500 test interface in an SOC is used to access multiple embedded cores. The IEEE 1500 core test wrapper, shown in Figure 7-6(a), has serial and parallel data ports, together with a wrapper instruction port (WIP). Internal registers include the wrapper instruction and wrapper bypass registers (WIR/WBY). Functional and test data can be driven through the wrapper parallel input/output (WPI/WPO) or wrapper serial input/output (WSI/WSO) buses and wrapper buffer registers (WBRs). Additional test data/control signals are also shown. The individual core TAMs can also be grouped together in the chip as shown in Figure 7-6(b).

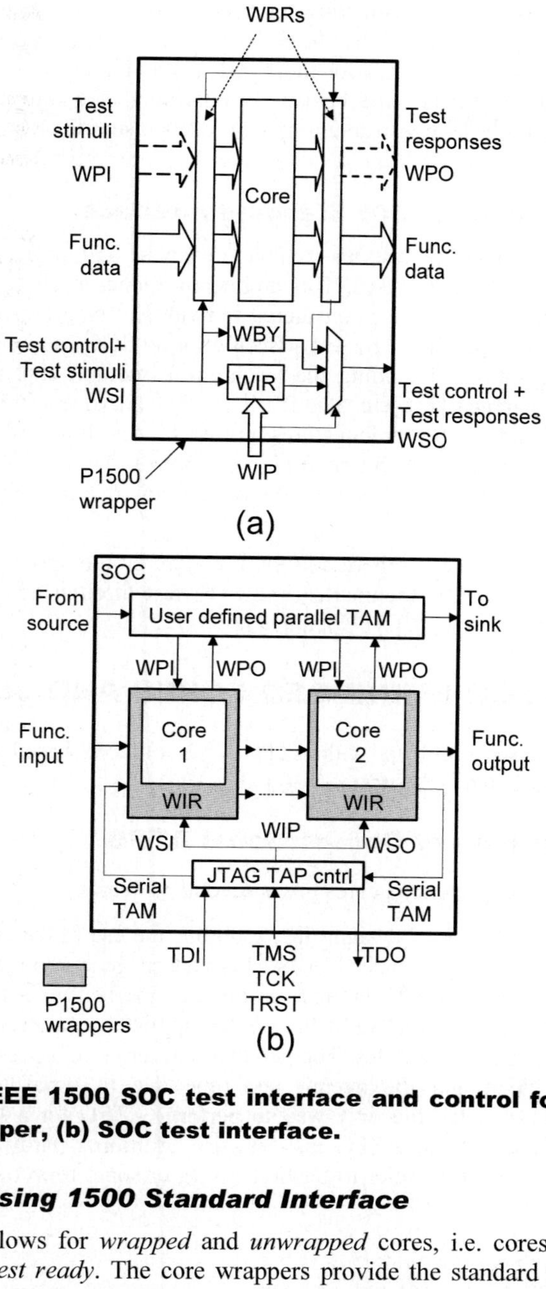

Figure 7-6: IEEE 1500 SOC test interface and control for core TAMs (a) core wrapper, (b) SOC test interface.

Core Test Using 1500 Standard Interface

The standard allows for *wrapped* and *unwrapped* cores, i.e. cores which are *test compliant* and *test ready*. The core wrappers provide the standard functionality of controllability at the inputs and observability at the outputs. The choice of which core I/Os to wrap is user defined based on the pins used for test and the tests to be applied.

The core test access mechanism and core tests themselves are described using the *Core Test Language* (CTL) which has been developed as part of this standardization effort [9]. The use of CTL models for individual cores, therefore, is helpful in two ways: (i) it provides a standard mechanism for sharing all the core test information and (ii) it enables the easy construction of test mechanisms for SOCs with several embedded cores.

SOC Test Using IEEE 1500 Standard Interface

Using the core CTL models, a test bus architecture can be built in the SOC to control the TAM ports for individual cores. This architecture lends itself to one of several test schedules, wherein the cores, considered individually or grouped together, can be tested serially or in parallel. The test schedule for an SOC can be prepared based on various considerations, including the tester infrastructure, chip power, type of tests, need for hierarchical test, etc. The IEEE 1500 standard test interface, together with CTL models and test bus architectures, presents a new framework for solutions to several interesting test scheduling problems. This is significant since the test time for SOCs can often be prohibitively large, and, therefore, its reduction is increasingly important[17].

Significant work has been reported in the literature on the implementation of the standard, creation of SOC benchmarks, and SOC test through this interface, with some important ones being [10], [11] and [12].

7.5 ATPG FOR EMBEDDED CORES AND SOCS

This section discusses problems with ATPG, and solutions based on well-known techniques, for SOCs with IP cores.

7.5.1 Limitations of Conventional ATPG

ATPG for SOCs with embedded cores poses several problems:

(a) The large size of an SOC design often prohibits the use of the conventional flat netlist based ATPG approach. Not only is pattern generation difficult, pattern compression is equally difficult and can often be ineffective. ATPG must hence be performed hierarchically, with ATPG applied separately to individual modules or groups of modules. The penalty is in terms of a possible increase in the pattern count and the overall test time due to serial test application. However, this is often the only way to perform ATPG for a large SOC. The adoption of hierarchical ATPG may require additional hardware support for scan and clock control, (refer to Section 7.3.3), or some form of automation for

[17] It may be noted that while the IEEE 1149.1 and IEEE 1500 test interface standards can co-exist, the test sequences used for one interface are not directly compatible with the other. Either interface can be used to access the other through suitable test data headers.

SOC *pattern instrumentation*, (*composition* or *packaging*), using individual core patterns[18].

(b) An IP core may just have a *black box* view, and new tests for it, therefore, cannot be created in the SOC. The core tests have to be re-used at the chip level. This is often only a pattern packaging problem; however, the core test modes must be carefully designed to permit pattern re-use and their parallel application for multiple such cores.

(c) A black-box IP core also impacts the test of the logic around it as it is a source of X (unknown) drivers at its outputs and a source of X receivers at its inputs[19]. The resulting coverage loss within the core and in the surrounding logic may be significant in a core dominated system. Section 7.5.2 describes a technique to provide an enhanced ATPG view of such an embedded core for testability improvement.

(d) The coverage of the SOC is not known until that of its individual cores and all other distributed logic is known. This may happen late in the design cycle, when any design changes for coverage improvement are infeasible. The method in Section 7.5.3 explains how this coverage can be estimated earlier.

7.5.2 Use of Scan Models

An IP core may offer only its simulatable view to the designer, and its structural view, necessary for ATPG, may not be available. Such an IP core, therefore, treats all its inputs as X drivers and, in turn, is itself an X driver at all its outputs. For more effective ATPG and for better coverage, the core model can be augmented in two ways:

(a) The core itself can be wrapped using boundary scan cells or *scan collar* elements. This provides the necessary controllability and observability to all the peripheral logic around the core. This approach is, however, costly due to extra logic and possible timing performance degradation at the core I/Os.

(b) A restricted *scan model* of the core can be constructed to include the core internal flip-flops, and selective logic at the primary inputs and primary outputs. Such a scan model enables ATPG of the logic surrounding the core without revealing the core IP itself. Figure 7-7(a) illustrates the construction of such a scan model. All the logic in dotted lines is not modeled. Only the scan chain and the logic required for its control are modeled.

---

[18] Patterns for a collection of IP cores and different test interfaces can be composed from individual sets using various techniques. This process is referred to as *pattern instrumentation*.

[19] As explained in Section 7.3.2 *Other DFT Requirements*, point (a) – some form of isolation is desirable for test and verification. Its absence can impact the fault coverage as well as correctness of verification.

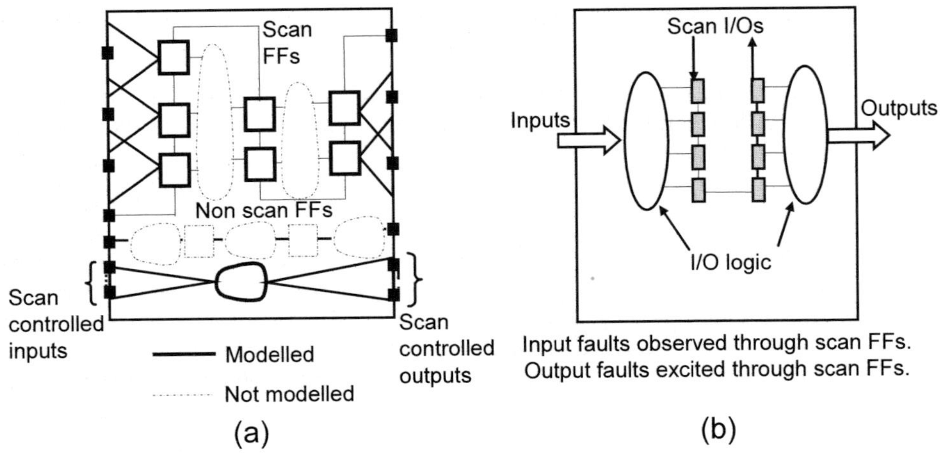

Figure 7-7: Scan model construction and module coverage estimation: (a) modelling details, (b) module and neighbours.

It may be noted that the IP core ATPG patterns and the core scan model must be generated by the IP core provider. In an SOC context, they can be used for obtaining coverage of the surrounding logic.

Scan Model Construction

The scan model can be generated with the following constraints:

(a) Only combinational logic between the primary inputs and nearest set of input flip-flops, and primary outputs and nearest set of output flip-flops is modeled.

(b) All internal flip-flops have their D inputs and Q outputs disconnected.

(c) All combinational logic between primary inputs and primary outputs is modeled.

(d) All test control logic is modeled too to put the core into the ATPG mode to correctly operate its scan chain.

It may be noted that the extent to which the internal logic of the core is modeled can be selected to arrive at a suitable compromise between the details included in the model and the achievable coverage. The generated scan model, in turn, can be verified in one of two ways, namely, (i) using formal techniques with appropriate constraints, or (ii) using simulation based techniques where the pattern generation and validation are carried out on alternate models.

Scan Model Simplification

Simplifications to this scan model are possible in the following ways:

(a) If all the core I/Os are directly bound as described in Section 7.3.2. (see *Other DFT Requirements*, point a), no specific combinational logic at the I/Os has to be modeled at all.

(b) If the core I/O flip-flops are grouped together in the scan chain, the scan chain inside the scan model need not include the internal scan flip-flops.

(c) Stitching all I/O bounding or nearest neighborhood flip-flops together also permits the scan model to be re-used across different core revisions, as long as the I/Os themselves do not change.

(d) The test control logic, which in complex processors controls various test and emulation operations through several data sequences over a standard JTAG TAP interface or otherwise, can also be simplified by using pseudo inputs to control the scan operation and scan clocks.

Scan Model Application

The IP core in an SOC is replaced by its scan model and patterns are generated using ATPG for the logic outside the core. These patterns can then be merged with the original core internal patterns to create the complete SOC test patterns set.

Another application of the scan model is in the generation of I_{DDQ} fault detection patterns using scan based ATPG techniques. I_{DDQ} current measurement is time consuming on the tester, and hence typically only a few, (in the range of ten), I_{DDQ} measurements are permitted. In an extreme case, with N cores, each having M I_{DDQ} ATPG patterns, up to $N \times M$ I_{DDQ} current measurements are required. Using the scan model, the individual core I_{DDQ} patterns can be combined to have only M such measurements with almost no loss of coverage.

7.5.3 SOC Test Coverage Estimation

The fault coverage of a core depends on the logic within it and the extent of controllability and observability offered through its interfaces. The type and number of cores accordingly influence the SOC test coverage. This coverage can be estimated by obtaining the core coverage bounds and information about its interconnection to other cores.

Obtaining Core Coverage Bounds

In the *unconstrained mode* of ATPG, all the I/Os of the core are assumed to be accessible. This gives the upper bound on the core coverage. In the *constrained* mode, none of the I/Os are assumed to be accessible, except those required for scan operation. This gives the lower bound on the core coverage. In an embedded context, the actual coverage for the core lies between these two limits. Using a simple framework, the impact of this coverage loss at the SOC level can be estimated. This helps in two ways: (i) modules, wherein coverage improvements must be driven, can be identified before they are integrated into the SOC; (ii) the maximum coverage that can be obtained for the SOC, when a particular set of cores is being integrated, can also be ascertained.

Basic Formulation

We give a simple formulation referring to Figure 7-7(b). The fault coverage for one module in the SOC context lies between the upper and lower bounds obtained using the unconstrained and constrained modes of ATPG operation, respectively. The potential coverage can be estimated by considering the neighbors of this module and ascertaining the controllability and observability provided by their inputs and outputs. Several possible formulations exist. An example is:

$$\mathrm{EDTC} = \frac{\sum_{J=1}^{N} \mathrm{EMTC}_J \times F_J}{\sum_{J=1}^{N} F_J}$$

where EDTC (EMTC) is the effective device (module) test coverage and F_J is the total number of faults in the Module J. Note that the EMTC lies between the constrained and unconstrained mode coverage for the module, based on the I/O controllability and observability as determined by its neighbors.

This formulation can also be extended to handle soft or firm cores, when the number of fault primitives therein are known. This information is useful at various stages in the SOC create process and the analysis can be repeated at different times for different fault models. Timely DFT architecture decisions can also be taken. For example, coverage loss at the boundary of the analog-digital module interface can be addressed by putting a suitable scan collar. The need for such a scan collar, (described in Section 7.5.2), can also be identified.

Several interesting related formulations have also been reported in the literature [13], [14].

7.6 SOC TEST MODES

This section explains various test modes and their impact on SOC test efficiency and quality.

7.6.1 Role of Test Modes

An SOC is tested under various test operation modes, depending upon the test environment and target of the test. Parameters driving this variety impact the type of tests, method of their application, data driven by these tests, etc. Examples of sources of this variety include:

(a) Types of testers, e.g. high end, low cost, burn-in, digital only, mixed signal, etc.

(b) Types of faults, e.g. stuck-at, I_{DDQ}, transition delay, path delay, bridging, etc.

(c) Purpose of the test, e.g. defect screening, speed binning, diagnosis, etc.

(d) Method of test application, e.g. functional tests through on-chip memory, scan based tests through scan interfaces and tester memory, etc.

(e) Volume of test data, e.g. from tester memory for scan based tests, through on-chip memory or control for BIST, etc.

(f) Type of modules being tested, e.g. core logic, peripheral logic, embedded memories, standalone memory modules, analog modules, I/O cells, etc.

(g) Type of tests, e.g. normal mode, stress mode, slow speed, at-speed, parametric, etc.

It is, therefore, necessary to provide support for various test operations set through these parameters. These, in turn, translate to different *test modes*. A rich variety of test modes provides the flexibility of creating different test programs and better quality tests for efficient and better defect screening, and is often a differentiating factor in the DFT architecture and implementation of complex SOCs.

7.6.2 Design and Categories of Test Modes

A detailed description of all test modes is given in the following. The design of such a gamut of test modes involves:

(a) Selection mechanism for specific test modes through appropriate decode logic or otherwise. (Refer to Section 7.6.4).

(b) Mechanism to drive and collect data for different test modes from the primary inputs and outputs. Appropriate pin assignment at the chip level for all test modes and test data (refer to Section 7.6.3).

(c) Support for different configurations, (through registers, etc.), within a test mode, e.g. BISTing a subset of all the memories.

(d) Integrating all test mode logic together with the individual cores to create the SOC.

(e) Performing verification, timing analysis, other electrical analysis, and DFT analysis on the test mode logic itself, just as for all other logic[20].

There are several categories of test modes. These are described in the following paragraphs.

Test Modes for ATPG

These include modes for scan based ATPG for different scan configurations and different fault models. While no additional control is required for stuck-at ATPG, contention control, as shown in Figure 7-3(b), and biasing for unbonded input pins in multi-package devices, must be provided for I_{DDQ} ATPG. Clock control, as shown in Figure 7-4, must be provided for transition delay fault and path delay fault ATPG. Note that all selection inputs to the multiplexers and other controls for fault model specific ATPG must be derived from the device test modes.

Test Modes for BIST

These include modes for configuring for self-test, running the tests, and exporting the results of the tests. Specific control is required for: (i) individual core modules to be logic BISTed, (ii) individual memories or groups of memories to be BISTed, (iii) BIST controller modes of operation, (iv) test versus debug and test versus repair modes, (v) checking the *signature register* at the end of logic BIST, (vi) checking for

[20] Consistent with general understanding, all DFT logic, including that for test mode support, must also be tested. Else, a fault in the test logic can erroneously cause another fault in the functional logic to be masked.

status and completion signals for memory BIST, (vii) internal or external clock application during BIST, etc.

Test Modes for Characterization

These include modes for I/O pin *characterization*, and characterization of analog blocks and interfaces. The characterization tests are further grouped into DC parametric measurements and AC parametric measurements[21]. These tests require different patterns, different access mechanisms and different extent of observability on a pre-defined set of device pins, which are often re-used across various test and functional modes of operation of the device. Dedicated test modes are, therefore, required to support this selection.

Analog Test Modes

Analog modules are screened using characterization tests which often require longer time and multiple dedicated pins for controllability and observability. Alternately, they can also be tested in *loop-back mode*, e.g. a DSP driving a DAC (digital-to-analog converter), which, in turn, drives an ADC (analog-to-digital converter), whose output is monitored by the DSP [15]. It is, therefore, necessary to have dedicated, isolation and concurrent test modes for them. Guidelines include: (i) digital logic in the analog modules, (e.g. filters), to lie on a separate scan chain, (ii) memories in analog modules to be BISTed separately, (iii) separate burn-in for analog modules and digital modules, (iv) concurrent application of analog characterization and digital scan/BIST tests, (v) isolation between analog and digital modules for better fault coverage, and (vi) different I_{DDQ} leakage current and noise measurement setups based on power supply and clock requirements [3], [15].

Test Modes for Testers

An SOC is tested under different conditions which require different tester platforms. Also, during its life cycle from early production to volume ramp-up, the tester platform can change based upon different manufacturing and volume requirements. Specific platforms of interest include: (i) high speed and high capacity testers, (ii) low cost testers optimized for a class of (usually structural) tests, (iii) burn-in testers, and (iv) mixed signal and analog testers. These platforms differ in their support for number of I/O pins, number of scan chains, number of clocks and clock frequencies, available memory per channel, ability to test multiple devices simultaneously, analog and digital instrumentation for parametric measurements, power supply distribution, leakage current measurement, voltage and temperature control, etc. A variety of test modes is, therefore, required to target the test of SOCs on a combination of these testers in different phases of the manufacturing test process.

---

[21] *Characterization* is the process of ascertaining the performance of analog functions using parametric tests and measurements. Examples include rise and fall times, and load carrying/driving capabilities of the device input/output buffers.

Test Modes for Emulation and Debug

There are modes which are non-functional and non-test modes, needed specifically for debug purposes. The requirements for these modes include the ability to use an alternate set of pins (e.g. different pin assignment or pin sharing for additional controllability or observability), clocks (e.g. burst mode versus step sequenced clocks), number of cycles of operation (e.g. test program halt for logic BIST debug), test schedules (e.g. serial versus parallel testing of individual modules), etc. Since the resources used and objects targeted for test and debug are often the same, it is important to also include these modes into the overall device test modes.

7.6.3 Test Pin Requirements

The test operations depend upon the device pin assignment and data transfer on these pins in various test modes. The selection of these pins and their assignment to specific test modes must be done considering the following:

(a) Input, output and bi-directional data transfer required by the pins.

(b) Shared usage of pins between functional and test modes, between different test modes, and dedicated test mode usage.

(c) Ability to handle high speed test data on the pins.

(d) Ease of connectivity of the pins to internal core I/Os through multiplexing logic and its performance impact.

(e) Location of pads on the die and pins in the packaged device for power distribution, signal quality, port grouping for tester controlled data transfer, multi-site testing, etc.

(f) Suitability of pins based on the I/O buffers (e.g. speed, voltage, etc.), driving them.

(g) Pins required to support concurrent test modes.

(h) Standard test interface requirements, e.g. bounding and compliance pins for IEEE 1149.1 test interface operation.

7.6.4 Test Mode Selection Mechanisms

Test modes for a device can be selected in one of three ways as depicted in Figure 7-8.

(a) Through primary input pins. There may be dedicated input pins for each of the several modes, or the mode selection is encoded using a smaller set of pins. It is important that the encoding and decoding be implemented to permit all the required combinations of test modes to be active. For example, if two modes TM1 and TM2 must also be active concurrently, then a third mode TM3, which is distinct from TM1 and TM2, must be added.

(b) Through JTAG TAP interface. This is a standard implementation, where one or more instructions are added to configure one or more data registers to select the appropriate test modes. The data registers may, in turn, have the encoded or decoded test mode information. The same considerations as in (a) before also apply.

(c) Through internal scan register. Here the test mode information is scanned into a dedicated register. The encoding can be one-hot to permit all combinations of test modes. This is a simple mechanism; however, this register must be on a separate scan chain, and the test mode selection must be updated based on primary input control, after the scan shift operation into this register is complete.

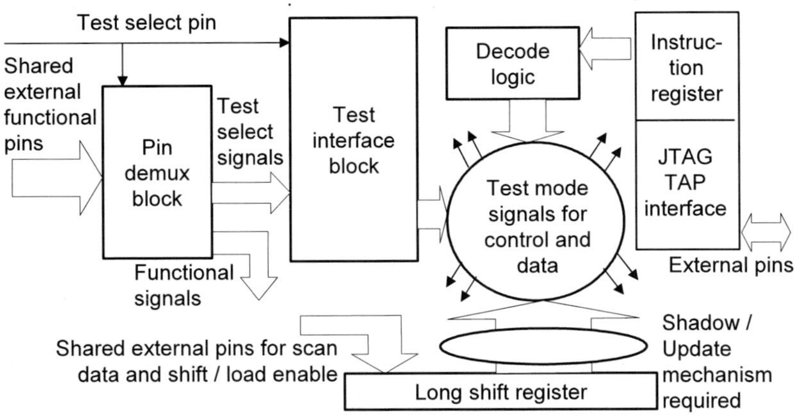

Figure 7-8: Test mode selection through different mechanisms.

7.6.5 Examples of Complex Test Modes

Examples of a few complex test modes, taken from a complex mixed signal SOC [3], are given below[22].

(a) Different scan ATPG test modes for parallel scan and serial scan across different scan groups, for different testers.

(b) Different at-speed ATPG test modes for simultaneous captures across all scan groups and sequenced captures across different scan groups.

(c) Different scan ATPG controls for concurrent and serial testing of IP cores.

(d) Different memory BIST operating modes, scan stuck-at ATPG, scan I_{DDQ} ATPG, burn-in, and functional modes of operation.

(e) Different burn-in modes, namely ATPG scan only, ATPG scan concurrently with memory BIST, ATPG scan concurrently with memory BIST and burn-in of analog modules, and static and dynamic burn-in for analog modules. Table 7-2 summarizes the burn-in test modes for this complex SOC.

---

[22] This design has twenty one test modes which are set using five external shared pins, resulting in possibly several different test programs targeted for different testers, different test and debug conditions, and different test data.

(f) Targeted test modes for selectable configurations of memory BIST and scan chains during burn-in.

(g) Concurrent test modes for simultaneous test of analog and digital modules, besides full chip scan mode for obtaining coverage on all the interface logic. Figure 7-9 illustrates such form of testing with three different sub-modes.

Burn-in modes	Digital modules		Analog modules	
	Logic	Memories	Digital part	Analog part
Static burn-in	Scan	Don't care	Scan	Don't care
	Don't care	Memory BIST	Memory BIST	Don't care
	Scan	Memory BIST	Scan and memory BIST	Don't care
Dynamic burn-in	Scan	Don't care	Loop-back tests	Loop-back tests
	Don't care	Memory BIST	Loop-back tests	Loop-back tests
	Scan	Memory BIST	Loop-back tests	Loop-back tests

Table 7-2: Burn-in configurations in a mixed signal SOC.

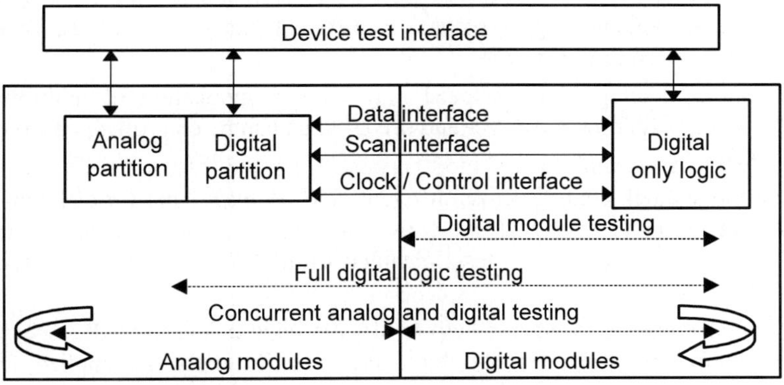

Figure 7-9: Concurrent testing of analog and digital modules.

7.7 DESIGN FOR AT-SPEED TESTING

This section discusses various DFT and test pattern generation techniques for at-speed testing, along with associated tradeoffs.

7.7.1 Need for At-speed Testing

Testing circuits at the rated operating conditions (*at-speed*) for detecting delay defects is important, especially for high performance designs in newer process technologies, due to new failure mechanisms, reduced performance margins and increased signal interference (due to coupling between high speed nets) in such designs. At-speed test techniques applied in practice fall into one of four categories, namely: (i) functional tests, (ii) scan ATPG tests for transition faults, (iii) scan ATPG tests for path delay faults and (iv) BIST for any delay fault model. Generation and application of structural at-speed tests impose several design restrictions, compliance to which significantly influences the impact of at-speed tests and the resulting test cost. At-speed tests are also being identified to substitute other tests, (for performance and stress test conditions), and hence their need is being increasingly felt.

7.7.2 Requirements for SOC At-speed Test

For an SOC with embedded cores, at-speed test requirements pose several test generation and test application problems, as listed below.

(a) Test application for embedded cores through device pins is now more complicated if some form of at-speed data transfer is required from the device pins.

(b) Different cores may have different operating frequencies. Hence additional DFT support is required to test them in parallel for delay defects.

(c) Apart from different clock frequencies, there may be multiple clock domains in an SOC, data transfer across which may be necessary for comprehensive testing.

(d) Depending upon the test methodology, separate treatment may be required for *false paths* and *multi-cycle* paths[23] in different components of the SOC, for correct at-speed test operation.

(e) Testing of interconnects at-speed may require application of multiple tests depending upon how the driver and driven logic can be controlled, observed and clocked.

(f) Test mode clock control across different clock domains and for different clock frequencies is very critical for correct and efficient at-speed test, since the pattern volume of at-speed tests obtained using ATPG techniques is typically large.

(g) While all these requirements help to test the logic inside the SOC, special AC characterization tests, through high speed testers, must be applied for I/O characterization.

It is apparent that at-speed testing for embedded cores and SOCs is another example where deployment of conventional DFT techniques may be rendered either inadequate or inefficient, in terms of test generation or test application [16]. Self-test techniques are, therefore, being increasingly used to overcome these inefficiencies.

---

[23] A structural path that is functionally never excited is a *false path*. (Note that a false-path is not a redundant path, since all the segments of this path belong to some other valid paths). A path that is required to operate slowly over multiple clock cycles, (as against one clock cycle), is a *multi-cycle* path.

7.7.3 Functional Tests for At-speed Testing

Functional tests have been traditionally used for at-speed testing. They represent the actual usage scenarios. However, tests generated for code or function coverage based on a high level verification plan are often inefficient, as their ability to target critical paths in the structural design is poor. Consider a functional path between two flip-flops as shown in Figure 7-10. There are *n* nodes (or gates), *r* registers and *lc* logic cones. Creating a transition and enabling its propagation along this path requires specific values at each node. Being able to construct a functional test to launch a transition at the start flip-flop and propagate it along the selected path to the end flip-flop can be extremely difficult. Also, it is necessary to simulate the test in timing mode under the required operating conditions to confirm the propagation of such a transition[24].

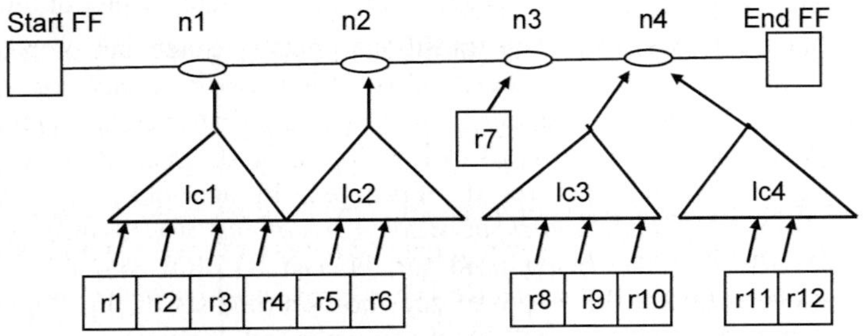

Figure 7-10: Critical path testing.

Functional tests typically run on different modules from internal memory. A typical test sequence consists of test vector load into memory, test execution, and end of test checks. A test may be *self-checking*, in which case only a pass/fail indication is obtained. Alternately the result of the test, e.g. as a memory dump, or the *signature* in a response compactor, (e.g. *parallel signature analyzer* – PSA or *multiple input signature register* – MISR), may be read out.

Functional tests become increasingly important in cases where the area or performance overhead of structural tests is unacceptable, or where the scope of the test is restricted to a well-defined functionality, or where an application test must be executed to check for the presence of unmodeled faults, (e.g. corresponding to functional fault models or electrical – as opposed to Boolean – fault models). Functional BIST is, therefore, important for application of such tests at different times and is discussed in Section 7.8.4.

[24] Structural tests generated using ATPG techniques, on the other hand, are more constrained. However, *robust* ATPG tests are guaranteed to meet such a condition, and no timing simulation is required. A robust ATPG test for delay faults cannot be invalidated under any timing conditions.

7.7.4 Scan Design and Scan Control

Various scan design considerations have been described in Section 7.3.3. As shown in Figure 7-4, scan groups may be created based on clock domains and clock frequencies for a suitable tester interface.

Scan based ATPG techniques using the transition and path delay fault models are being increasingly used. Test patterns for these models are classified as *robust*, *non-robust* or *functionally sensitizable*, depending upon how the controlling and non-controlling inputs along the path that is being sensitized for fault propagation are handled. A test for a delay fault, (or a set of delay faults along a path), is *robust* if the result of the test application does not depend on other delays in the circuit. The result for a *non-robust* test depends on the absence of conflicting delays along other paths. The result of a *functionally sensitizable* test is dependent on the presence of enabling delays involving multiple paths. (Formal definitions can be found in [17]). A detailed discussion of delay testing can be found in a separate chapter of the book.

Two techniques exist to launch transitions on the desired net or path. In the *launch off capture clock* technique, (also called *clock launch* or *functional justification*), two at-speed clocks are applied, with the first one causing a transition at the output of one flip-flop (one set of flip-flops) and the second one capturing it into another flip-flop (another set of flip-flops). In the *launch off shift clock* technique, (also called *shift launch* or *scan justification*), two at-speed clocks are applied, with the first one being used for the last bit shift into the scan chain resulting in a transition at the output of one flip-flop (one set of flip-flops), and the second one capturing it into another flip-flop (another set of flip-flops).

Launch off shift clock patterns are often more efficient, resulting in higher transition fault coverage for a lower pattern count. However, in this case, the scan enable control signal must be de-asserted at-speed between the two fast clocks. Such a transition cannot be forced on a primary input pin through the tester. And hence an internal pipeline stage is often used to register the scan enable control signal. Note that a separate pipeline register is required for each scan group with a different operating frequency, i.e. different capture clock, because this register must be also clocked using the corresponding capture clock. The scan enable signal may fan out as a tree and multiple pipeline stages may be required depending upon how many flip-flops are being driven by each branch of the tree[25].

7.7.5 Clock Control for At-speed Testing

Multiple clock domains and clock groups must be handled separately for at-speed test application. The two at-speed clock pulses can be sourced from either an external source or through an internal phase locked loop (PLL) driven clock generator. The number of such clock pulses can also be greater than two. For multiple clock frequency domains, such pulses must be generated separately for each clock domain.

Various clock application schemes are possible. (Refer to Figure 7-4).

---

[25] The same considerations must also addressed during Logic BIST based on the STUMPS architecture. This is discussed further in Section 7.8.3.

(a) Load the three scan groups and apply at-speed clocks to each of the three groups *simultaneously* (*in parallel*). This is the most efficient technique. However, it requires that inter-clock domain paths be handled appropriately. (This is discussed in Section 7.7.6).

(b) Load the three scan groups and apply at-speed clocks *staggered* in a particular order. Here the flip-flop contents in an earlier group can also influence the coverage in a subsequent group. This requires intelligent pattern generation and fault simulation, and is tied to the ATPG tool's ability to handle *named clock captures* or *custom clock captures*[26].

(c) Load the three scan groups and apply at-speed clocks to only one group. This is the simplest technique. However, pattern generation has to be repeated (*sequenced*) every time a new group is being clocked.

Several variations in the above techniques are possible, namely: (i) at-speed clocks being applied to a set of groups as against all groups, (ii) more than two at-speed clocks being applied for functional coverage, (iii) combinations of launch off shift and launch off capture clock patterns, (iv) scan groups operated individually and top-up ATPG patterns generated later for coverage of faults in logic between these groups, (v) different forms of capture clock alignment, etc.

The timing diagrams of Figure 7-11 illustrate the different clocking schemes for parallel, staggered and sequenced captures referring to Figure 7-4. Several techniques have been reported in the literature for adequate scan and clock control for the generation of at-speed ATPG patterns, and for their application through standard tester interfaces [18], [19].

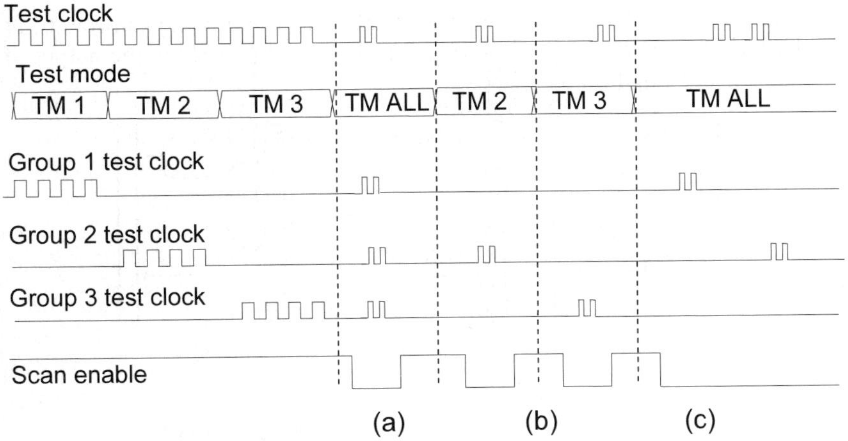

(a) Parallel captures: Simultaneous captures – G1, G2 and G3
(b) Sequenced captures: Separate test modes – Capture in G2 -> G3
(c) Staggered captures: Same test mode – Capture in G1 -> G2 -> ...

Figure 7-11: At-speed clocking schemes with different capture mechanisms.

[26] Sequential ATPG may also be performed depending upon the launch technique used.

7.7.6 Handling Violating Paths

A well cited problem with structural at-speed tests is their inability to handle *false paths* and *multi-cycle* paths, and also the difficulty in handling paths between logic blocks lying in different clock domains or operating at different clock frequencies. (This is due to the delay insensitive models used for test generation by ATPG tools). Techniques to handle such *violating* paths, i.e. paths not meeting the timing constraints, are listed below.

(a) False paths and multi-cycle paths can be eliminated through re-design in the form of: (i) logic partitioning leading to path splitting, (ii) breaking *long* paths in the test mode using intermediate flip-flops, (iii) re-synthesizing these paths so as to run at a faster clock speed, or (iv) masking the outputs at these paths through blocking logic in the test mode.

(b) Inter-clock domain paths can be similarly handled. As illustrations, consider the cases shown in Figure 7-12. In (a), all the inputs and outputs of the logic under test lie in the same clock domain. There are no violating paths. In (b), all the inputs lie in one domain and the outputs in another. The paths have to be synthesized so as to operate at-speed with respect to the two clock domains. This is possible if the same test clock is used to drive the two clock domains; the clocks therein will have a specific edge alignment relation. In (c), the inputs and outputs lie across two clock domains. The path synthesis requirements are similar to the case shown in (b). However, for the validity of the test, it is essential, that the inputs across different domains are set simultaneously and are held constant till the outputs are captured[27].

(c) Forcing the ATPG tool to handle these paths either by forcing constraints, or by masking the values in the capturing flip-flops, or by masking the output of an entire scan group in some cases[28].

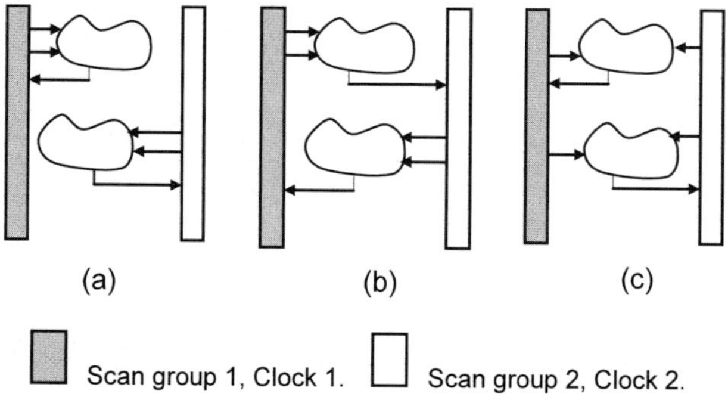

Figure 7-12: Combinations of inter-clock domain paths (a) independent clock domains, (b) separate I/O clock domains and (c) merged I/O clock domains.

---

27 This form of at-speed testing can also be generalized across more than two clock domains.

28 This is, however, not applicable for BIST techniques, wherein a deterministic value must be captured in all the flip-flops for every clock. Alternately, they may keep holding a present (constant) state.

7.7.7 Test Control Through I/Os

As explained in Section 7.7.4, at-speed test control is difficult through the chip I/Os, since they may not operate at high speeds. An example is the control of the scan enable signal. Another example is the application of two at-speed clocks through the JTAG TAP interface. This is not possible unless the TAP is in *Run_Test/Idle* state, as against the conventional *Capture* state. Observing memory BIST status signals in every cycle for diagnosis through the chip outputs also may not be possible for the same reason.

While techniques have evolved for at-speed testing of logic internal to the chip, they are still evolving for that of the chip I/Os. This is due to the parametric nature of the tests involved and also the need to contact the I/Os themselves. Recently, [20] and [21] have reported use of BIST methods with dedicated logic structures and clock control to observe data transfer at the I/Os.

7.7.8 Pattern Generation Techniques

With the background of techniques given in the previous sections, a few important recommendations for pattern generation and selection for at-speed testing are:

(a) Launch off shift clock patterns are generally more efficient than launch off capture clock patterns, resulting in a higher transition fault coverage with a lower pattern count. In this case, stuck-at fault scan ATPG patterns can also be used for transition fault detection.

(b) Selection of the right set of critical paths for path delay fault ATPG is crucial. It is not enough to select the top very few critical paths as available from the *static timing analysis* (STA) tool timing report. In fact, a large number of near critical paths must be selected for three reasons. (i) As there are a larger number of paths with reduced timing slack in high performance designs, a delay defect of a certain size can cause a larger number of paths to fail. (ii) The path delay fault coverage through ATPG techniques is typically low, (often below 10%). Hence, the larger the set of paths selected for ATPG, the more the number of paths that can be covered. (iii) For multiple clock domain designs, it is necessary that paths are selected across different clock domains, rather than just from those in the highest frequency clock domain.

(c) It is important to make a judicious choice of tests to be applied for at-speed testing. Transition fault ATPG patterns are considered suitable for detection of gross delay defects; however, path delay fault ATPG patterns are more suitable for speed binning, since the latter can detect more subtle delay defects. Functional tests are often difficult to grade for their fault detection capabilities, (for reasons mentioned in Section 7.7.3). However, they may help to isolate defects under different use conditions through functional sensitization.

(d) At-speed testing for designs with multiple clock domains requires one or more of the following: (i) sophisticated clock control; (ii) some form of over-design; (iii) increased test time due to sequential captures; (iv) masking of captures into some flip-flops[29].

7.8 DESIGN FOR MEMORY AND LOGIC BIST

This section describes various techniques, together with design and implementation issues, for Built-In Self-Test (BIST) of memories and logic.

7.8.1 BIST Overview

BIST is a DFT technique wherein extra hardware is added for generating the test stimuli for the modules being tested and capturing their response to this stimuli. As the name suggests, built-in self-test indicates the capability to perform the operation of testing through internal resources. The only external control operation required is the ability to start the BIST operation, and to check the pass/fail status upon its completion. This capability can be achieved through different means depending upon the kind of stimuli required, type of module being tested, and mechanism used for capturing the response. BIST techniques are described in [22], [23], [24] and [25].

BIST, in the general sense, has several variations. Of specific interest in an SOC context are BIST of individual modules, BIST of the entire device, BIST using dedicated test resources, BIST using shared test resources, serial BIST operation, parallel BIST operation, and BIST optimizations for area overhead, test time, test data volume, test power, fault coverage, and fault detection latency. The coverage in this chapter is restricted to BIST techniques for the two main components in today's SOCs, namely memories and logic[30].

Advantages and Limitations of BIST

BIST methodology offers several advantages. The important ones include: (i) ability to test without dependency on external (costly) tester infrastructure, leading to the ability to perform periodic testing, (which can also be extended to *concurrent testing* and *on-line testing*), and the ability to perform in-system testing, (ii) better quality of tests to detect unmodeled faults with a non-minimal test set, (iii) ease of test program generation due to minimal test pin requirements and minimal test data transfer across the device I/O pins, (iv) scope of test time reduction due to increasing high speed internal test application and decreasing low speed external data transfer, and test parallelism, (v) ability to provide IP cores together with their test sets in capsule

---

[29] Though option (iv) is often identified as the preferred option since it has no impact on the design, its impact on test pattern generation, validation and application, and overall test time together with increased overhead of tester transactions, is largely ignored. Moreover, this option is not applicable to designs with BIST. Hence, in practice, it is equally important to consider alternatives (i), (ii) and (iii).

[30] Several EDA tools are available for memory BIST and logic BIST implementations. As a result, BIST deployment for memories and logic is widespread. (BIST for ADCs, DACs, PLLs, I/Os and interconnects has also been proposed).

form for ease of re-use and for hierarchical or distributed test, as well as for IP protection, (vi) often re-use of the same stimuli generated by the internal pattern generator for different fault models, and (vii) ease of testing embedded IP cores through restricted set of device pins, and thereby, re-definition of the SOC test problem to a test integration problem (just as the SOC design paradigm is based on IP core integration).

However, an embedded test or self-test methodology has some attendant limitations too. The important ones include: (i) need for deterministic patterns for improving coverage and/or pattern efficiency beyond a threshold requirement, (ii) need to support conventional external test interface and test data transfer mechanisms for such improvement, (iii) increase in the on-chip hardware to support internal test and self-test, and attendant increase in the design effort since such BIST hardware can often be intrusive to logical and physical design, (iv) additional such overhead in the form of test points, pattern storage, etc. if the need of external deterministic patterns has to be minimized, (v) difficulty of debug due to restricted controllability and observability (except in the case of specific debug test modes), (vi) intolerance to BIST specific DFT non-compliance (e.g. X generators, timing violations, etc.), in IP cores and chips rendering the design process more stringent, and (vii) inability to modify tests applied through hard-wired (as opposed to programmable), controllers, e.g. as in memory BIST.

The adoption of BIST techniques is increasing, as the overall test cost is increasing in proportion to the design cost, and as the cost of external test is increasing even faster. However, the need for external tests and testers has not yet been alleviated.

7.8.2 Design Techniques for Memory BIST

Memories can be tested in three ways:

(a) Through a parallel test interface, where the address, read and write buses, and memory control signals can be directly controlled through an external parallel bus interface. While this approach provides high flexibility, it is infeasible in complex SOCs due to the large number of memory cores present, their high speed of operation, and the restricted I/O test bandwidth available.

(b) Through a programmable internal interface, e.g. a CPU or an equivalent peripheral, which can make the required read and write accesses to the memory with a choice of appropriate data. This is a traditional embedded functional test approach often requiring no new additional test resource. However, the programmability may be limited depending upon the type of the interfacing peripheral, the memory bus architecture and their communication protocol.

(c) Through a dedicated, (usually hardwired), BIST interface, where the address and write buses are driven with appropriate data, and the read bus contents are checked at the appropriate cycles, through canonical operations performed by a finite state machine (FSM) controller. This is an autonomous approach ideally suited for BIST. However, the algorithmic read and write sequences are built into the controller and cannot be changed.

A combination of techniques are in use today, depending upon the types of memories, the types of failures targeted in a particular process technology, and the memory configuration and memory/peripheral interface organization in the SOC. While the classical CPU based test and test through dedicated BIST controllers are widely used, there is an increasing need for run-time programmable algorithms for better defect screening[31]. Design considerations and tradeoffs in the implementation of memory BIST are described in the following sub-sections.

Memory BIST Architecture

Consider the memory BIST block diagram in Figure 7-13. It has four important components, namely, the BIST controller, BIST collar, data-logger and interconnects between the memory core and BIST logic[32].

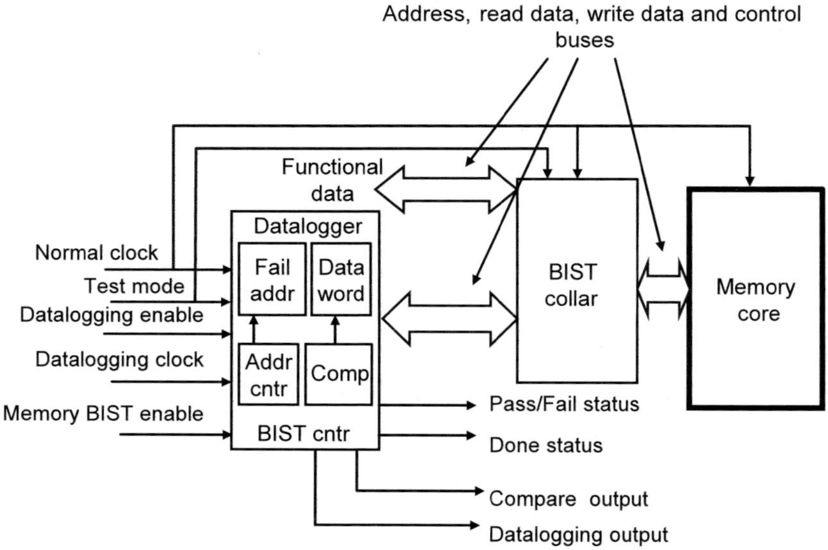

Figure 7-13: Typical memory BIST implementation.

The controller provides the stimuli to the memory through algorithmically generated read/write sequences, and compares the actual response with the expected one at pre-determined cycles. The collar selects the appropriate test or functional data. The choice of a memory BIST architecture for an SOC is based on several optimality and feasibility considerations. These include[33]:

---

[31] A complete review of these techniques is beyond the scope of this chapter. The interested reader may refer to [22], [23], [24] and [25].

[32] The *data-logger* is optional as it is required for memory failure diagnosis and repair, and is not required for go/no-go testing.

[33] The detailed construction and operation of memory BIST logic is available in EDA tool manuals and will not be repeated here.

(a) Physical association of controllers to individual memory cores, leading to different tradeoffs in the BIST logic and integration overhead.

(b) Logical association of controllers to individual memory cores, leading to multiple configurations, e.g. parallel test operation for defect screening in memories, serial test operation for their diagnosis and repair, controller re-use across different memory types, putting smaller memory cores into larger groups for ease of test, etc.

(c) Handling different memory types, e.g. (i) RAMs, ROMs and register files, for different algorithmic test sequences, (ii) repairable and non-repairable memories, for different diagnosis mechanisms, (iii) single ported and multiple ported memories, for different controller access mechanisms, etc.

(d) Test of embedded memories inside IP cores, with or without dedicated controllers, and their integration at the device level.

Different Collar and Controller Operating Modes

In an SOC, four operating modes exist for the memory BIST controller and collar.

(a) *Functional mode*—The controller is inactive or its outputs are ignored. Functional data is driven into memory through the collar.

(b) *BIST mode*—The controller communicates with the memory, through the collar.

(c) *Scan test mode*—(i) For stuck-at testing, the BIST logic is entirely on scan. The memory is bypassed through the collar. (ii) For I_{DDQ} testing, the controller is left out of scan to enable background states to be set and retained into the memory for various I_{DDQ} measurements. (iii) For at-speed testing, the BIST logic is normally on scan. However, in case functional paths are targeted through the memory (e.g. through sequential ATPG), the collar again selects functional data over the BIST controller data.

(d) *Burn-in mode*—To enable simultaneous burn-in of logic and memories, the BIST logic is again left out of scan, to enable parallel memory BIST operation and application of scan patterns.

Optimizations in the memory BIST operating sequence include:

(a) Running all controllers and their associated memories in parallel (to the extent possible), and capturing the status of failing memories. Thereafter, BIST is run sequentially only on the failing memories, to log the failing addresses and data.

(b) Performing memory I_{DDQ} and retention tests together with logic I_{DDQ} tests, using the same set of background patterns, (e.g. checkerboard, inverse checkerboard, etc.).

(c) Allowing for multiple failures in the data-logging mode through the use of internal registers in first-in first-out (FIFO) configuration, and running BIST and data-logging scan operations asynchronously.

Integration and Verification of Memory BIST Logic

Memory BIST in an SOC must meet the varying requirements of implementation, integration, verification and test pattern generation [26]. These include:

(a) Support for at-speed operation through appropriate insertion of pipeline stages in the BIST data path and control, (necessitating changes in the controller FSM itself), and testability of the BIST logic through additional scan in the controller and observe points around the collar logic wrapping the memory core.

(b) Device level integration of individual memory BIST controllers and export of their individual and combined status across various hierarchies corresponding to embedded cores, and assignment of device test modes and pins for start and stop sequences, with and without data-logging,

(c) Verification through comprehensive *fault injection* in different memory cores in different addresses and bit locations, to ensure coverage for all modeled faults through correct controller and collar integration[34].

(d) Export of failure data logs, processing them for redundancy allocation (conventionally done off-line, or on-line in case of *built-in self-repair*—BISR), and support for fuse programming, in the case of repairable memories.

(e) Creation of different test pattern sets, termed as *TDL pattern sets* (or TDLs in brief, TDL: Tester Description Language), for manufacturing tests, for the various operations of memory BIST described earlier. For example, a combination of N controllers, each addressing M repairable memory cores, may result in up to N TDLs for fault detection, and $N \times M$ TDLs for diagnosis and repair purposes.

7.8.3 Design Techniques for Logic BIST

Memories, being regular in structure, lend themselves easily to FSM-based algorithmically generated tests. Random logic does not. This has led to the development of (pseudo-) random pattern BIST with two types of tests, namely, *test-per-clock* and *test-per-scan* [22]. Design considerations and tradeoffs in the implementation of logic BIST are described in the following sub-sections.

Logic BIST Architecture

Linear Feedback Shift Registers (LFSRs) are used as pattern generators (*pseudo-random pattern generator* – PRPG) and response compactors (*multiple-input signature register* – MISR or *parallel signature analyzer* – PSA). The rest of the logic is grouped into sets of scan chains, data into which is loaded through the PRPG and captured into the MISR, resulting in the popular *STUMPS* (Self-Test Using MISR and PRPG Structures) architecture [22] This is a test-per-scan type of extension to scan design and is depicted in Figure 7-14. The PRPG shifts data into the various scan flip-flops grouped into the STUMPS channels. The *phase shifter* is

[34] The objective of fault injection here is the verification of the complete BIST implementation, and not that of the controller itself. This should not be confused with memory fault simulation.

used to break any correlation when one bit of the PRPG drives multiple STUMPS channels. (A similar cloud of XOR gates can drive the MISR). At the same time, the response of the scan flip-flops is shifted out into the MISR. This operation is repeated for each pattern. The shift counter and pattern counter control the number of shifts and the total number of patterns, respectively. (The STUMPS inputs and outputs are also independently accessible to facilitate generation of additional patterns using ATPG if required).

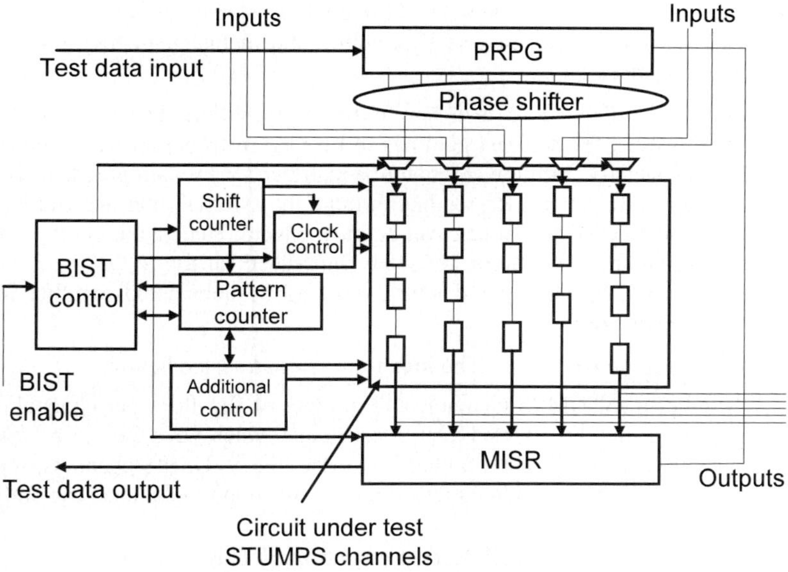

Figure 7-14: Logic BIST implementation based on STUMPS architecture.

Several implementations and variations have been proposed to the STUMPS architecture. These include: (i) deterministic BIST with periodic re-seeding either from the tester or through internal storage [27], [28], (ii) OPMISR (On-Product MISR) with only inputs from ATE [29], (iii) scan compression BIST with de-compressors on chip [30], (iv) X-tolerant BIST for X bypass [31], and (v) test-per-clock type circular BIST [22].

Tradeoffs in Implementation

Scan implementation for logic BIST requires a more diverse set of considerations to be met in the design, as compared to conventional scan implementation. The important considerations are listed below. They are explained with reference to the standard STUMPS based implementation outlined in Figure 7-14.

(a) *STUMPS architecture*—This includes the definition, i.e. number and sizes of the PRPG and MISR, number of STUMPS channels, and number of flip-flops per channel. Various tradeoffs can be exploited in test application time (due to test concurrency, test pattern efficiency, and faster shift speed), test coverage across multiple clock domains, area overhead due to additional logic and routing congestion, and test time power.

(b) *Test points*—Addition of observe and control test points is necessary for improving the test coverage of pseudo-random patterns, as also their efficiency. Synthesis for timing closure can change the netlist structure and hence, in turn, impact the selection and suitability of a given set of optimal test points. Test point insertion, therefore, may cause timing closure iterations, and is often cited as an important barrier to adoption of logic BIST on large designs. One solution for rapid timing convergence is to preclude the inclusion of test points on a window of critical paths.

(c) *Periodic re-seeding*—This is a less intrusive technique as compared to test points. However, the self-test goal has to be relaxed to permit new seeds to be shifted periodically from the external test interface, or pre-computed seeds have to be stored in the device. Re-seeding through the external interface and internal shift into the STUMPS channels can be independent. Hence, the optimal rate of re-seeding, number of patterns per seed, shift speed for the STUMPS channels, and number of flip-flops per channel, must all be considered together for an efficient architecture.

(d) *Programmable parameters*—The important ones are listed below.

- Number of STUMPS channels and number of flip-flops per channel: New flip-flops may be inserted due to test points which have then to be stitched into the existing STUMPS channels. Also, the STUMPS channels must be partitioned into individual clock domains, to support efficient scan shift and capture clocking.
- Number of PRPG/MISR pairs per controller: This is determined based on individual pairs operating in tandem or independently.
- Number of patterns: This is controlled by the pattern counter inside the controller. The number of patterns can be very large in self-test mode or small for specific debug operations, etc.
- Number of shifts: This is controlled by the shift counter, based upon the number of flip-flops (including those due to test points), in the STUMPS channels.
- Number of run phases: To enable effective and efficient coverage using test points, BIST runs can be divided into phases, with a set of test points being effective in each phase. The specific advantages of this technique are explained in [32].
- Number of test points: The number and location of test points can be controlled to maximize coverage with minimum (or acceptable) physical design overhead and number of timing closure iterations.

(e) *STUMPS clocking*: Based on the STUMPS channel grouping and the shift and capture mechanisms in the STUMPS flip-flops, a variety of clocking schemes is possible (refer to Section 7.7.5 and Figure 7-11). These mechanisms can be employed for concurrent test across different clock domains, application of launch-off-shift and launch-off-capture transition fault patterns, test power reduction through slower shift or selective shifts and captures, obtaining coverage for logic in inter-clock domain paths, etc. Synthesis of the clock network for efficient logic BIST is, therefore, an important design consideration.

(f) *Additional design support*: Several forms of design support are required for successful logic BIST implementation. These include (i) suppression of all X sources, e.g. due to embedded memories, unwrapped cores, non-scan flip-flops, multiple bus drivers, etc., either by elimination or masking, (ii) elimination of all timing violations, e.g. due to long false paths, multi-cycle paths, etc., to enable BIST to run at-speed, either by over-synthesis to meet timing, or by path segment splitting, or by path end-point masking, (iii) support for normal ATPG through STUMPS channels re-stitching into long scan chains for coverage improvement and debug, (iv) support for applying device internal clocks during shift as well as capture, and (v) BIST start, stop, and debug control mechanisms. For these reasons, logic BIST is often considered design intrusive, as the constraints for both DFT insertion and physical integration are more stringent.

The above considerations also provide space for various optimizations, based on the resulting tradeoffs in test quality, test time and DFT logic overhead. Interesting implementation case studies are reported in [33] and [34].

Verification and Debug for Logic BIST

The computation of the signature for every design change and establishing its correctness via simulation is the verification problem. On the other hand, resolving the cause of signature mis-match between simulation and silicon results is the debug problem. Both verification and debug of logic BIST implementations are difficult. This is because the only indication of the successful completion of self-test mode logic BIST is the correctness of the signature, which is obtained at the end of the BIST operation. The signature may be incorrect due to: (i) simulation model mis-matches, (ii) timing mode simulation failures, and (iii) detected modeled and unmodeled faults in silicon. Situation (i) will cause good devices to fail. Situation (ii) requires a design fix. Situation (iii) may require failure analysis for yield improvement. In all these cases, there is need for a sophisticated debug methodology, based on BIST hookup checks, stop-and-resume checks, and PRPG/MISR control checks. The problem of identifying a failing flip-flop, given a pattern count and final signature value, is computationally complex. Multiple errors can also result in the same signature. This form of *signature aliasing* impacts fault detection as well diagnosis. Practical debug techniques involve automation for running a subset of failing patterns and creating fault dictionaries, followed by intelligent searches.

7.8.4 Functional BIST

Using functional tests for manufacturing time testing continues to be one of the alternate test methodologies for different modules of a design. This topic was reviewed in Sections 7.7.3 and 7.8.2, with respect to at-speed testing and memory testing, respectively. The programmability offered by a CPU-like control interface, through its instructions, and similarity of these tests to actual application tests, make their adoption attractive. Creation of functional tests can be difficult. However, for regular structures like memories and data paths, they are relatively easy to create for cases where the CPU can be used to access these structures. These tests run off the internal memory, and hence are also a form of BIST. Design considerations and tradeoffs in such forms of functional BIST are reviewed in the following sub-sections.

Processor BIST

Figure 7-15 illustrates a functional BIST implementation using the CPU core, and its data and instruction memories [35], [36]. Based on the high level description of the different functional modules, which are addressable using the CPU, instruction routines are created and pre-loaded into the internal memory for execution. While regular data paths are easy to test, control logic is harder. This is because creation of functional tests to exercise or set the control signals may require several sets of internal registers to take specific values over multiple cycles of operation. (This has been depicted in Figure 7-10). However, through an appropriate process of information extraction and instruction selection (including operators and operands), it is possible to construct instruction routines to test different modules.

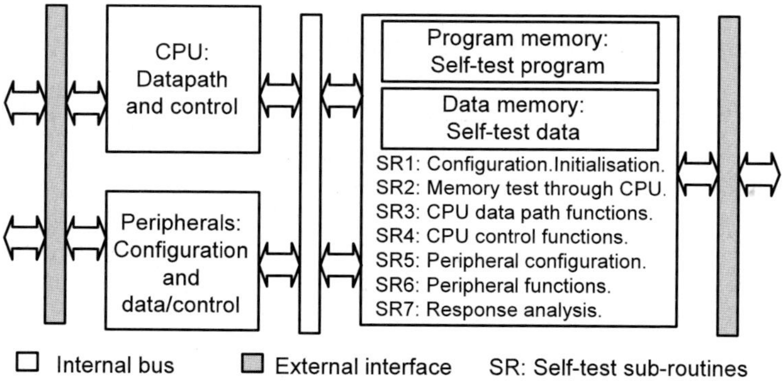

Figure 7-15: Processor BIST architecture for self-test.

This process of test case generation can be formulated for different processors, data path and control modules, and associated peripherals, to facilitate their automatic generation. While this idea was proposed long ago using the register transfer language of microprocessors, it has regained attention on account of the difficulties and rigor associated with structural BIST implementations. (Refer to Section 7.8.3, *Tradeoffs in Implementation*).

Tradeoffs Against Conventional BIST

It may be noted that functional BIST offers unique advantages, on and above conventional BIST, in the form of representing actual use conditions, no overhead either in terms of extra design time, performance or area, and more directed coverage than achievable with pseudo-random patterns, etc. BIST with pseudo-random patterns is known to detect unmodeled faults, and functional BIST helps to restrict these classes of faults to an acceptable subset. This approach is also highly amenable to some form of periodic in-system testing, since dependency on the external tester interface is eliminated. A few important considerations must, however, be noted with functional BIST.

(a) Though it is well-known that though some hard to detect faults can be uncovered using functional patterns, some easy to excite faults may also require non-functional inputs. Hence the instruction set routines must be carefully constructed.

(b) Different SOC architectures may lend themselves differently for test through one or more processor cores. Based on the choice and location of various peripheral modules, CPU access may also be denied in the normal mode of operation.

(c) Due to the inherent latency in the fault detection process, care must be exercised to ensure that faults excited in one cycle are not masked till they result in at least one unique bit write into the memory.

Significant work has been reported in the literature on functional BIST, covering processors with different fault models, and the test case creation, pattern count reduction and fault simulation techniques [25], [35], [36], [37] and [38]. These together highlight situations where functional BIST can effectively complement structural BIST tests.

7.8.5 SOC BIST Architecture

The above variety in BIST of logic and memories, using dedicated controllers or functional mode of operation, must be considered together with additional forms of BIST, namely for analog blocks such as PLLs, ADCs, DACs, as well as for I/Os, etc. At the SOC level, therefore, multiple configurations can emerge. Special care must be taken for the test modes, test pins, and test clocking to enable the integration of individual controllers. The various operating modes at the SOC level include: (i) group-wise series or parallel operation, (ii) concurrent BIST with normal operation on different partitions, (iii) distributed BIST and hierarchical BIST across different sub-systems, and (iv) BIST test and debug modes of operation. Such an architecture provides further tradeoffs in test scheduling, test time and test quality, together with ease of physical integration[35].

7.9 CONCLUSION

In this chapter, a description of various DFT, ATPG and BIST techniques for testing embedded cores and SOCs has been presented. The various design and test considerations, and tradeoffs in their implementation have been highlighted. It will be evident to the reader that the problem of test within this new design paradigm is a complex one, with several optimization possibilities depending upon these tradeoffs. A comprehensive coverage of such optimizations is beyond the scope of this chapter.

---

[35] A popular example is the impact of BIST operation on the IR drop across the power grid in the device. Tradeoffs here include running BIST modularly, or running BIST slowly, versus over-designing the power grid to handle peak BIST mode switching requirements.

However, for the sake of completeness, the important considerations for enabling them, are listed below.

- *DFT architecture*: Choices exist in terms of the test methodology, (namely, ATPG or BIST or functional tests, etc.), test access mechanisms, and deployment of test automation. These lead to several tradeoffs in the test logic overhead and test concurrency possible at the SOC level, in turn, giving rise to various implementation options and test schedules.
- *Pattern selection*: Selecting the right pattern mix is important since multiple pattern sets have an overlap in the detected faults and have varying effectiveness for different fault classes. Also, the selection of the patterns determines the effectiveness of the coverage for defect screening and ease of debug. These choices lead to tradeoffs in test generation time, test application time and cost, and test volume.
- *Test quality*: It is well understood that the test quality can be improved through better selection of tests or better design. Examples of the former include tests for detection of bridging faults and multiple detection (*N-detect*) of given faults. Examples of the latter include over-design to prevent marginality failures. This improvement in quality has an associated cost, and cost-quality tradeoffs (in terms of design/test time) must be well understood. For example, catalogue devices and devices with low DPPM[36] have different quality requirements, which in turn, can drive different implementations.
- *SOC design process*: The SOC design paradigm, leading to rapid construction of chips incorporating system functions, is based on the integration of pre-designed IP cores. The SOC test paradigm must similarly evolve into that of test integration of these cores. Such a framework requires the development of efficient mechanisms to define, implement, verify and integrate the core test logic at the device level. Since the core development can often happen concurrently with that of the SOC, different forms of concurrent engineering and test synthesis can be employed. Illustrative examples include: (i) creation of a distributed BIST architecture, (e.g. with variable number of pipeline stages on the data path), (ii) optimal pin assignment for functional and test modes in multiple package devices, (iii) timing constraints for minimal number of design iterations, (iv) power grid design for peak test power, (v) creation of a library of standard DFT IP cores, (e.g. memory BIST and logic BIST controllers), (vi) adoption of generic formal techniques to verify all DFT IP and integration logic, (vii) concurrent module and device level DFT compliance checks and coverage analysis, (viii) rapid integration of design fixes, (e.g. timing related), in the physical implementation, and (ix) creation of a cost function based SOC test schedule[37].

---

[36] Defective Parts Per Million.

[37] In the experience of the author, based on several designs done at Texas Instruments, test in the form of DFT, DFM and DFY, (design for testability, manufacturability and yield), significantly

- *Design automation*: The techniques described in previous sections can be largely deployed through the use of several existing design automation aids, in the form of tools and flows. A detailed summary is outside the scope of this chapter. However, it will suffice to state that several new forms of optimizations will be powered by tool based automation leading to the synthesis of suitable test architectures for SOCs.

Today's high performance designs are constrained by the dual problems of increasing test cost and unpredictable test quality. Addressing these problems in the context of large system-chips built using a collection of embedded cores, (through effective DFT architectures, efficient ATPG, and increased adoption of self-test techniques), continues to be challenging, thereby attracting researchers to the development of newer standards, newer implementation techniques, and newer optimizations [39], [40] and [41].

ACKNOWLEDGEMENTS

The author acknowledges the help of his several colleagues in Texas Instruments, notably Jais Abraham, Srinivasa Chakravarthy, Nikila K. and Satish Panigatti, (TI Bangalore, India), and Ken Butler, (TI Dallas, U.S.A.), for various discussions and collaborative efforts, which have helped develop several ideas and methodologies described in this chapter.

REFERENCES

[1] Y. Zorian, E. J. Marinissen and S. Dey, "Testing Embedded Core based System Chips", Intl. Test Conf., 1998, pp. 130-143.

[2] J. Abraham, N. Prasad, S. Chakravarthy B. S., A. Bagwe and R. Parekhji, "A Framework to Evaluate Test Tradeoffs in Embedded Core Based Systems - Case Study on TI's TMS320C27xx", Intl. Test Conf., 2000, pp. 417-425.

[3] Nikila K. and R. Parekhji, "DFT for Test Optimisations in a Complex Mixed-Signal SOC – Case Study on TI's TNETD7300 ADSL Modem Device", Intl. Test Conf., 2004, pp. 773-782.

[4] IEEE Computer Society, *IEEE Standard Test Access Port and Boundary Scan Architecture, IEEE Std. 1149.1*, 2001.

[5] IEEE P1500 web site, http://grouper.ieee.org/groups/1500.

[6] K. P. Parker, *The Boundary-Scan Handbook*, 3rd edition, Kluwer Academic Publishers, 2003, ISBN 1-4020-7496-4.

[7] L. Whetsel, "An IEEE 1149.1 Based Test Access Architecture for ICs with Embedded Cores", Intl. Test Conf., 1997, pp. 69-78.

[8] D. Bhattacharya, "Hierarchical Test Access Architecture for Embedded Cores in an Integrated Circuit", VLSI Test Symp., 1998, pp. 8-14.

[9] R. Kapur, *CTL for Test Information of Digital ICs*, Kluwer Academic Publishers, 2002, ISBN 1-4020-7293-7.

---

influences the SOC design process, and in many cases, has been amongst the most important tasks in the entire SOC create process.

[10] F. DaSilva, Y. Zorian, L. Whetsel, K. Arabi and R. Kapur, "Overview of the IEEE P1500 Standard", Intl. Test Conf., 2003, pp. 988-997.

[11] E. Marinissen, V. Iyengar and K. Chakrabarty, "A Set of Benchmarks for Modular Testing of SOCs", Intl. Test Conf., 2002, pp. 519-528.

[12] K. Chakrabarty, *SOC (System-on-a-Chip) Testing for Plug and Play Test Automation*, Kluwer Academic Publishers, 2002.

[13] M. Nicolaidis, R. A. Parekhji and M. Boudjit, "E-Groups: A New Technique for Fast Backward Propagation in System Level Test Generation", Asian Test Symp., 1996, pp. 34-41.

[14] I. Pomeranz and Y. Zorian, "On Testing of Non-Isolated Embedded Legacy Cores and Their Surrounding Logic", VLSI Test Symp., 1999, pp. 41-48.

[15] M. Burns and G. Roberts, *An Introduction to Mixed-Signal IC Test and Measurement*, Oxford University Press, 2001.

[16] B. Nadeau-Dostie, *Design for At-speed Test, Diagnosis and Measurement*, Kluwer Academic Publishers, 1999.

[17] M. L. Bushnell and V. D. Agrawal, *Essentials of Electronic Testing for Digital, Memory and Mixed-Signal VLSI Circuits*, Kluwer Academic Publishers, 2000.

[18] N. Tendolkar, R. Raina, R. Woltenberg, X. Lin, B. Swanson and G. Aldrich, "Novel Techniques for Achieving High At-speed Transition Fault Test Coverage for Motorola's Microprocessors Based on PowerPC Instruction Set Architecture", VLSI Test Symp., 2002, pp. 3-8.

[19] J. Saxena, K. Butler, J. Gatt,R. Raghuraman, S. P. Kumar, S. Basu, D. J. Campbell and J.Berech, "Scan Based Transition Fault Testing - Implementation and Low Cost Test Challenges", Intl. Test Conf., 2002, pp. 1120-1129.

[20] S. Sunter and B. Nadeau-Dostie, "Complete Contactless I/O Testing - Reaching the Boundary in Minimizing Digital IC Testing Cost", Intl. Test. Conf., 2002, pp. 446-455.

[21] M. Tripp, T. Mak and A. Meixner, "Elimination of Traditional Functional Testing of Interface Timings at Intel", Intl. Test Conf., 2003, pp. 1014-1022.

[22] C. Stroud, *A Designer's Guide to Built-In Self-Test*, Kluwer Academic Publishers, 2002.

[23] A.Van de Goor, *Testing Semiconductor Memories - Theory and Practice*, ComTex Publishing, 1998.

[24] R. D. Adams, *High Performance Memory Testing: Design Principles, Fault Modelling and Self-Test*, Kluwer Academic Publishers, 2002.

[25] D. Gizopoulos, A. Paschalis and Y. Zorian, *Embedded Processor-Based Self-Test*, Kluwer Academic Publishers, 2004.

[26] T. Powell, W-T. Cheng, J. Rayhawk, O. Samman, P. Policke and S. Lai, "BIST for Deep Submicron ASIC Memories with High Performance Applications", Intl. Test Conf., 2003, pp. 386-392.

[27] P. Wohl, J. Waicukauski, S. Patel and M. Amin, "Efficient Compression and Application of Deterministic Patterns in a Logic BIST Architecture", Design Automation Conf., 2003, pp. 566-569.

[28] P.Wohl, J.Waicukauski, S.Patel and M.Amin, "X-Tolerant Compression and Application of Scan ATPG Patterns in a BIST Architecture", Intl. Test Conf., 2003, pp. 727-736.

[29] C. Barnhart, V. Brunkhorst, F. Distler, O. Farnsworth, B. Keller and B. Koenemann, "OPMISR - The Foundation for Compressed ATPG Vectors", Intl. Test Conf., 2001, pp. 748-757.

[30] J. Rajski, J. Tyszer, M. Kassab, N. Mukherjee, R. Thompson, K. H. Tsai, A. Hertwig, N. Tamarapalli, G. Mrugalski, G. Eide, J. Qian, "Embedded Deterministic Test for Low Cost Manufacturing Test", Intl. Test Conf., 2002, 301-310.

[31] S. Mitra and K. S. Kim, "X-Compact: An Efficient Response Compaction Technique for Test Cost Reduction", Intl. Test Conf., 2002, pp. 311-320.

[32] N. Tamarapalli and J. Rajski, "Constructive Multi-Phase Test Point Insertion for Scan-Based BIST", Intl. Test Conf., 1996, pp. 649-658.

[33] G. Hetherington, T. Fryars, N. Tamarapalli, M. Kassab, A. Hassan and J. Rajski, "Logic BIST for Large Industrial Designs: Real Issues and Case Studies", Intl. Test Conf., 1999, pp. 358-367.

[34] M. Kusko, B. J. Robbins, T. J. Koprowski and W. V. Huott, "99% AC Test Coverage Using Only Logic BIST on the 1 GHz IBM S/390 zSeries 900 Microprocessor", Intl. Test Conf., 2001, pp. 586-592.

[35] N. Kranitis, G. Xenoulis, A. Paschalis, D. Gizopoulos and Y. Zorian, "Application and Analysis of RT-Level Software-Based Self-Testing for Embedded Processor Cores", Intl. Test Conf., 2003, pp. 431-440.

[36] J. Abraham and J. Shen, "Native Mode Functional Test Generation for Processors with Applications to Self-Test and Design Validation", Intl. Test Conf., 1998, pp. 990-999.

[37] L. Chen, X. Bai and S. Dey, "Testing for Interconnect Crosstalk Defects Using On-Chip Embedded Processor Cores", JETTA, Vol. 18, No. 4/5, 2002, pp. 529-538.

[38] P. Parvathala, K. Maneparambil, W. Lindsay, "FRITS – A Microprocessor Functional BIST Method", Intl. Test Conf., 2002, pp. 590-598.

[39] E. Volkerink, A. Khoche, J. Rivoir and K. D. Hilliges, "Modern Test Techniques: Tradeoffs, Synergies and Scalable Benefits", JETTA, Vol. 19, No. 2, 2003, pp. 125-135.

[40] R. Parekhji, "How (In)Adequate Is One Time Testing?", Intl. Test Conf., 2003, pp. 1279.

[41] R. Madge, B. Benware, R. Turakhia, R. Daasch, C. Schuermyer and J. Ruffler, "In Search of the Optimum Test Set - Adaptive Test Methods for Maximum Defect Coverage and Lowest Test Cost", Intl. Test Conf., 2004, pp. 203-212.

Chapter 8

Embedded Memory Testing

R. Dean Adams

8.1 INTRODUCTION

Memories are everywhere in today's semiconductor chips. Memories dominate the layout of virtually all chips. Memories likewise dominate all of the impacts on the yield and quality of these chips. Test of memories is probably the single most critical piece in current and future testing related to semiconductor chips.

There are a vast number of stand-alone memory chips but the amount of embedded memory on chips is worth examining. It can easily be said that embedded memories occupy 50% or more of the silicon area today and many believe that this number will reach nearly 95% in the foreseeable future [1].

Since memories occupy the bulk of silicon area there should be greater concern about this memory area and assurance that a correspondingly greater design and test effort also be applied. Not only are memories occupying the vast majority of chip area but they are also the densest of circuitry. The distance between adjacent structures is smaller in memory regions than in other areas of the chip. The design ground rules are modified for memory portions by the foundries to allow tighter

D. Gizopoulos (ed.), Advances in Electronic Testing: Challenges and Methodologies, 263-300.

packing of the circuitry. There are long metal runs, often with minimum dimension and minimum spacing as well as polysilicon shapes and diffusion shapes closely spaced. This is a recipe for being sensitive to smaller defects than in the logic and mixed-signal portions of the chip. Digital portions of the chip have full rail voltage differences to distinguish between a "1" and a "0". Memories have only small signal swings to differentiate these levels [2]. Analog portions of chips do not utilize minimum design dimensions in critical areas. Memories, however, have analog portions in the areas with minimum design dimensions. So memories have dense circuitry which is more sensitive than anything else on semiconductor chips. As a result, defects are far more prone to impact memory operation than anything else on a chip. Because of these factors careful testing of the memories will do more to impact the overall chip quality than anything else.

Because memories are sensitive in ways that other on-chip circuitry is not, memories need to be tested more thoroughly. Many companies have tested their memories in a haphazard ad-hoc fashion and not seen problems. Not detecting a problem is not the same as not having a problem, however. When chips contain a small amount of memory and when technologies are sufficiently robust, other portions of the chip receive the lion's share of attention for defects and testing. Companies which have produced a reasonable number of chips with a reasonable amount of on-chip memory understand all too well the need for thorough memory testing. Many people first started to realize the predominance of *subtle* defects at the 130nm technology node. With 90nm, 65nm, and beyond the sheer amount of memory and the susceptibility of the tiny circuitry to subtle defects drive the memory test paradigm through a required change.

Because memories are so dense and are so susceptible to defects, they are the only chip portions that are fully anticipated to include failures and still be shipped. Redundancy is typically utilized to implement spare memory elements in order to enhance memory yield. An average 32 Mb stand-alone SRAM is expected to have between three and four failures on it [3]. Since this is an average chip, more than four redundant elements are required to enhance yield for "worse than average" chips. A recent chip report stated that the yield for on-chip logic was 54% and the yield for on-chip memories was 31% [4]. This chip, with redundancy implemented, yielded the same memories at 71%. Clearly, redundancy is critical to memories, given that the slim profit margins on most chips is less than the difference between these redundancy and non-redundancy implemented yield numbers and the memory area required for redundancy is typically well less than 10% [5] with the chip area impact being even smaller. With the amount of memory on chip growing, the need for redundancy is continuing to grow as well [6]. Figure 8-1 shows the long term growth in the amount of memory on chip, which is Moore's law as applied to embedded static memory. The second (bottom) line is a trend for the amount of on chip memory which can be included without requiring redundancy to achieve adequate yields. Historically the number has grown from memories having 1K bits, through 64K bits and 256K bits to the current number of approximately 2M bits. Most foundries require that chips with more memory than 2M bits have redundancy implemented to ensure that chips have adequate yield. If more memory than the redundancy limit is included on chip then yield will be reduced and if sufficient memory is included without having redundancy then there will be no chip yield at all. Foundries have actually seen this happen.

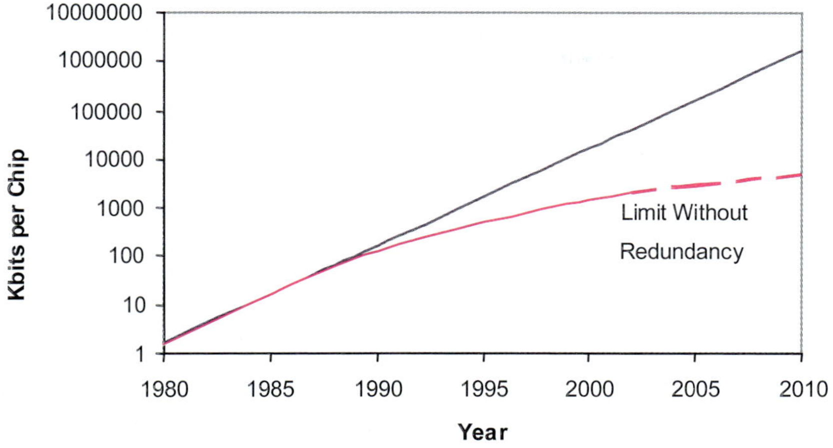

Figure 8-1: On-chip memory and redundancy growth trend.

If all of the on-chip memory is grouped into a single unit, testing and redundancy handling is vastly simplified. Instead, the number of unique on-chip memories is growing very rapidly.

Figure 8-2 shows the growth trend for the number of embedded memories on a single chip. Many people think of the number of on-chip memories in terms of instruction caches, data caches, and the like. This was mostly true up until the early 1990s. Since then chip designers have seen the advantage of having numerous small storage elements scattered throughout the chip. These register files and even small static RAMs have driven the number of individual memories through the roof. It is typical to see chips with several hundred memories on them with several cases at the time of this writing having between *one and two thousand separate memories on a chip*. By the end of the decade it is anticipated that nearly 10,000 unique memories may reside on a single chip.

Memory testing, given the challenges described in the chapter so far, is a daunting task. Memories are embedded within a chip and therefore require embedded test[1]. Most System-on-Chip (SOC) semiconductors are tested with logic testers, oriented towards testing the random logic. The memory portions of the chip are not well suited to being tested by these logic testers.

Stand-alone memory chips are tested by memory testers. These memory testers are algorithm oriented and are not well suited to testing logic, where large amounts of data storage are required to apply the various Automatic Test Pattern Generation (ATPG) vectors to the random logic. Thus a different solution must be provided for testing embedded memories than that utilized for stand-alone memories, that solution being *Built-In Self-Test* (BIST).

[1] Test application and response collection performed on-chip.

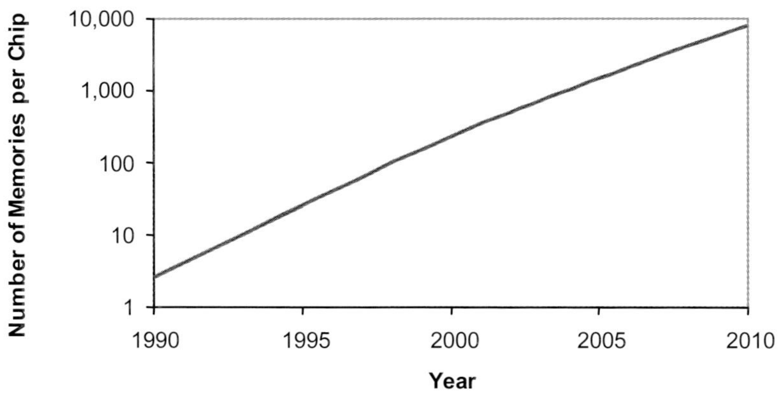

Figure 8-2: Growth trend for the number of on-chip memories.

A memory BIST architecture provides stimuli to the memory circuitry and a compression or comparison means for the data being read from the memory. A BIST which incorporates redundancy calculation is often referred to as *Built-In Redundancy Analysis* (BIRA) or *Built-In Self-Repair* (BISR). The BIRA term is normally utilized when the redundancy calculation results are for subsequent laser or electrically blown fuse repair [7]. The BISR term is used when the redundancy calculation results are stored on chip and automatically utilized to implement the needed redundancy repair. BISR is used in the field at each power up to determine the needed redundancy implementation.

There are many reasons that BIST is utilized for embedded memory. BIST is needed because the memory I/O do not usually reach chip I/O. A BIST test can be done at speed, whereas a pattern applied by scan or by multiplexing the inputs to the memories cannot be done at speed. The higher speed test of BIST is a key factor that allows BIST testing to be of superior quality than embedded memory testing via a through-the-pins method. The higher speed test improves the quality of test and of course reduces test time. Although most SOC area is occupied by memories, most SOC testing time is dominated by logic test or by mixed-signal test. The memory test is typically quite short, since the BIST execution requires a limited number of back-to-back cycles applied at high speed. Only *retention testing*[2] increases memory test time considerably and for static memories, which are the vast majority of embedded memories, this time is still inconsequential when compared to the overall chip test time. By using BIST, the test development is pushed into the design phase of the project, where a better test strategy can be planned. If the right BIST is inserted as part of the design then the test is better overall. If, however, a poor BIST is inserted into the design then the manufacturing test will be poor and continue to be that way until a chip re-design is performed. Built-in self-test can be re-used at wafer, component, card, system, and field test. In this manner the test need only be developed once and, since it is developed at the same time as the chip design, the test can be better.

---

[2] Retention testing involves holding the memory contents constant, with no reads or writes, to ensure that the stored data does not leak away or disturb on its own.

The BIST can be custom developed for the memory it is testing but this is problematic when realizing the number of memories included on chip, considering Figure 8-2. For a single memory or even for a few memories, a BIST solution could be custom designed. Custom BIST design for a small number of memories is still not really desirable when evaluating the design effort involved. If memories are custom designed and BIST is custom designed then the effort involved in doing the BIST design can easily reach 25% of the effort required to design the memories themselves. This is a significant number and repeatedly surprises development managers who anticipate putting in a small amount of register transfer level (RTL) logic to provide a simple test. The BIST effort balloons from an anticipated brief job of a few months by one person part time into a continuous full-time job by more people than ever expected. BIST design simply cannot be tucked with other efforts. Part of this is because a complete understanding of memory circuitry is required along with a complete understanding of test coupled with a detailed knowledge of RTL design. This combination of skills is rare to find in a small team, let alone any one individual. If one of the skills is lacking then a poor quality BIST and poor quality test results. The result of poor quality testing is that chips which are defective pass test and go to the customer, which is certainly undesirable.

The task of doing memory BIST development should appear to be daunting even without considering the large number of memories involved. Once the number of memories is considered only an *automated* approach can be utilized. Electronic design automation (EDA) allows for the generation and insertion of memory BIST in a chip through Design-for-Test (DFT) processes. DFT is not the calculation of a test but is instead the modification of a customer's design to insert the needed control and observation to facilitate test. Memory BIST is a key portion of DFT because it provides means for testing on-chip memories. DFT, since it is modifying a design, belongs as a fully integrated piece of design automation. When thousands of memories are contemplated, the generation and insertion of the memory BIST pervades throughout the chip design. In order to accomplish a small and efficient BIST the whole chip design needs to be considered. The larger memories are often fixed very early in the chip design development. The smaller memories can change in size and even number virtually up to the point of chip tape out. The BIST needs to change as the chip design changes and therefore the EDA DFT solution needs to modify the BIST portions of the chip in a fluid fashion considering the entire chip design. That said, the test of the memories needs to still be of an exceedingly high quality and needs to factor in the design topology of the memory under test. A small arbitrary BIST is insufficient as the chip quality would suffer horrendously.

In this chapter the memory design factors which influence test are covered followed by the faults encountered in memories. Typical memory test patterns are then covered followed by BIST and BISR concepts. Finally, newer technologies and memories as well as their test-related considerations are reviewed.

8.2 THE MEMORY DESIGN UNDER TEST

As stated in the introduction, good testing of embedded memories requires understanding the memory that is being tested. Different memories have different designs and require different tests. There is no magic pattern that works for all memories nor even works well for a large sub-set of memories. The pattern applied to test the memory must take into account the memory topology or schematic as well as the layout or GDSII[3]. People often think of memories as being black boxes which contain ones and zeros. Figure 8-3 shows a typical static random access memory (SRAM) and a typical dynamic random access memory (DRAM) cell. By simple examination the SRAM cell shows six transistors and includes a *feedback path* that retains data as long as the power is maintained to the chip. By contrast the DRAM cell contains only a single transistor with no feedback method. The DRAM cell must frequently be *refreshed* or the data that it contains will be lost. These two memory cells are radically different and yet many people try testing these with identical methods. The design is essentially different and therefore the test should be different.

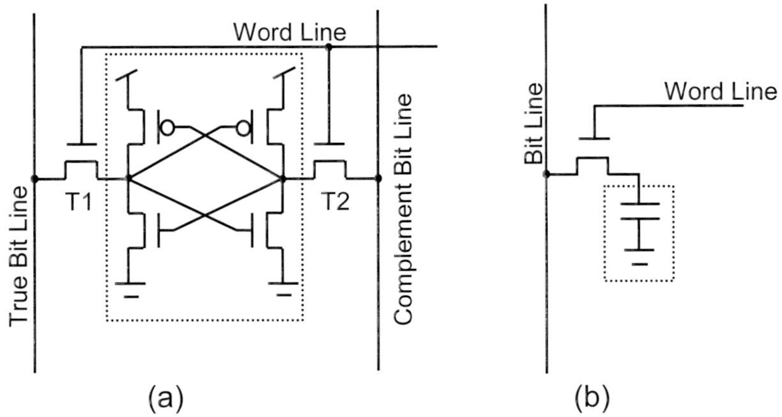

Figure 8-3: Schematics for (a) SRAM and (b) DRAM cells.

Most embedded memories are static in nature and therefore most of the focus in this chapter is on this type. There is sufficient discussion of dynamic memories as well as non-volatile memories to grasp the complexity of the problem and understand some of the key test and BIST implications.

8.21 Static Memory

The simplest of memories is a single port static memory with the cell as shown in Figure 8-3(a). This type of memory has been used for decades in stand alone memory chips and is the most commonly used embedded memory. It is quite interesting to note that this memory is a solid stalwart as compared to dynamic memories, where more subtle defects can cause damaging cell leakage, or as compared to newer non-volatile memories, where the fault models are still little understood [8]. Even with the vast level of experience in the industry using static memories new fault models are introduced when this type of memory is implemented in more exotic technologies such as *silicon on insulator* (SOI) or *strained silicon* [9]. Even in the more pedestrian bulk silicon technology, as the

[3] Graphic Design System II: industry standard format used to capture physical-design data.

channels have become shorter new and bizarre fault models have become evident to those who produce enough memory or those who just look closely enough at the failures which do occur.

Sub-threshold leakage has grown radically to the point that standby currents mask almost any defect that could previously have been detected with quiescent current (I_{DDQ}) testing. Transfer device (T1 and T2 in Figure 8-3(a)) leakage has grown to the point of requiring stricter limits on the maximum number of cells along a bit line and has an impact on the types of data patterns that need to be applied during test. The static memory cells have become so small that their inherent internal capacitance is trivial and their ability to retain data is compromised by *soft errors* caused by cosmic rays or alpha particles. This type of event drives a *soft error rate* (SER) that is much higher now for SRAMs than DRAMs. Even though the SRAM has a feedback mechanism inside the cell, the SER for SRAMs is much higher because SRAM cells have less internal capacitance. All this is stated as a preface for needing to understand the memory designs in order to understand the needed testing.

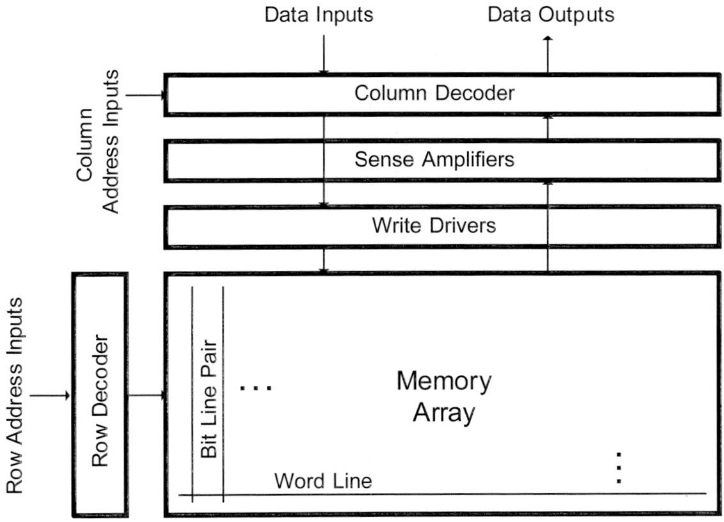

Figure 8-4: Block diagram of random access memory.

A simple description of an SRAM is an array of bits in the form of rows and columns. The specific location being addressed is selected by a row decoder and a column decoder, as illustrated in the random access memory block diagram shown in Figure 8-4. When data is written into the memory, the information coming from the data inputs is applied to the bit line pairs through a write driver. When data is read from the memory, the signal is read into the sense amplifier prior to be sent to the data outputs.

Each column has a pair of bit lines. The cells are accessed by this bit line pair where one of the bit lines is the "true" side while the other is the "complement" side. Both bit lines are utilized for reading the column of cells, as well as for writing the column. Multiple arrays can be assembled together in the form of banks or blocks with associated addressing.

The cells along a row are each contacted by a word line, where the contacts for a single SRAM cell are shown in Figure 8-3(a). To read or write a cell the word line is brought high, turning on the transfer devices. To write a cell a "1" is driven to one of the bit lines while a "0" is applied to the other bit line in the pair. The transfer devices are very effective at driving a "0" since they are NFETs but are not effective at driving a "1". As voltage scales downwards the NFETs' capability to drive a "1" is lessened further. Thus to write a "0", zero volts is driven onto the true bit line whereas to write a "1", zero volts is driven onto the complement bit line. When a word line is brought high it accesses all the cells in that row. Based on the bit address selected, the cell information from a specific word is sent to the outside of the memory. The intervening cells, not selected by the bit address, are ignored and the bit lines are restored when the read operation is complete.

The difference between an SRAM and a *register file* (discussed in the next subsection) is the presence of a sense amplifier on the SRAM. Because an SRAM has a sense amplifier the memory can be much larger and still have cell contents read very rapidly.

A sense amplifier is employed so that only a very small amount of signal can be developed across the bit line pair before a determination is made as to whether the cell being read has a "1" or a "0" in it. Typically 100mV or less of signal is required before making this determination and this number is becoming smaller with technology scaling (for example in 90nm technology this value can go as low as 60mV). The small signal development is required to give SRAMs the high speed access for which they are known. Even with this small signal development, the single largest portion of the access time is in allowing this signal to develop, as opposed to the time required for word address decoding, bit address decoding, and driving the data output of the memory.

The bit lines are both pre-charged to the high power supply level (V_{DD}). During a write, one of the bit lines fully transitions. During a read, one of the bit lines transitions only slightly. For example, during a read of a "1" the true bit line will stay at V_{DD} while the complement bit line will transition to V_{DD}–100 mV. In between each read or write the bit lines are all restored to their V_{DD} pre-charge condition.

The layout of an SRAM cell can vary in numerous ways. There are three primary layouts that make up the categories as shown diagrammatically in Figure 8-5. Figure 8-5(a) shows the power supply contacts internal or inboard to the cell. Figure 8-5(b) shows the power supply contacts external or outboard to the cell. Figure 8-5(c) shows the split word line configuration of the cell [10]. Each of these cell layouts fail in different ways and these ways impact the type of testing that should be applied as well as the redundancy calculation that needs to be performed for repair. Cells with inboard power supplies (Figure 8-5(a)) have more common-mode noise sources in the presence of defects and therefore a different cell disturb defect sensitivity exists. A defect in the V_{DD} contact for this cell will impact the cell shown and the one mirrored above it which shares the same V_{DD} contact. A resistive defect in one of the bit-line contacts will impact that cell and the one mirrored below it, which shares the bit line contacts, affecting only one data type for this pair of cells. A resistive defect in a bit line contact first impacts the write operation and only a larger defective resistance affects the read operation because a write involves a full rail voltage swing and requires overcoming the storage latch inside the cell [11].

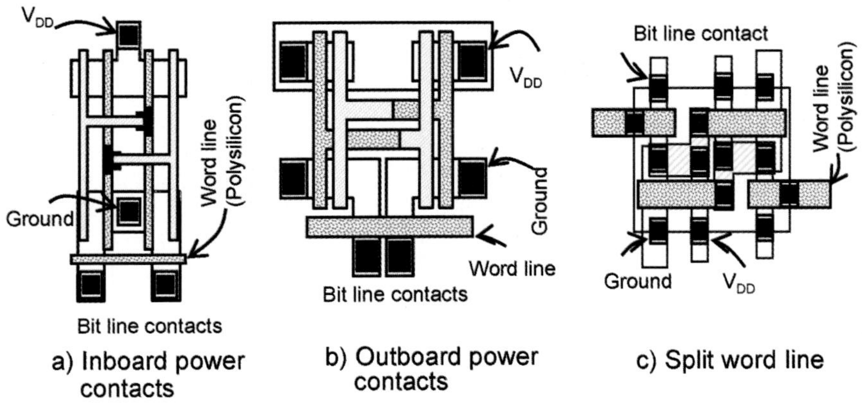

Figure 8-5: Three cell layout configurations.

Cells with outboard voltage contacts (Figure 8-5(b)) can share these contacts with three other cells. A defect in one of these contacts can cause a failure in the next cell vertically and the pair horizontally adjacent to these.

A split word-line cell (Figure 8-5(c)) is currently preferred by foundries as it is easier to make manufacturable, especially for deep sub-micron technologies. The shapes in this cell's layout have eliminated non-90 degree angles. All of the layouts shown are simplified representations and the other two layouts typically have a number of 45 degree angles in them. This cell is easier to fabricate, especially when considering the transfer device and controlling its I_{on} to I_{off} currents. It should be noted that there are far more contacts in the split word-line cell and therefore there are more locations that a defect can impact V_{DD}, ground, or bit line contacts to the cell.

8.2.2 Register Files

A typical register file is similar to an SRAM cell in that it has six devices for writing and storing the information. It does, however, have two additional devices for reading data from the cell, as shown in Figure 8-6. The extra devices and associated metal cause the register file to be larger but also faster than a single port SRAM cell. This is a two-port register file and there are numerous other possible configurations. One port is for reading while the other port is for writing where both operations can be performed on the register file simultaneously or asynchronously. There are restrictions on the combinations of operations that can be performed to the same cell because a read and a write operation to the same location could easily produce unpredictable or "X" results. There are two word lines, one associated with each port. Similarly there is a bit line pair associated with writing the cell and a single bit line or data line associated with reading the cell.

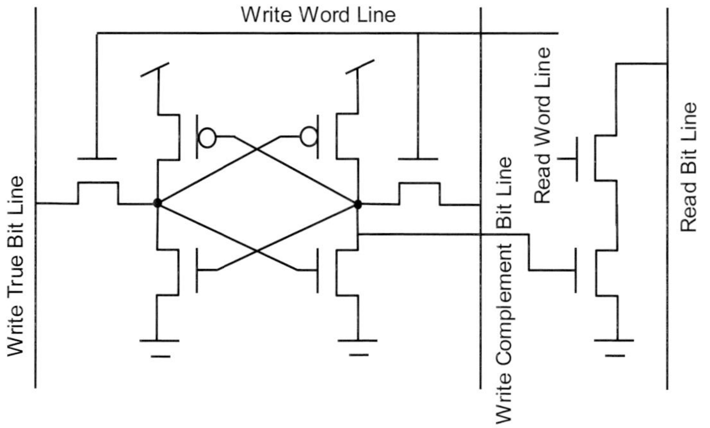

Figure 8-6: Register file cell schematic.

Register files are considered much closer to logic than any other form of memory since the read bit line makes almost a full rail voltage swing and no differential sense amplifier is required at the end of the bit line to accomplish the read. A skewed inverter, latch, or simple single ended sense amplifier may be employed to detect the "1" or "0" being read from the register file cell. The read bit line must be pre-charged to V_{DD} and the write bit lines typically are as well.

A register file, while considered closer to logic and having a more robust read operation, nonetheless has some greater sensitivities to defects than SRAMs. An SRAM cell has a single word line and a pair of bit lines contacting it. A two-port register file has two word lines and three bit lines which contact it. Obviously more metalization is covering the cell than that of an SRAM. An SRAM's size is constrained by diffusion and polysilicon whereas a register file's size is constrained by metal lines and is thus larger. The objective in memory design is to pack the cells in very tight proximity in order to increase the number of bits on a chip. Thus the metal lines are often narrower and have less space between them for register files. Metal short defects need to be of greater concern. There can be register files with many more ports, with a pair of bit lines added for each write port and a read bit line added for each read port. Each addition provides further possible defect sites where a short can occur between the added bit lines or between the added word lines. These shorts cannot be detected via a stuck-at fault model pattern. More will be discussed about multi-port memory test patterns later.

8. 2.3 Dual Port Memories

A dual port memory contains two read-write ports. Each port can be utilized to read or to write the memory, with the ports operating simultaneously or asynchronously from one another. There are certain restrictions in dual port memories since writing and reading the same cell normally produces unpredictable results as does simultaneously writing the same cell with both ports[4].

It can be seen in Figure 8-7 that each dual port cell has two word lines and four bit lines contacting it. Dual port memories are wiring bound, forcing the metalization to be in very tight proximity. The bit lines are all pre-charged to V_{DD}. The bit lines are more sensitive than those of a register file since the bit lines have sense amplifiers to detect the small signal differential that develops between the true and complement bit lines during a read operation. A write causes one bit line to make a rapid full rail swing in order to apply zero volts onto either the true or complement side of the cell, to store a "0" or a "1", respectively.

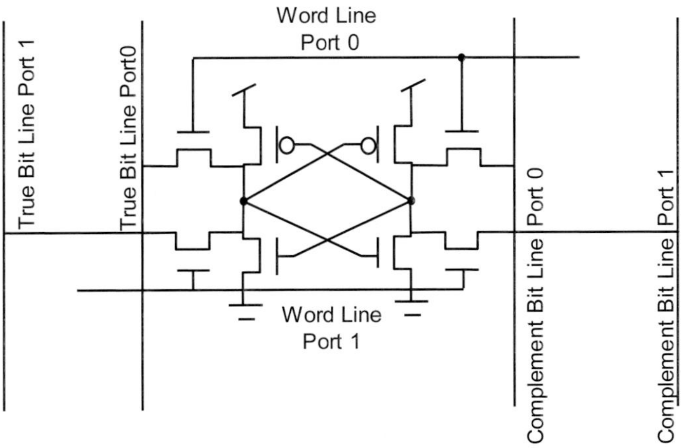

Figure 8-7: Dual port cell schematic.

Shorts or coupling defects between bit lines are of significant concern in dual port memories, especially due to the sensitivity of the read operation sense amplifiers and since the metal lines are so tightly packed. Often, isolation power lines are placed between the bit line pairs to avoid coupling. This helps avoid shorting type defects but often results in metal lines being at minimum dimension with minimum spacing. The word lines are also packed tightly together which increases the opportunity for manufacturing defects causing shorts between them. Because the read operations are more sensitive it is important to test for word line shorts by performing read operations. Word line shorts can exist, depending on the memory design, between word lines accessing the two ports on one cell or between word lines accessing vertically adjacent cells.

---

[4] The simultaneously read and write of a single address is usually specified as not allowed but certain designs can allow it by controlling the design timing. This can be accomplished by performing a write through from the data input to the data output or by forcing timing to have the "simultaneous" read and write occur in a specific order inside the memory array.

8.2.4 Content Addressable Memories

A *content addressable memory* (CAM) stores *cell data* and *match data*. Rather than having the cell data examined solely on the basis of an address, match data is provided to the input of the memory [12]. All the CAM's entries are examined in parallel to determine if the match data corresponds to that stored in one of the memory locations. If it does then that memory location's cell data is provided to the output of the memory. Thus, the design of a CAM includes both cell storage information, as in the case of an SRAM, and compare logic to check the match data correspondence [13].

A *ternary CAM* (TCAM) includes the capability of doing a "don't care" match selectively on a per bit basis. The complexity of a TCAM "cell" requires the storage of two bits as shown in Figure 8-8, in order to contain the values of "1", "0", and "don't care" states. The fourth state, which is possible for these two bits, is an invalid functional condition but is still needed for a thorough memory test, a fact that complicates TCAM testing. A TCAM can also be composed of dynamic cells [14], in which case only two transistors and two capacitors compose the TCAM cell; all of the leakage issues associated with dynamic memories need be tested.

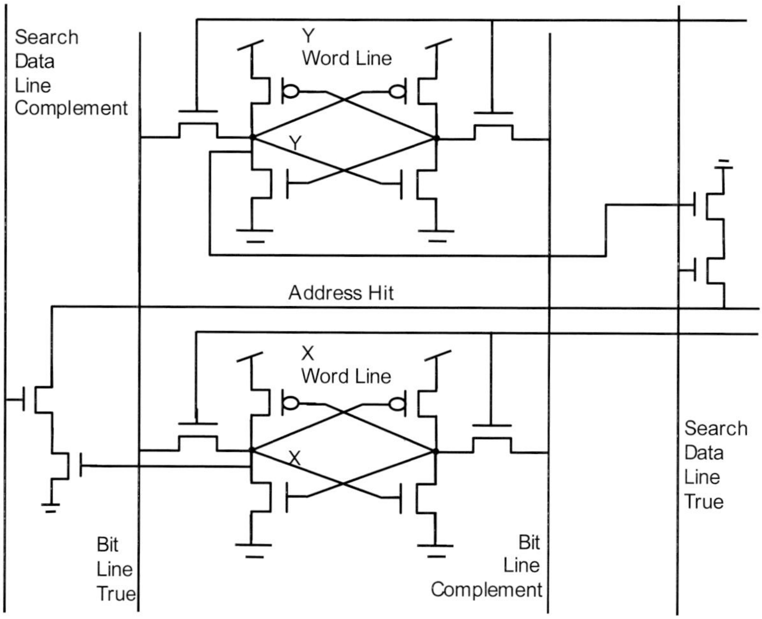

Figure 8-8: TCAM cell schematic.

8.2.5 Dynamic Random Access Memories

A DRAM is the simplest structure at the cell level, as shown earlier in Figure 8-3(b). Each cell is composed of a single transistor and a single capacitor with a diagrammatic representation of a cell cross section shown in Figure 8-9. The trench aspect ratio is much larger with the depth being far greater than shown in the Figure.

The trench capacitor has an exceedingly high quality dielectric, being produced very early in the fabrication process.

While the DRAM is the simplest at the cell schematic level, the capacitor provides significant complexity at the functional level. At the system level the required regular refresh of all the cells provides very real overhead and logistical complexity. Since there is no feedback path in a DRAM the cell's data charge slowly bleeds off. Occasionally, each cell location in the DRAM needs to be read and the data re-written into the location. This is referred to as a *refresh operation*. Embedded DRAMs can be designed where the refresh operation is hidden from the user and therefore the user does not need to handle the refresh logistics. Even so, there are a limited number of allowed read operations in a certain period of time and a limited time where the hidden refresh is performed resulting in some functional operation limitations. Much complexity is added to the memory to accomplish this ease of use. This complexity masks access to the memory in many ways and can make testing quite difficult since the freedom to apply patterns and operations directly to the memory is hampered.

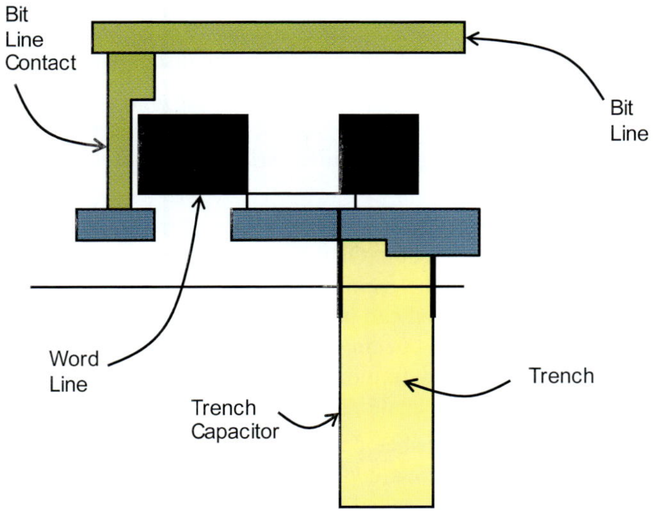

Figure 8-9: DRAM cell diagram.

Whenever a DRAM cell is read, the data in the cell is destroyed[5] and must therefore be written back into the cell. This is referred to as a *write back* operation. Since an SRAM's read operation is not destructive, a read is completed by simply restoring the bit lines. A DRAM's read operation, however, is only completed after performing a write back and then restoring the bit lines to a pre-charge condition[6].

[5] DRAM cells have destructive reads because there is no feedback within the cell. To read the cell the charge on the storage capacitor is transferred to the bit line. Once the charge is transferred the cell's data is lost and the read is therefore termed destructive.

[6] For this reason DRAMs require longer cycle times than SRAMs.

DRAM bit lines can be restored to differing levels, depending on the design, but the most common pre-charge value is $V_{DD}/2$.

Because an entire DRAM read operation is far more complex than a typical SRAM it is critical that the DRAM test is executed at a high cycle rate. High speed testing is advantageous for SRAMs, and allows for better quality testing, but high speed testing is paramount for DRAM memories.

8.3 MEMORY FAULTS

Logic testing is primarily geared toward stuck-at faults. There is some testing that is oriented toward transition and path delay faults[7] but relatively little fault modeling is done when considering random logic. For memories, the fault model set that needs consideration is far larger. Anyone who tests memory considering only stuck-at faults is foolish in the extreme. Because memories are so dense and have analog operations, they are "defect magnets." A defect which would cause no problem in logic can cause subtle defective operation in memories. Subtle defects are not subtle from a customer perspective but instead cause intermittent problems that are difficult to diagnose. In this section, first, the four simple standard fault models will be reviewed and then the more subtle fault models will be discussed.

The most basic memory fault model is the *stuck-at fault model* where a given memory cell can only contain a "1" or a "0". It has a strong corollary to the logic stuck-at fault model. Any test pattern which writes and reads both zeros and ones will catch this type defect.

Next is the *transition fault model* where a cell can store either a "1" or a "0" but it can only transition in one direction. Once the cell has transitioned it can no longer go back to its previous value. The memory behavior is much like that of a stuck-at fault, except that the defective cell can power up in either state. Testing for this type of fault model is simple. It should be noted that the transition fault model is different from the transition fault model definition used in testing in logic where a transition in a logic line is expected to propagate.

A *coupling fault model* describes defective operation where one cell causes an error in its neighboring cell. There are various forms of coupling fault models where a state or a transition of one cell, called the *aggressor*, causes defective operation in the *victim* cell. A state coupling fault is where the state of a given cell causes a change in operation of the victim cell. A transition coupling fault is where a transition in an aggressor cell causes erroneous operation in the victim cell. There can be a single aggressor or multiple aggressor cells.

Lastly, a *neighborhood pattern sensitive fault* (NPSF) is one where the operation of one cell is dependent on the neighborhood in which it resides. A neighborhood can be the eight cells immediately surrounding the base cell under test or it can be all the cells in the same row or in the same column as that of the base cell. Historically, only DRAMs needed to be considered for neighborhood pattern sensitive faults but with the advent of more recent advanced technologies, such as silicon on insulator

---

[7] See other chapters of the book for defect-oriented testing and delay testing issues.

(SOI) or even bulk 90 nm silicon, SRAMs also need to consider neighborhoods, especially those along the column in which the base cell resides.

These four fault models compose the traditional fault models considered in memory testing. Numerous other subtle fault models must be analyzed, however, in order to achieve the needed quality levels that customers are demanding. For general purposes, memory cells are dynamic, static, or non-volatile. A dynamic cell must be refreshed and can be used in a DRAM, a dynamic TCAM, or other similar type structure. A static cell retains data indefinitely whether in a quiescent state or whether there are an infinite number of read operations as long as power is maintained. Static cells can be used in a standard SRAM, static CAM memories, or other similar type memories. Non-volatile memory cells retain their data even if powered off; examples are Flash memory, EEPROM, and even ROMs. The fault models must differ as the memory cell type differs.

Subtle fault models are best described at a transistor level of abstraction. Since semiconductor memories are actual physical circuits they fail due to actual physical defects which can be described electrically. The first broad category of subtle defects is covered by the *retention* or *disturb fault model*. A defect free cell retains the "1" or "0" which is stored in it but a defective cell does not. In the case of a DRAM cell this means that the cell does not retain data for the duration between refresh cycles. For an SRAM cell it means that data is not retained until new data is written into the cell. For a non-volatile memory cell a disturb is considered to be any change in state without intentionally writing to it [15], [16].

A read disturb fault model describes defective operation where a cell is read and the data is lost. This can occur in a DRAM by simply having the write back operation fail after the read of the cell is completed. For an SRAM a read disturb defect can be a high resistance pull-down path in one side of the cell. A pull-down resistive defect can be as simple as a resistive ground contact or a defectively longer NFET pull-down device. A DRAM cell can have numerous physical defect sites that cause charge to leak from the cell capacitor more rapidly than designed and cause the cell to lose data over time, including a poor dielectric or a leaking transfer FET [17]. Figure 8-10 shows example sites for some of these defects. The defect on the DRAM cell shown in Figure 8-10(b) is simple. There are normal leakage paths within a DRAM cell but the defect in this Figure indicates a leakage that is beyond the tolerable specification. For an SRAM cell the possibilities are more definitive and will now be discussed.

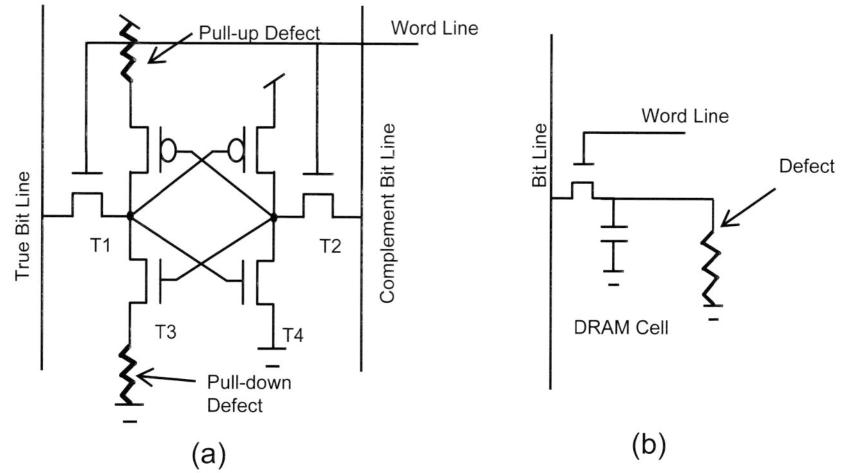

Figure 8-10: Example defect sites for (a) SRAM and (b) DRAM memories.

Within an SRAM cell, the ratio of the pull-down strength to the transfer strength is referred to as the beta ratio. The pull-down devices are T3 and T4 shown in Figure 8-10(a) whereas the transfer devices are T1 and T2. A small cell beta ratio of 1.0 means that the transfer device is the same strength as the pull-down device, causing the cell to lose its data whenever it is read. A large cell beta ratio of 2.0 or higher means that the cell is very stable and will not disturb during a read operation. The beta ratio can be sufficiently high, though, to make a cell so stable that it cannot be written. The presence of defects can change the effective beta ratio. A pull down defect, as shown in Figure 8-10(a) reduces the beta ratio by weakening the pull-down capability of the true side of a cell. The cell will have trouble retaining a "0" state during a read operation. It is important to note that a pull-up defect does not impact the beta ratio or the cell stability during a read operation since the bit lines are pre-charged to a high state. Performing a read in the presence of a pull-up defect simply reinforces the high state for that side of the cell.

Figure 8-11 describes the amount of resistance, for a pull-down defect such as the one shown in Figure 8-10(a), that causes a read disturb type defect. During a read, starting with the bit lines pre-charged to V_{DD}, the word line for a given address goes high turning on the transfer devices. The cell node which is low pulls down the corresponding bit line. The other bit line remains at its high state due to the large amount of capacitance it has. When the first bit line is pulled low there is an effective divider network between the transfer and pull-down devices. In a defect free cell, the low node rises a small amount due to the divider network operation. The small amount is less than an NFET threshold voltage, which assures that the opposite pull-down device remains off. When there is a resistive pull-down defect, the low node rises to a higher than normal voltage. With sufficient resistance the low node rises far enough to turn on the opposite pull-down device. At this point the cell becomes unstable and will easily flip [18]. The low node need only rise to an NFET threshold voltage in order to cause the cell to be unstable.

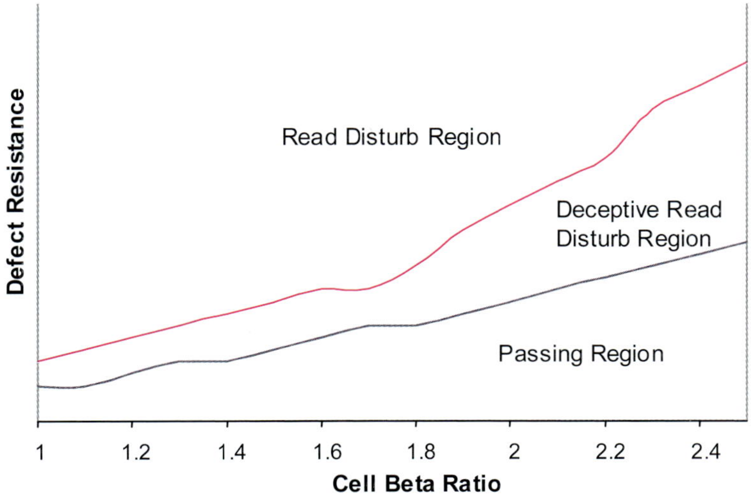

Figure 8-11: Graph of defective pull-down resistance and resulting operation.

Figure 8-11 shows that a range of defective resistances exists which allow normal operation to continue, shown as the passing region. It should also be noted that the effect of the resistance depends on the cell beta ratio with a more stable cell (a higher cell beta ratio) requiring a greater defective resistance to cause a problem. The upper line in the graph indicates when the cell disturbs and is detected on the first read. This is what is typically thought of as a read disturb defect. In between the top and bottom lines is a range of defective resistance where, depending on internal timings designed into the SRAM, a cell can disturb and contain incorrect data but the correct data reaches the output of the memory. This deceptive erroneous operation occurs due to the sense amplifier detecting the value of the cell *prior* to the cell being disturbed. With this intermediate amount of defective resistance, it takes longer for the cell to disturb. The timings are such that the correct bit line has discharged and the sense amplifier sets prior to the cell disturbing.

For bulk type silicon SRAMs there are only two possible defect sites per cell for this type of disturb defect, a pull-down defective resistance can exist for either the true or complement side. For more exotic technologies, such as silicon on insulator, there are numerous other defect sites. There are four additional possible defect sites which can abnormally strengthen the transfer devices, two added sites for the pull-down devices, and even two sites on pull-up devices that can cause read disturb type defects.

For dual port memories there is also a greater concern for read disturb type defects. Dual port memories have two transfer devices on each side of the memory cell, as shown earlier in Figure 8-7. Because each port is independent of the other, the cell needs to be designed so that both ports can read from that cell at the same time. When both ports are reading, twice the current is being discharged into the low node. Because of the required extra current, the beta ratio must be designed to be higher on dual port memories. By examining Figure 8-11, it can be seen that memories with high beta ratios also have a larger range of resistance which allow for the deceptive read disturb operation to occur. Since dual port memories are becoming a larger portion of the memories in the field more of these type defects can be expected.

In addition to pull-down defects which can cause retention concerns in SRAMs, pull-up defects can cause retention problems as well. A defect, such as that shown earlier in Figure 8-10(a), can cause a quiesced cell to lose its data over time. Since reading a cell involves having both bit lines pre-charged high and bringing the word line high, a resistive pull-up defect cannot be activated by performing a read. A drooping high cell node can in fact be restored when a read is performed, due to the high bit line potential and the miller capacitance coupling from the word line being activated. In order to detect the presence of a resistive pull-up defect either a long time must transpire for the cell to lose its data or the power supply can be bumped to try to disturb the cell. Another detection possibility is to have a special mode designed into the memory to pre-charge the bit lines low and perform dummy reads which attempt to disturb the cells in a similar fashion to that used to examine for resistive pull-down defects [19].

Beyond cell defects, problems can exist in the peripheral circuitry of the memory as shown in Figure 8-4. There can be problems in write circuitry, the sense amplifiers, the decoders, and elsewhere. Decoder circuitry can have problems where the memory points to the wrong location, points to no location, or points to multiple locations. All of these operations are catastrophic and therefore relatively easy to detect. One subtle type decoder defect involves opens in the decoder tree. Some decoders include dynamic logic where a pre-charge operation occurs prior to each evaluation. When a dynamic decoder is utilized, an open defect is again easy to detect. However, when a static decoder is utilized, an open defect such as that shown in Figure 8-12 is difficult to detect. During normal march patterns[8] the address order is either incremented or decremented through the address space. Since the middle PFET never has to uniquely pull the output of the three-input NAND gate high, the output maintains its high level due to internal capacitance and defective operation is not detected. Only through an addressing sequence that requires each PFET to uniquely pull the NAND output high will allow static decoder open defects like this to be detected.

[8] We discuss memory test patterns in detail in subsequent sections.

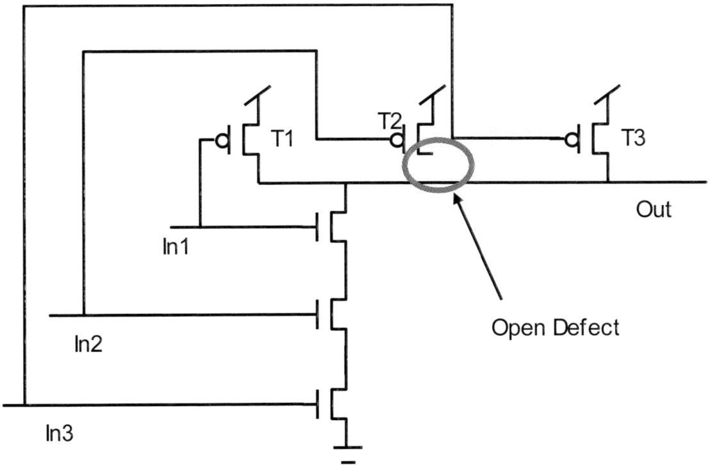

Figure 8-12: Static decoder defect and sequential sequence operation.

Memories with multiple ports can also exhibit defects that are not easily detected [20]. Some people approach testing a multi-port memory by applying a test to each port sequentially but never exercising multiple ports in parallel. This type of test does not look for any port-to-port interactions that are only activated by exercising both ports simultaneously, often requiring different addresses [21]. There are intra-port faults where two ports of the same type but with different addresses interact. There are inter-port faults where two different ports interact with one another [22]. When an inter-port fault exists it can be either an intra-address or an inter-address fault. An inter-port inter-address fault is one where two different ports from two different addresses interact. An inter-port intra-address fault is one where two different ports from the same address cause an erroneous operation. An example of an inter-port intra-address fault is shown in Figure 8-13 where one word line is driven high and a second word line from the same address and a different port is defectively pulled high. Since a word line is tied to the NFETs, which comprise the transfer devices in the cells, the word line need only be pulled up by an NFET threshold voltage in order to activate the transfer devices. Defects on the lower cell in Figure 8-13 can impact operations on the upper cell, especially when the upper cell experiences a read. Word lines on dual port memories are in tight proximity to one another, as stated earlier in the description of the dual port memory design. Due to this tight proximity and due to the fact that the word line need only be pulled up by a threshold voltage, these types of multi-port faults can occur frequently [23].

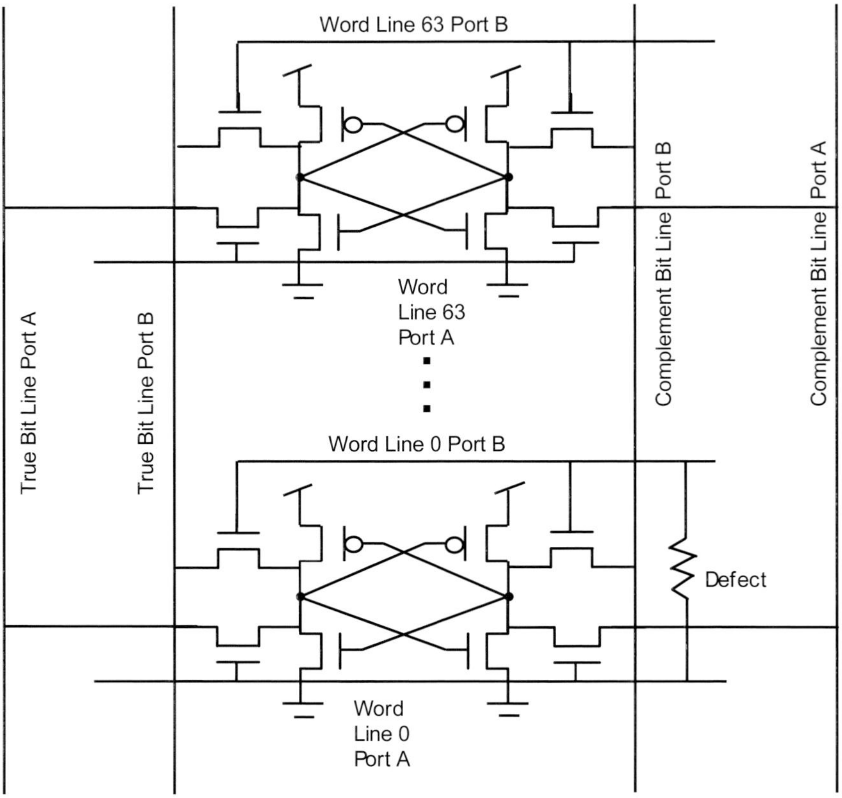

Figure 8-13: Inter-port intra-address defect.

The coverage of subtle faults certainly is not complete based on this brief description. The types of faults discussed here do, however, provide a glimpse into the defects that can cause erroneous operation and show the need for understanding the actual memory design, in order to determine the impact. New memory designs and new technologies will require the test engineer to be ever vigilant in order to maintain the correct set of fault models and therefore the correct tests. Without these fault models, quality embedded memories cannot be produced. For a detailed description of subtle faults in memories the reader may refer to [2].

8.4 MEMORY TEST PATTERNS

Memory test patterns are the key for providing quality embedded memory test [24]. In order to select the best patterns or to develop the appropriate ones, first the memory design must be understood along with the technology that is being used for fabrication. Then the appropriate fault models need be selected and subsequently the correct patterns defined. There is no single magic pattern which can be used to test memories. The patterns must be carefully selected or developed to detect the ways that real manufacturing defects can make the memory circuits fail to operate correctly. A host of memory test patterns have been defined over the decades. The most important are described in this chapter to provide an understanding of the components that comprise a good test pattern and how to analyze the value of these patterns. This analysis will also provide understanding of the appropriate techniques used in developing patterns needed to detect defects in new memories and new technologies.

8.4.1 Pattern Nomenclature

There are a number of terms which are used repeatedly in describing memory test patterns and which benefit from explanation. The term *base cell* has been used loosely so far in this chapter but it is important to establish a clearer description. A base cell is one which is the focus (the *cell under test*) of a specific test pattern. A base cell is surrounded by a neighborhood of cells immediately north, south, east, west, and diagonally adjacent, as shown in Figure 8-14. This Figure shows the base cell, B, and the surrounding neighborhood cells, N, in the context of an eight by eight memory array. There are therefore a total of nine cells in the neighborhood: the base cell and the eight surrounding it. Sometimes a *deleted neighborhood* is referred to when describing only the surrounding cells, not including the base cell. A *word* is composed of data contained in multiple cells, with all cells being accessed by a single address. The address which is the focus of a given test sequence operation is composed of base cells (cells that are tested by the test sequence) and is therefore referred to as the *base address*. The *order* of a pattern indicates the number of cycles involved in testing the memory. A 4N pattern indicates that each location in the memory has been accessed four times. N is the number of address locations in the memory so the total number of cycles for this 4N pattern is four times the number of memory address locations. A small "n" refers to the number of bits in a memory and is useful when evaluating bit-oriented memories. Memories can be either *bit-oriented* memories (BOM) or *word-oriented* memories (WOM). A bit-oriented memory contains a single bit per address whereas word-oriented memories have more than a single bit in each word or address location [25]. From a practical point of view, virtually all memories have word widths of greater than a single bit and therefore testing should be focused on word-oriented memory testing.

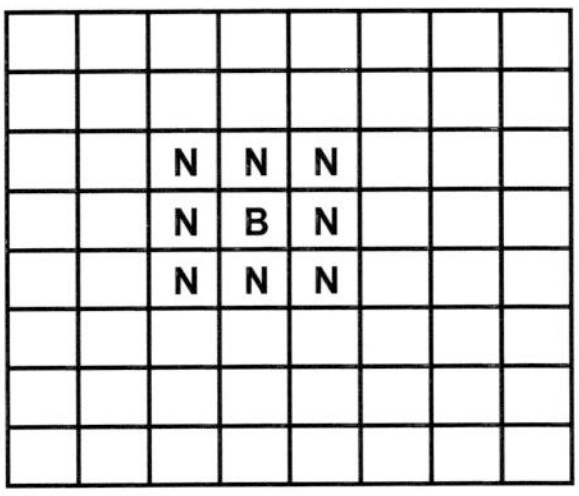

Figure 8-14: Base and neighborhood test cell diagram.

Address sequencing is very critical in memory test effectiveness. There are three address sequence types: *marching*, *walking*, and *galloping*. There are also random and pseudorandom sequences but they are of little value and will be discussed later in the BIST section. Marching patterns are the most frequently used. The address space is progressed through sequentially in either an incrementing or decrementing fashion. Walking patterns also progress through the address space sequentially. The difference between marching and walking patterns has to do with the data state before and after the base cell or address has completed being accessed. With a marching pattern, after a given sequence, the base cell is changed to the opposite state from when the sequence was started. As the memory address space is progressed through, more and more of the cells contain the opposite value from that which was in the memory prior to the sweep. At the end of a sweep all of the data values have been changed in a marching test pattern.

In a walking pattern the data state is returned to the original value that was in the base address. At any given time only a single address location can have different data from that contained at the beginning of the sweep. At the end of the sweep, the data state in the memory is as it was at the beginning of a walking pattern sweep.

A galloping pattern involves many more cycles than marching or walking patterns. For each base address, every other address is accessed before or after the base address. This makes all possible address transitions occur in a galloping pattern test and therefore the test is of the order N^2. Galloping patterns are very thorough but their length makes them impractical for normal manufacturing tests.

8.4.2 Key March Patterns

A few selected patterns will now be discussed. There is a whole series of lettered memory test patterns starting with a pattern named March A. The most commonly described memory test pattern is the March C– pattern. There is a March C pattern which has a redundant test operation which, once eliminated was renamed March C–. The March C– pattern can be described as "W0 ⇕; R0, W1 ⇑; R1, W0 ⇑; R0, W1 ⇓; R1, W0 ⇓; R0 ⇕" with a ";" separating each of the sweeps[9]. A tabular representation is show in Table 8-1. The first sweep involves proceeding through the memory address space while writing zeros. There are a total of six sweeps and March C– is a 10N pattern. The address space is incremented, in sweeps two and three, and decremented, in sweeps four and five, through for each data type. Sweeps one and six can be implemented as either incrementing or decrementing. March C– is a very valuable basic memory test pattern and is a useful starting point in developing further tests. The March C– pattern covers stuck-at, transition faults along with some coupling and some address decoder faults.

[9] W0 means to write a "0 ," W1 means to write a " 1", R0 means to read a "0", and R1 means to read a "1". Operations separated by commas happen on different cycles. Incrementing through the address space is indicated by ⇑ while decrementing is indicated by ⇓. A ⇕ symbol means to sequence the addresses but the order is not specified.

1	W0 ⇕
2	R0, W1 ⇑
3	R1, W0 ⇑
4	R0, W1 ⇓
5	R1, W0 ⇓
6	R0 ⇕

Table 8-1: March C– pattern description.

An enhancement to the March C– pattern is referred to by the name *Partial MOVing Inversion* (PMOVI). It involves an additional read operation at the end of each sweep in the March C– pattern along with a minor address sequence difference. The PMOVI pattern is shown in Table 8-2 and is a 13N pattern. Similar patterns are sometimes referred to as March 13N or as a three step, since each primary sweep includes three steps, unique address pattern. Please note that unique address patterns are called such because they ensure decoder operation uniquely exercises a single address location. The PMOVI pattern provides significant coverage for any memory. The PMOVI pattern covers stuck-at and transition faults along with some additional coupling and address decoder faults. To detect defects in a memory with a differential sense amplifier it is useful to include another write at the end of each three-step sequence sweep. This write puts the same value back into the cell that already exists, an example being sweep 2 becoming "R0, W1, R1, W1 ⇑". By including this final write, at each address transition a write of one data type is immediately followed by a read of the opposite data type [26]. A write causes one of the bit lines to discharge fully while a read of the opposite data type requires that the bit lines fully restore and the opposite bit line discharge a small amount. Any defect impacting the bit line pre-charge operation will be easily detected by the addition of the write to the end of each sweep. When the write is added, the pattern is referred to as the March 18N, four step unique address, or enhanced March C– pattern and is shown in Table 8-3. When using this pattern, addressing order becomes very critical. It is important to modify the row addresses most frequently so that stepping from one address to the next remains in the same column, since pre-charge defects impact on a column basis and are easiest to detect that way. The March 18N pattern covers stuck-at and transition faults along with some additional coupling and address decoder faults but is especially useful for detecting pre-charge faults.

1	W0 ⇓
2	R0, W1, R1 ⇑
3	R1, W0, R0 ⇑
4	R0, W1, R1 ⇓
5	R1, W0, R0 ⇓

Table 8-2: Partial moving inversion (PMOVI) pattern description.

1	W0 ⇕
2	R0, W1, R1, W1 ⇑
3	R1, W0, R0, W0 ⇑
4	R0, W1, R1, W1 ⇓
5	R1, W0, R0, W0 ⇓
6	R0 ⇕

Table 8-3: March 18N pattern description.

An additional lettered pattern is the March LR pattern as shown in Table 8-4 [27]. This is an interesting pattern in that some of the individual sweeps are marching operations while other sweeps are walking operations. March LR is a 14N pattern. The walking sweeps are very useful for detecting cumulative cell leakage or other problems that impact a whole column. By having only a single bit in a column with the opposite state, the lone cell being read must overcome all of the other cells' leakage pulling down the opposite bit line. Leakage currents for off cells can become significant in advanced technologies and verifying that a single cell can still be accurately read in presence of this leakage is important. The March LR pattern covers stuck-at and transition faults along with even more coupling and address decoder faults.

1	W0 ⇕
2	R0, W1 ⇓
3	R1, W0, R0, W1 ⇑
4	R1, W0 ⇑
5	R0, W1, R1, W0 ⇑
6	R0 ⇑

Table 8-4: March LR pattern description.

The last lettered pattern to be discussed in this section is the March G pattern and it is shown in Table 8-5 as being 23N plus two pauses. This pattern is important to discuss since it includes the pause portions that are critical to retention defect testing [28]. The required pause duration is dependent on the technology in which the memory is fabricated. The pause can also be combined with a power supply bump. The duration of the bump can be quite short from the external tester, including only power supply settling times. This duration is quite long from the memory's perspective, however, since settling times are essentially DC operations from a memory circuit's vantage point. Pauses can be included in other patterns but the March G pattern is described here because it explicitly includes pauses which help capture retention defects. The March G pattern covers stuck-at and transition faults along with some additional coupling and address decoder faults but is especially useful for detecting retention faults.

1	W0 ⇕
2	R0, W1, R1, W0, R0, W1 ⇑
3	R1, W0, W1 ⇑
4	R1, W0, W1, W0 ⇓
5	R0, W1, W0 ⇓
6	Pause
7	R0, W1, R1 ⇕
8	Pause
9	R1, W0, R0 ⇕

Table 8-5: March G pattern description.

The patterns which have been discussed in this chapter are summarized in Table 8-6. There are at least another dozen lettered patterns described in memory testing literature. Each year more patterns are added to the art, mostly driven by new memory designs and new memory technologies. Many of the new patterns have been added for testing multi-port memories since most of the port-to-port defective interactions have only begun to be understood in the last few years. The volume of material required to adequately cover the possible patterns cannot be described in this brief chapter and the reader is referred to other referenced literature for further information, in particular [2], [20] and [24].

Pattern Name	Order	Faults Covered	Reference
March C–	10N	Stuck-at, transition, some coupling, some decoder	[24]
PMOVI	13N	Stuck-at, transition, some coupling, some decoder	[28]
March 18N	18N	Stuck-at, transition, some coupling, some decoder, pre-charge	[2]
March LR	14N	Stuck-at, transition, some coupling, some decoder	[27]
March G	23N	Stuck-at, transition, some coupling, some decoder, retention	[28]

Table 8-6: Memory test patterns summary.

8.4.3 Memory Data Backgrounds

When considering word-oriented memories, data-background patterns[10] become critical to testing since each bit cannot be individually controlled. There are *blanket background patterns* where the memory contains either all zeros or all ones. Then there are *checkerboard background patterns* where the memory contains either a physical checkerboard or inverse-checkerboard pattern. It is important to implement

[10] Data stored at the beginning of a test pattern.

a physical checkerboard pattern because applying a logically addressed checkerboard pattern does not yield a physical checkerboard in the memory itself and therefore does not stress adjacent structures in the manner that a checkerboard pattern should. Further data backgrounds are possible based on the number of bits in each word. The number of possible data-background patterns is $\log_2(m)+1$, where m is the number of bits in a word. Blanket and checkerboard patterns are sufficient for most static memories but dynamic memories require a richer set of data backgrounds[11].

It is useful to consider the impact of the data background patterns on the physical structures in an embedded memory. For a dynamic memory the data contained in the cell is indicative of the state of that cell's structure relative to its neighbors. For a static memory, where there are both true and complement portions inside each cell, the impact on adjacent structures is not so obvious. Each static cell can be represented by a true and a complement side, which will be identified as "T" and "C", respectively. In this manner, if three cells are sequentially stepped they would be shown as TC-TC-TC or simply TCTCTC. Intervening cells are often mirrored in order to pack the cells in tighter proximity. The same three cells, with the middle cell being mirrored about the y axis would be represented as TC-CT-TC or simply TCCTTC. If the cells are stepped then a blanket background, of either all zeros or all ones, provides the greatest stress between adjacent structures. This is because a "000" background in these three cells becomes, when considering the "T" and the "C" in the cell, 01-01-01 or 010101. If the cells are mirrored then the adjacencies are represented as 01-10-01 or 011001. Each "T" and "C" represent a structure inside a cell. Each "1" and "0" represent the potential of their respective structures. Clearly the blanket pattern applies more stress between adjacent structures when the cells are simply stepped. Expanding on this concept, a nine-cell neighborhood is shown in Figure 8-15 with both blanket and checkerboard patterns being shown for memories with cells that are mirrored and those which are stepped. Between the application of blanket zeros and ones along with the application of checkerboard and inverse checkerboard, all of the adjacent structures are stressed, both vertically and horizontally.

	Blanket						Checkerboard			Inverse		
	Zeros			Ones								
Stepped:	01	01	01	10	10	10	01	10	01	10	01	10
	01	01	01	10	10	10	10	01	10	01	10	01
	01	01	01	10	10	10	01	10	01	10	01	10
Mirrored:	01	10	01	10	01	10	01	01	01	10	10	10
	01	10	01	10	01	10	10	10	10	01	01	01
	01	10	01	10	01	10	01	01	01	10	10	10

Figure 8-15: Data representation for cell stepping and mirroring.

[11] DRAMs need richer data backgrounds for testing based on empirical results from manufacturing test. Their susceptibility to cell-to-cell leakage and other cell interactions which are detected only with specific data types easily explain the need for the different background patterns.

8.4.4 CAM Test Patterns

A content addressable memory is a good example of a memory which needs more than just memory patterns to adequately test the circuitry. CAMs contain match logic which checks for a comparison between the match data in the CAM and that applied to the match inputs of the memory. Patterns must be applied to detect faults in the compare logic in addition to the memory test algorithms. Figure 8-16 contains a simple pattern which facilitates test of the compare logic inside the memory. The example TCAM has a data width of eight bits. This Figure shows an all zeros match followed by walking a "1" across the compare bits. The complement is also implemented in the latter half of the pattern shown in the Figure. This simple pattern covers 100% of all the stuck-at faults in the XOR-OR tree that composes the compare logic [29]. It is also helpful to realize that the pattern includes single-bit mismatches, which provide the least pull down capability in the dynamic logic normally comprising the CAM's compare circuit. In addition to this pattern, it would also be useful to have an all bit mismatch followed immediately by a match to ensure that the dynamic logic pre-charge circuitry operates correctly.

	D0	D1	D2	D3	D4	D5	D6	D7	
	With array at all 0s								
1	0	0	0	0	0	0	0	0	match
2	1	0	0	0	0	0	0	0	mis-match by 1
3	0	1	0	0	0	0	0	0	mis-match by 1
4	0	0	1	0	0	0	0	0	mis-match by 1
5	0	0	0	1	0	0	0	0	mis-match by 1
6	0	0	0	0	1	0	0	0	mis-match by 1
7	0	0	0	0	0	1	0	0	mis-match by 1
8	0	0	0	0	0	0	1	0	mis-match by 1
9	0	0	0	0	0	0	0	1	mis-match by 1
	With array at all 1s								
10	1	1	1	1	1	1	1	1	match
11	0	1	1	1	1	1	1	1	mis-match by 1
12	1	0	1	1	1	1	1	1	mis-match by 1
13	1	1	0	1	1	1	1	1	mis-match by 1
14	1	1	1	0	1	1	1	1	mis-match by 1
15	1	1	1	1	0	1	1	1	mis-match by 1
16	1	1	1	1	1	0	1	1	mis-match by 1
17	1	1	1	1	1	1	0	1	mis-match by 1
18	1	1	1	1	1	1	1	0	mis-match by 1

Figure 8-16: CAM compare circuit test pattern.

A ternary content addressable memory has two bits per cell as shown earlier in Figure 8-8. There are also only three valid combinations for the data stored in these two bits: a "0", a "1", and a "don't care." While these three states include all the functional possibilities, they will not suffice to cover the possible defects inside the TCAM. Exercising all four data possibilities is required since otherwise subtle faults, which would impact the TCAM's operation in the field, would go undetected at the time of manufacturing test. Figure 8-17 shows the result of placing blanket zeros and ones as well as checkerboard and inverse checkerboard patterns in the TCAM bits.

Desired Blanket:

	x	0	0	0	x	1	1	1
	y	0	0	0	y	1	1	1
	x	0	0	0	x	1	1	1
	y	0	0	0	y	1	1	1
	x	0	0	0	x	1	1	1
	y	0	0	0	y	1	1	1
Result:	Masked				Invalid			

Desired Checkerboard:

	x	0	1	0	x	1	0	1
	y	1	0	1	y	0	1	0
	x	0	1	0	x	1	0	1
	y	1	0	1	y	0	1	0
	x	0	1	0	x	1	0	1
	y	1	0	1	y	0	1	0
Result:		0	1	0		1	0	1
		0	1	0		1	0	1
		0	1	0		1	0	1

Figure 8-17: Background pattern impacts on TCAMs.

Addressing order is critical to provide high quality memory testing. Since a memory has physical circuits that can fail, it is important to be aware of the addressing in order to detect these defects. Row addresses are separate from column addresses. It is easier to detect errors in the row decoder by rippling rows most frequently, as it also useful to ripple columns most frequently to detect errors in the column decoder. These address sequences are sometimes referred to as fast-x and fast-y patterns. Including a galloping row address pattern, where a single column is exercised while ping ponging through the row addresses is quite useful for detecting row decoder defects. A galloping column pattern is likewise very effective for exercising the column decoder. Beyond testing the decoders, address order is critical for exercising defects that impact a column. For example, rippling rows most frequently allows an address sequence that proceeds along a column which is useful for detecting pre-charge and other column oriented defects. In addition to the data and address pattern sequences discussed in this section are key environmental issues. The power supply state, temperature, and timings can all be modified to exacerbate certain defects. Power supplies can be bumped up or down, while the memory is quiesced, to bring out retention defects. Temperature can be increased to increase leakage. All of the possible test tools and capabilities must be brought to bear on pulling out defects to make testing the most effective. The objective is to eliminate defects so that the memories, and chips they reside on, can have very high quality when used in products.

8.5 SELF TEST

Built-in self-test is the vehicle for testing embedded memories. The chapter thus far has described memory faults, test patterns, and designs. All this information is useful for stand alone as well as embedded memories. When it comes to Design-for-Test and BIST, circuitry is designed on the chip to facilitate testing. The proper testing needs to be developed but the BIST development is what allows embedded memories to actually be tested. It is critical to have a good test strategy and a good BIST in order to have a high quality memory as a result.

Self test of logic usually involves the application of pseudorandom stimuli [30], [31]. Testing of random logic can be adequately performed with pseudorandom patterns. Memories, however, are regular structures and require regular patterns. The pseudorandom patterns applied with Logic BIST (LBIST) are designed to detect stuck-at defects and little more. As already mentioned, stuck-at testing is insufficient for memories. An extensive fault model set is required to be tested in order to detect many of the memory defects. Patterns specifically oriented to these fault models are required to provide a high quality test. Whenever pseudorandom patterns are implemented and uniquely detect failing memories there is weakness in the test strategy that needs to be identified and eliminated. Catching memory defects with pseudorandom patterns indicates a deficiency in the remainder of the test strategy.

Since the memories are embedded, an embedded test is required. The memory inputs and outputs do not reach the chip boundaries and therefore a self test must be provided for the memories. This BIST must provide both stimulus and result comparison. To provide stimulus, address, data, and read/write control inputs must be provided to the memory. To perform output comparison the memory outputs must be either compressed or they must be compared with expected values calculated by the BIST. The key pieces of such a memory BIST are shown in Figure 8-18. These are not drawn to scale since the central memory dwarfs the size of the memory BIST that is testing it. The primary address components are an address counter and an address compare circuit. These allow for up and down counting in rippling either rows or columns most frequently. The data generation logic provides data inputs to impose blanket zeros, blanket ones, checkerboard, inverse checkerboard, and other data backgrounds as desired. The read/write control generation logic provides read enables, write enables, and any bit write type control capabilities required by the memory. For multi-port memories, CAM memories, and other complex memory types the address, data, and read/write control is more complex and the BIST needs to be correspondingly more complex. For the data out comparison shown in Figure 8-18 the data generation logic provides expected values which can be compared against the output values coming from the memory.

All of the components that comprise a memory BIST are based on the developed patterns and required memory test strategy. It is easy to have a mediocre BIST with poor defect coverage. Having a high quality BIST requires understanding the memories being tested, the technology they are being produced in, and the types of defects which are possible during fabrication.

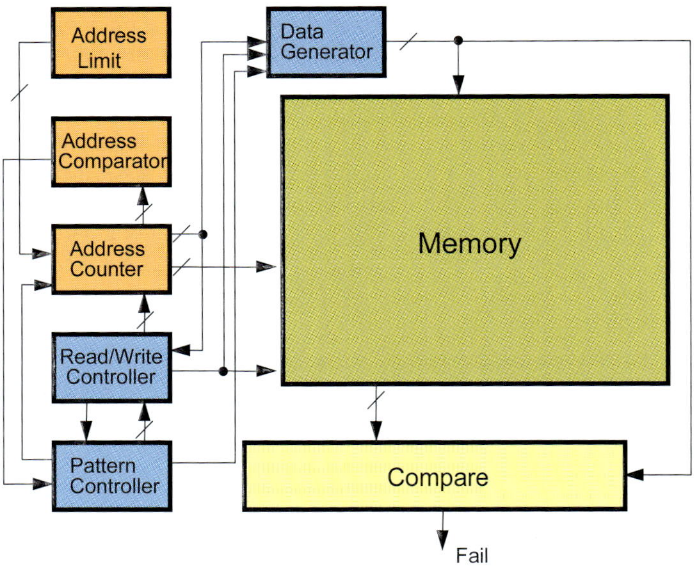

Figure 8-18: Memory built-in self-test components.

For medium to large sized memories, the redundant elements employed by BISR or BIRA are included in the memory. The redundancy calculation that determines which redundant elements to implement in place of which defective portions, is included in the BIST logic. Sometimes this is still just referred to as BIST but often when the redundancy calculation is utilized for *hard redundancy*, implemented by electronic or electrical fuses, the function is referred to as built-in redundancy-analysis (BIRA). When the redundancy calculation is automatically implemented during or at the conclusion of the BIST execution, the term most frequently used is built-in self-repair (BISR). When BISR is utilized the redundant elements are normally implemented using *soft redundancy*. Hard redundancy means that electrical or laser fuses (or anti-fuses) have been blown to permanently set the desired redundancy implementation. Hard redundancy can also be implemented in a type of flash memory technology. When soft redundancy is utilized the redundant elements are only implemented temporarily. At each power-up the BISR must be re-executed and the redundancy calculation performed. The soft calculated redundancy result is then stored as long as power is applied to the chip. A combination of soft and hard redundancy can be implemented in a composite fashion. In this case, at time of fabrication, a hard redundancy solution is implemented through a BIRA calculation to fix any known permanent manufacturing defects. After system installation, BISR is performed at each power up. This composite of hard and soft redundancy can allow reliability defects to be repaired via BISR on top of the defects detected at time of manufacture via BIRA.

The redundancy calculation needs to be performed and the intermediate result retained during BIST execution. The redundancy calculation logic works with the output compare logic to store information about failing elements as fails are detected. Therefore the redundancy calculation logic is typically placed in close proximity to the output compare logic. Redundancy calculation logic requires a fair amount of space due to the amount of information which needs to be retained.

Memories can have different types of redundancy. There can be spare rows, spare columns, spare I/Os, or spare blocks, representing four different dimensions of redundancy. The type of redundancy is optimally determined after examining defects from a specific technology. If it is determined that the predominant defect mechanism is going to cause column failures then column redundancy should be included, at a minimum. The relationship between the dimension affected most by defects and the dimension of the redundancy is obvious. Each dimension can have a single spare, such as one spare row, or multiple spare elements in that dimension. A single redundant element can be used for repair, such as a single column, or a group can be utilized such as a group of eight adjacent columns. When a group is replaced then a larger defect can be repaired. Granularity of the redundancy should be understood, however, since a group of redundant elements cannot be replaced arbitrarily. For example, a group of four redundant columns may be replaced starting at the zeroth, fourth, eighth, etc. column. If there is a short between the third and fourth columns, even though only two columns need repair, the group of four redundant columns cannot perform the repair due to the granularity. If there were two groups of four redundant columns then both groups of four would be required to perform the repair. One group would replace columns zero through three and the other would replace columns four through seven. This arrangement, even though eight redundant columns are utilized, would repair the short between the two columns number three and number four.

Multiple dimensions of redundancy can be utilized, such as the combination of both rows and columns. One example of a memory with this combination of redundancy is illustrated in Figure 8-19. A single cell fail can be replaced by either a spare row or a spare column. It is, however, very critical to optimally replace failing locations. A pair of vertically adjacent fails can be repaired by either a single column or by two rows. Implementing a non-optimal redundancy will result in some repairable chips not having a correct redundancy solution determined and the chip discarded. The BIST logic for calculating the redundancy must include a carefully selected algorithm for choosing the redundant elements, in order to provide the most optimal replacement and resulting in the highest yield.

There are redundancy calculation approaches which try all possible solution paths. There are also those which make assumptions about the most likely defect types and select a best all-around redundancy implementation algorithm. One method is to count all the fails along each row and all the fails along each column. Then a determination is made when a column must be fixed and when a row must be fixed. Following this, the remaining fails are allocated to the remaining available redundant elements. Another common technique is to detect when more than one bit is failing in a row. When a multi-bit failure is detected then a row is automatically allocated for redundancy repair. Most redundancy calculations take a long time and require a large amount of storage for the redundancy implementation result in addition to scratch space for storing intermediate calculations. There are a very small number of redundancy calculation techniques, however, which are ideally suited to BIST implementations and require a small amount of hardware overhead.

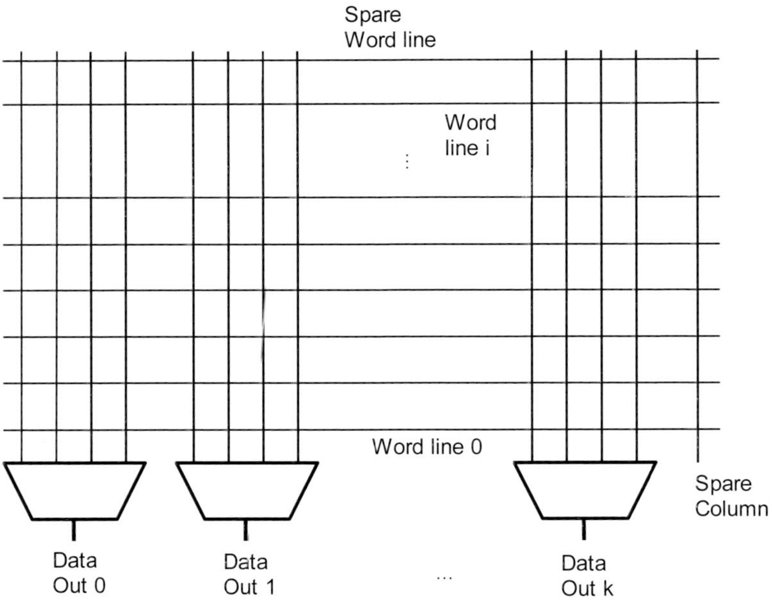

Figure 8-19: Example memory with two dimensional redundancy.

The BIST logic is typically designed in RTL using either VHDL or Verilog. This logic is then integrated into the chip design and synthesized with the remainder of the logic. As the number of memories grows according to Figure 8-1, the amount of memory BIST logic which must be designed and integrated becomes highly restrictive, without the proper electronic design automation (EDA) tooling. Even with EDA tooling integrating the BIST into the chip design in a manner that considers the actual physical placement and routing is very unusual. Figure 8-20 shows a proposed EDA flow which considers all of the relevant information to implement a memory BIST and the other DFT portions into a design considering the process, layout, and other relevant factors. Such an EDA methodology helps to implement the BIST in close proximity to the memory with minimal impact on the functional path around the memories. It should be noted that the EDA tooling needs to allow designing an advanced chip with a large number of memories. Just the ease of integration of these memories and memory BIST is not the goal but doing so with a very high quality memory test that understands the ways that real memories fail with real manufacturing defects.

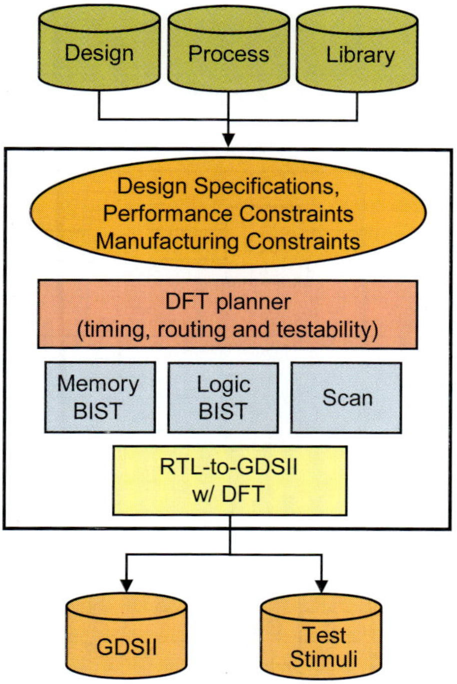

Figure 8-20: EDA design for test flow.

8.6 ADVANCED MEMORIES & TECHNOLOGIES

New memory designs are generated and memories are embedded onto chips fabricated in new technologies each year. Channels are driven to ever shorter lengths, increasing the sub-threshold leakages and reducing the I_{on} to I_{off} ratio. These changes drive new defect mechanisms and sensitivities which need to be tested to ensure high quality. Low K insulators allow metal lines to be packed more tightly together. The porous and soft nature of these insulators has produced a plethora of manufacturing problems. High K dielectrics reduce gate leakage currents but there is a challenge to maintain carrier mobilities at a high level. These new processes are only developed further through encountering fabrication problems and this requires better testing to ensure that parts which pass test are indeed good, especially when dimension scaling is involved. In memory manufacturing there are, however, more unusual technologies than those which are just generated through process scaling.

Silicon on insulator (SOI) technology provides performance advantage through the reduction in source and drain capacitance and by taking very little gate drive to turn on the FETs. SOI has the unique characteristic that FET operation is a function of the recent sequence of operations performed, the so-called *history effect* [32]. This history effect can cause any given transistor to have a very low threshold voltage and therefore a higher sub-threshold leakage current. Also characteristic to SOI is a floating body inside the FET. Since the body floats, a parasitic bipolar transistor is formed which may be turned on when, for example, the source of an NFET is pulled low.

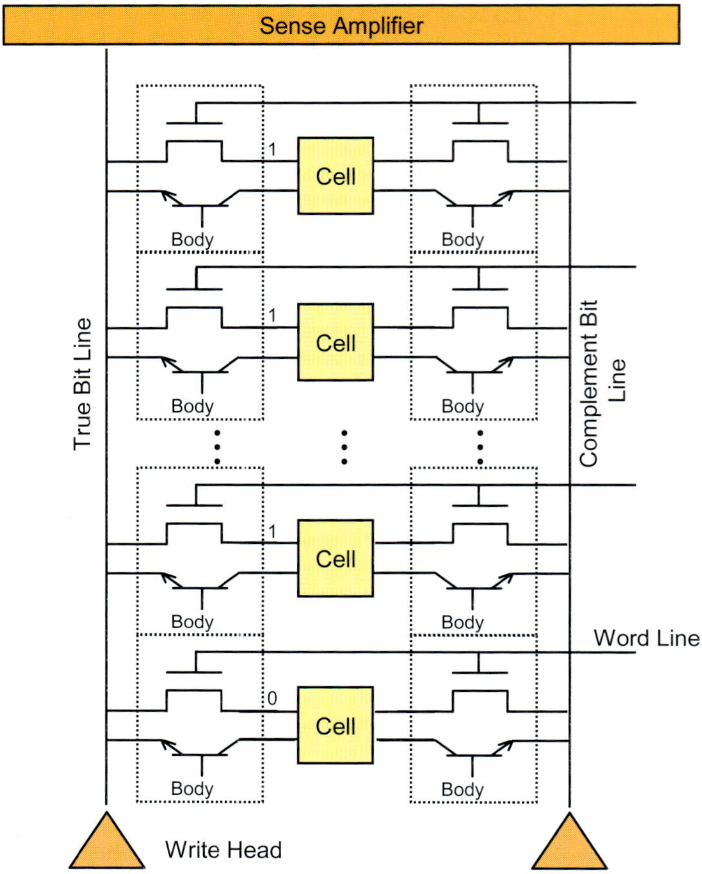

Figure 8-21: SOI memory column.

Figure 8-21 shows an SRAM column implemented in SOI technology with the two transfer devices shown for each cell (the two NFETs). Shown in parallel with the transfer FET is the parasitic bipolar transistor. Due to the bit lines being pre-charged to a high state it is possible to have all of the bodies of the FETs, attached to one of the bit lines, to float high, depending on the data type stored in the cells along the column. When the bit line needs to be pulled low, such as when writing the bottom cell to a "0" state, the bodies are only pulled low by turning on the parasitic bipolar device. This unusual current flow must be accounted for in the design and in the test. It is recommended that a combination of pauses, in concert with a walking pattern, be utilized for SOI memory testing [33].

Strained silicon is a new technology which stretches the crystal lattice to increase the mobility of the carriers. High quality silicon is especially critical with strained silicon due to concerns about dislocations. The increased mobility can generate a 15-20% drive current advantage which would yield desired significant performance improvement. Strained silicon can be utilized with bulk silicon or it can be implemented in a silicon on insulator technology. The unique aspects of strained silicon mean that careful testing need be implemented starting with a minimum of the test recommendations associated with SOI.

Three dimensional gates are also a new technology addition that provide for reduced I_{off} currents without the need for high K dielectrics. Normal FETs are planar structures but a three dimensional gate allows multiple surfaces around the channel to have gates. More mask levels and processing steps are required resulting in more opportunities for defects. In addition, with a multi-sided gate it is possible to have only one side of the gate fail. When this happens the drive capability through the FET is reduced and delay defects or reduced signal into a sense amplifier structure can be the result. Thorough and careful testing needs to be performed as this type of technology is approached.

Non-volatile memories have filled a critical niche in memory applications. Flash and EEPROM have proved to be very useful memories, albeit with restricted numbers of operations. These non-volatile memories have only had limited application in embedded memory environments and therefore not had as much focus for BIST solutions. New non-volatile memories including FeRAM[12], MRAM[13], and ovonic[14] memories provide solutions that retain data even when powered off but are not restricted in the number of operation cycles [34], [35], [36]. Manufacturing problems exist with each but these problems are being solved over time. One of these memories will likely replace all other forms of non-volatile memory within the next five years. The advantages of non-volatile memories are tremendous with the obvious benefit of data retention with the power off but also of reduced standby power and improved density. The non-volatile memories are poised to take over a much larger share of the entire memory market including the embedded space. If a low cost fabrication process is developed that is compatible with logic technology processes then only the small, high speed SRAMs and register files could remain beyond the reach of non-volatile memories. Certainly embedded DRAM would easily be replaced by non-volatile memories. That said, the defect types of these non-volatile memories need to be understood. The MRAM, with its free magnetic layer could easily be sensitive to disturb type defects, with the elevated currents required during write operations [37]. The FeRAM could have polarization defects within the ferroelectric capacitor, either making it hard to properly polarize or being difficult to maintain a stored polarization. The phase changes within the ovonic or OUM

[12] FeRAM stands for Ferroelectric Random Access Memory. Information is stored on material in the cell node which is electrically polarizable and retains its polarization once voltage is removed.

[13] MRAM stands for Magnetoresistive Random Access Memory. The cell state is stored in a material which changes resistance based on the magnetic orientation of the surrounding materials.

[14] Ovonic memories are known by several names including Ovonic Unified Memories (OUM), chalcogenide memories, and Phase-change Random Access Memories (PRAM). Data is stored based on the phase state of the material which is heated and then allowed to cool either relatively slowly or rapidly. The two phases have differing resistances allowing two states to be stored in the cell.

memories could be difficult to predict, given the presence of a defect. It is critical that the proper fault modeling be performed on whatever non-volatile memory ultimately dominates. This effort can then result in the proper pattern development and test strategy.

8.7 CONCLUSIONS

Semiconductor memories are vastly complex structures with very high density. Diffusions, polysilicon, metalization, and all fabricated structures that compose memories are in very tight proximity to one another. Memories also have analog functions with very precise timing. Given all of these factors, memories are far more susceptible to defects than any other portion of chip electronics. This susceptibility drives a greater need for very precise and accurate fault modeling. The fault models need to reflect the ways that real defects cause actual physical circuits to behave erroneously. These fault models must be far more comprehensive than are ever used in the testing of on-chip logic.

Once the proper set of fault models for a given memory design and fabrication technology have been determined, the appropriate test patterns and test strategy can be developed. Only when all of these factors work together can high quality testing and therefore high quality memories result. Given the amount of memory on the chips, the quality of the chips is governed by the quality of the embedded memories on them. The yield for these memories is determined by the suitable available redundant elements and the redundancy calculation that decides where to use the various redundant elements.

All of the testing and redundancy calculation must be performed by built-in self-test logic embedded on chip around the memory structures. The BIST must be implemented and integrated on the chip through the use of EDA tools which understand the memories, the process, and the physical constraints of the chip.

Future memory designs and technologies will require new fault models and better testing. Only through the most advanced BIST capabilities can the needed chip quality and yield be obtained.

REFERENCES

[1] International Technology Roadmap for Semiconductors, 2003 Edition, http://public.itrs.net/Files/2003ITRS/Home2003.htm

[2] R.D. Adams, *High Performance Memory Testing: Design Principles, Fault modeling, and Self-Test,* Kluwer, 2002, ISBN 1402072554.

[3] T. Barnett, A. Singh, "Relating yield models to burn-in fall-out in time," IEEE International Test Conference 2003, pp. 77-84.

[4] S. Subramanian, et al., "A practical experience of implementing memory repair in COT," IEEE VLSI Test Symposium 2004, presentation.

[5] M. Kume, et al., "Programmable at-speed array and functional BIST for embedded DRAM LSI," IEEE International Test Conference 2004, pp. 988-96.

[6] R.D. Adams, et al., "An integrated memory self test and EDA solution," IEEE Memory, Technology Design, and Test Workshop 2004, pp. 92-5.

[7] M. Ouellette, et al., "On-chip repair and an ATE independent fusing methodology," IEEE International Test Conference 2002, pp. 178-86.

[8] M. Mohammad, K.K. Saluja, "Flash memory disturbances: modeling and test," IEEE VLSI Test Symposium 2001, pp. 218-24.

[9] K. Bernstein, N. Rohrer, *SOI Circuit Design Concepts*, Kluwer, 2000, ISBN 0792377621.

[10] K. Osada, "Universal-V_{DD} 0.65-2.0 V 32-kB cache using a voltage-adapted timing-generation scheme and a lithographically symmetrical cell," IEEE Journal of Solid-State Circuits Vol. 36, No. 11, 11/2001, pp. 1738-43.

[11] H. Pilo, R.D. Adams, et al., "Bitline Contacts in High-Density SRAMS: design for testability and stressability," IEEE International Test Conference 2001, pp. 776-82.

[12] H. Miyatake, M. Tanaka, Y. Mori, "A design for high-speed low-power CMOS fully parallel content-addressable memory macros," IEEE Journal of Solid-State Circuits, Vol. 36, No. 6, 6/2001, pp. 956-68.

[13] H. Kimura, et al., "Complementary ferroelectric-capacitor logic for low-power logic-in-memory VLSI," IEEE Journal of Solid-State Circuits Vol. 39, No. 6, 6/2004, pp. 919-26.

[14] V. Lines, et al., "66MHz 2.3M ternary dynamic content addressable memory," IEEE Memory Technology, Design, and Test Workshop 2000, pp. 101-5.

[15] L. Dilillo, et al., "Data retention fault in SRAM memories: Analysis and detection procedures," IEEE VLSI Test Symposium 2005, pp. 218-24.

[16] P. Pavan, L. Larcher, A. Marmiroli, *Floating Gate Devices: Operation and Compact Modeling*, Kluwer, 2003, ISBN 1402077319.

[17] Z. Al-Ars, et al., "Simulation based analysis of temperature effect on the faulty behavior of embedded DRAMs," IEEE International Test Conference 2002, pp. 783-92.

[18] R.D. Adams, E.S. Cooley, "Analysis of a deceptive destructive read memory fault model and recommended testing," IEEE North Atlantic Test Workshop 1996, pp. 27-32.

[19] J. Brauch, J. Fleischman, "Design of cache test hardware on the HP PA8500," IEEE International Test Conference 1997, pp. 286-93.

[20] S. Hamdioui, *Testing Static Random Access Memories*, Kluwer, 2004, ISBN 1402077521.

[21] J. Zhao, et al., "Detection of inter-port faults in multi-port static RAMs," IEEE VLSI Test Symposium 2000, pp. 297-302.

[22] S. Hamdioui, "Testing multi-port memories: Theory and practice," Ph.D. Dissertation, Delft University, 2001.

[23] R. Gibbins, R.D. Adams, et al., "Design and test of a 9-port SRAM for a 100Gb/s STS-1 switch," IEEE Memory Technology, Design, and Test Workshop 2002.

[24] A.J. van de Goor, *Testing Semiconductor Memories: Theory and Practice*, ComTex Publishing, Gouda, The Netherlands, 1998, ISBN 9080427616.

[25] A.J. van de Goor, I.B.S. Tlili, "March tests for word-oriented memories," IEEE Design, Automation and Test in Europe Conference 1998, pp. 501-8.

[26] A.J. van de Goor, S. Hamdioui, R. Wadsworth, "Detecting faults in the peripheral circuits and an evalution of SRAM tests," IEEE International Test Conference 2004, pp. 114-23.

[27] A.J. van de Goor, et al., "March LR: A test for realistic linked faults," IEEE VLSI Test Symposium 1996, pp. 272-80.

[28] A.J. van de Goor, "Using march tests to test SRAMs," IEEE Design & Test of Computers, 3/1993, pp. 8-13.

[29] K. Lin, C. Wu, "Functional testing of content-addressable memories," IEEE Memory Technology, Design, and Test Workshop 1998, pp. 70-5.

[30] P.H. Bardell, W.H. McAnney, J. Savir, *Built-In Test for VLSI*, Wiley, 1987, ISBN 0471624632.

[31] C.E.Stroud, *A Designer's Guide to Built-In Self-Test*, Kluwer Academic Publishers, 2002, ISBN 1402070500.

[32] E. MacDonald, N.A. Touba, "Delay testing of SOI circuits: Challenges with the history effect," IEEE International Test Conference 1999, pp. 269-75.

[33] R.D. Adams, P. Shephard III, "Silicon on insulator technology impacts on SRAM testing," IEEE VLSI Test Symposium 2000, pp. 43-47.

[34] G. Braun, et al., "A robust 8F2 ferroelectric RAM cell with depletion device (DeFeRAM)," IEEE Journal of Solid-State Circuits, Vol. 35, No. 5, 5/2000, pp. 691-6.

[35] S. Tehrani, et al., "Progress and outlook for MRAM technology," IEEE Transactions on Magnetics, Vol. 35, No. 5, 9/1999, pp. 2814-9.

[36] M. Gill, T. Lowrey, J. Park, "Ovonic unified memory – a high-performance nonvolatile memory technology for stand-alone memory and embedded applications," IEEE International Solid State Circuits Conference 2002, pp. 202-3.

[37] C.L. Su, R.F. Huang, C.W. Wu, "MRAM defect analysis and fault modeling," IEEE International Test Conference 2004, pp. 124-133.

Chapter 9

Mixed-Signal Testing and DfT

Stephen Sunter

There is a long-running debate: some engineers say that all circuits are digital but some circuits can't make up their minds; others say that all circuits are analog but some are more non-linear than others. The irony is that analog designers maximize performance of circuits by making them more digital (e.g., sigma-delta converters), and digital designers maximize performance of circuits by making them more analog (e.g., differential logic gates). Thus, knowledge of both analog and digital test techniques is useful for *all* circuit testing.

The key difference between digital and mixed-signal testing is the measurement of continuous variables. Although the variables can include analog parameters of digital circuits and mixed-signal boards or systems, this chapter focuses on testing analog and mixed analog/digital integrated circuits (ICs). The parameter values of interest have the following characteristics:

- they are non-deterministic and can only be described statistically, for example, average value, standard deviation, and upper and lower test limits;
- their value is affected by many conditions, for example, power supply voltage, temperature, and manufacturing process variations;

D. Gizopoulos (ed.), Advances in Electronic Testing: Challenges and Methodologies, 301-336.

- they may be considered in multiple domains, for example, time or frequency domain.

The practical implications of these characteristics are that circuit simulation is much more computation-intensive than it is for purely digital circuits; fault models are more nebulous, and test time is much longer.

According to various sources, about 60% of all new IC designs in 2004 included mixed-signal circuitry, and the percentage is growing. Ever-shrinking IC process technologies permit more of a system to be incorporated into a single IC, and since most systems must interact with the "real" world, which is analog, system-level ICs must include analog circuitry. A growing number of analog intellectual property (IP) providers have been anticipating this growth and provide a variety of mixed-signal circuit blocks to permit digital design teams to add a mixed-signal circuit block to their IC without having to design the block. The most commonly used mixed-signal functions are the Phase-Locked Loop (PLL), Analog-to-Digital converter (ADC or A/D), Digital-to-Analog converter (DAC or D/A), comparator, operational amplifier (op-amp), and filter. Unfortunately, most mixed-signal IP blocks do not include an accompanying test solution.

Other chapters in this book discuss industry's tidal shift to structural testing of digital circuits as a way to facilitate automated DfT, test generation, and (more recently) fault diagnosis. The functional versus structural testing debate is still on-going for mixed-signal (and gigahertz digital) circuits. At one extreme, measuring individual resistor and capacitor values of a filter is not only impractically complex to do, it simply does not prove that the filter will meet functional specifications because of the many interdependencies within the circuit. At the other extreme, measuring every functional specification of a filter requires an uneconomical amount of time. In practice, test engineers usually characterize representative sample devices of a new design and determine which measurements are highly correlated, and therefore redundant, and which are not. Based on experience, knowledge of the particular design, available time, and tester capabilities, a test engineer chooses a combination of purely functional tests, structure-based functional tests, function-based structural tests, and purely structural tests. This chapter will explore these methodologies, as they affect testing, DfT, and fault modeling.

9.1 A BRIEF HISTORY

There are many ways to implement most circuit functions. The same test approach is rarely applicable to all implementations of a function. Function or specification-based tests apply signals implied by the circuit's data sheet specifications, and are based on the intended application of the circuit. Structural tests may apply signals that are not typical of the signals seen in the application, and are focused on verifying that the circuit's structure has not been altered by manufacturing defects. The range of functional and structural tests for mixed-signal circuits is far wider than for digital circuits, and the debate about which is the best methodology is a long one.

9.1.1 Functional vs. Structural Test

First, consider some common examples of function-based structural tests for the ADCs whose simplified schematics are shown in Figure 9-1. A sigma-delta ADC uses a large number of logic gates in its digital filter, so the logic gates can be first tested with a scan-based test, in a few milliseconds, before performing conventional sine wave-based testing, which often requires 500 ms or more. A pipelined ADC or Successive Approximation Register (SAR) ADC can benefit from an analog test that only measures voltages for major bit transitions. A flash ADC, that uses a large number of resistors in series, requires testing of every bit-transition using a linear ramp and histogram-based analysis.

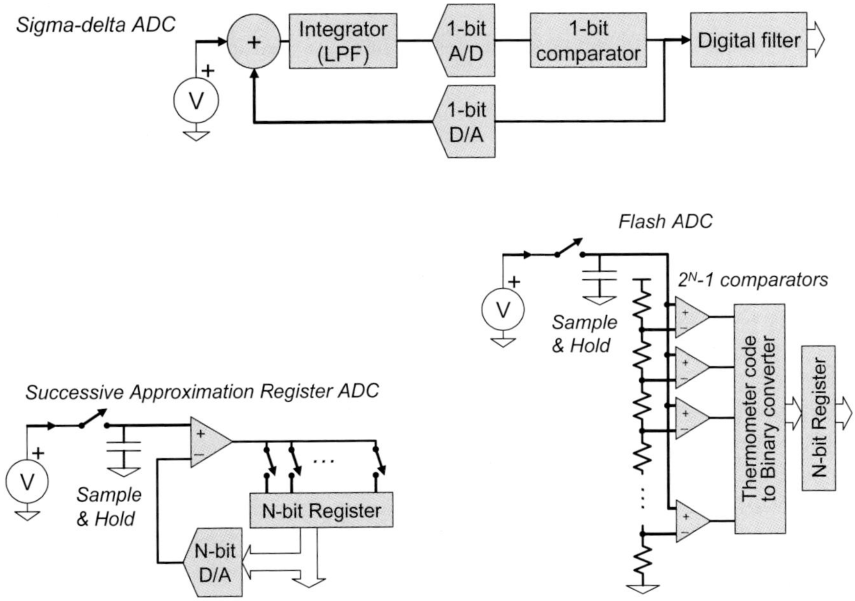

Figure 9-1: Different types of ADCs.

Structure-based function tests for an ADC include measuring its frequency response while applying a multi-tone sine wave – its response to *all* in-band frequencies is not measured but, considering the structure, it is almost certainly fault-free if the response is acceptable for some well-chosen discrete tone frequencies. In fact, all test programs rely upon correlation between untested parameters and tested parameters – even for a simple filter it is impractical (and unnecessary) to measure frequency response at all frequencies for all test conditions. Another example of where ADC structure affects its functional test is self-calibration circuitry – it is often essential to alter this structure during test to prove the calibration circuitry is able to compensate for all values within the supply voltage and temperature range, and to prove that the uncalibrated ADC is not excessively inaccurate.

Identifying under-tested portions of a circuit is the primary goal of fault modeling and fault simulation. Parametric fault modeling is the subject of many research papers, but is not widely used in industry due to its computation-intensive nature and the difficulty of corroborating the analysis with manufacturing results.

Although many academic studies have concluded that mixed-signal circuits should be structurally tested, as concluded for digital circuits, industrial test engineers have been justifiably skeptical. One insightful paper [1] described how both approaches were used to test approximately two thousand devices. The study concluded, in effect, that a defect-oriented structural test is an economic and acceptable alternative for circuit parameters that have lots of margin relative to their test limits, but is not acceptable when there is little margin because too many defective devices pass the test. This has proven to be generally true.

9.1.2 Testing

The development of mixed-signal IC test techniques was driven initially mostly by telecommunications. It was the first application where the complexity of digital and analog circuitry on a single IC could be justified because of the need for high volume production and high reliability, at reasonably low frequencies. Most of the early mixed-signal testing publications from the mid-to-late 1970's discuss telephony: voice-band frequencies, mu-law/A-law codecs, etc.

In the 1980's, the need for more repeatable, accurate, and faster analog testing became paramount and this led to the development of the Digital Signal Processor (DSP)-based mixed-signal tester[1] [2]. Companies such as LTX and Teradyne produced large and infamously expensive testers that were cost effective because they were so fast and repeatably accurate. Whereas a spectrum analyzer can sweep the 4 kHz audio spectrum in a few seconds, a DSP-based tester could perform an equivalent test in one tenth of the time – a multi-tone sine wave stimulus consisting of four or five discrete frequencies replaced a linear sweep of all in-band frequencies. Nevertheless, the number of DSP-based tests needed still meant total test times of 5 to 15 seconds, whereas tests of digital circuitry required only a second or two because there were less than a few thousand logic gates on each IC.

As technology permits more and more (now hundreds of millions) transistors on an IC, the test time for the logic gates has been reduced by higher clock rates and the greater parallelism permitted by scan-based tests, but the analog test times per function have not reduced comparably. Many publications report, consistent with anecdotal evidence, that test times for analog circuitry on a mixed-signal IC are the majority of the total test time, even though the vast majority of the IC is digital. Reasons for this include the following:

- the stimulus rate is limited by the analog circuit-under-test (CUT) bandwidth, not by the tester;
- analog signal levels are decreasing due to decreasing supply voltages, but thermal noise is not, therefore, more measurement averaging is needed;

[1] A DSP-based tester captures analog signals with ADCs, performs analyses with digital, matrix calculations (e.g., Fourier transforms), and generates analog signals with DACs.

- analog bandwidths are increasing, which increases noise levels because thermal noise power is proportional to bandwidth, therefore, more measurement averaging is needed;
- access to core circuits is limited by the surrounding circuitry, which complicates parallel testing of circuit blocks;
- the number of high quality analog tester channels is usually four or less.

Some of these limitations can be overcome by Design-for-Test (DfT).

9.1.3 Design-for-Test

Design-for-Test (DfT) has been practiced by IC designers for longer than they realize. Since the early days of IC design, designers have placed small probe pads on various circuit nodes and provided special function modes to permit diagnostic access. A designer once cynically remarked, "If I provide diagnostic access to a signal, it will work perfectly – the only circuit nodes that will fail will be those without diagnostic access." These access techniques were often not documented for the test engineers because that would lead to additional specifications and design constraints, and the access was not robust enough for production testing. Examples of these ad hoc design-for-diagnosability techniques, which have evolved into bona fide DfT techniques, include *analog test buses*, *loop-around*, and *dedicated digital scan paths*, illustrated in Figure 9-2. An *analog test bus* would typically be an interconnect wire to which a dozen or more selected signals were connected by CMOS transmission gates. The bus would be connected to a bond pad directly or via an existing analog buffer that provided a low-priority function. *Loop-around* refers to, for example, connecting the output of a DAC to an ADC on the same IC to permit a digital input stimulus and a digital output response. *Dedicated scan paths* are relatively short scan paths that only provide data to, for example, a single DAC, so that the input word to the N-bit DAC can be updated every N clock cycles, thus providing a reasonably fast analog output rate – variations on this technique continue to be published.

The first paper [3] to describe the use of systematic mixed-signal DfT proposed using conventional digital scan to control the transmission gates connected to an analog bus wire, and discussed the benefits of monitoring or injecting analog signals at various points along an analog signal's path. In fact, this was more or less what many designers were already doing in an ad hoc way, but the impact of the bus capacitance on the monitored signal was a key problem[2].

[2] Test bus capacitance is usually comparable to or larger than an accessed node's capacitance, therefore, circuit performance may change significantly when the node is accessed via the bus.

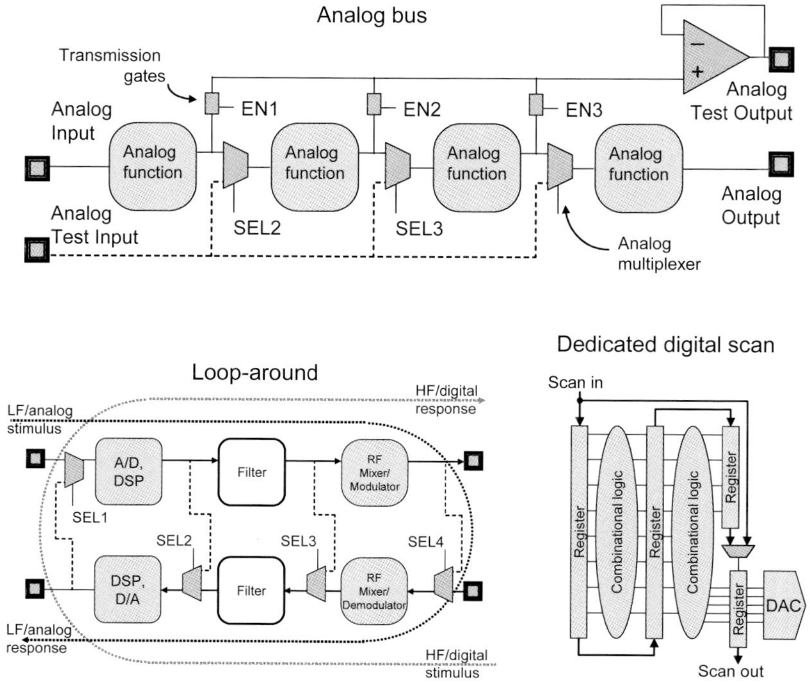

Figure 9-2: Different mixed-signal DfT techniques.

Not many mixed-signal DfT papers were published until the advent of mixed-signal Built-In Self-Test (BIST) in 1995. The first paper [4] to describe a BIST for mixed-signal functions proposed connecting a DAC output to an ADC input (assuming the IC had both) and stimulating the DAC with a pseudo-random pattern generator (PRPG), as done in digital BIST. The output of the ADC was connected to a multiple-input signature register (MISR) to generate a digital signature. In the ideal case of noise-free signals, exactly equal signal ranges, and no distortion, this arrangement might work. However, a single bit error would cause a faulty signature because there was no averaging. The paper suggested that the least significant bit (LSB) should not be included in the test to reduce the impact of noise, but that is exactly where testing is most needed because the LSBs are the bits most sensitive to manufacturing process variations.

Many researchers have proposed using a PRPG as the stimulus for mixed-signal BIST. Such a stimulus has very attractive properties: it has been used very successfully for digital circuit testing, the generation circuit uses very little IC area, it is easy to design, and it accurately generates a wide, uniformly-distributed frequency spectrum. Cross-correlating the output of a linear CUT with its random input signal produces its impulse response [5]. The difficulty has proven to be in the output response analysis: reliably separating a good from a bad response, especially for noise-related specifications. This method also does not allow measurement of distortion because output distortion frequencies overlap the signal frequencies. Using a digitally-generated impulse as the stimulus does not help: an amplitude-limited impulse typically has too little energy to produce a good signal-to-noise ratio at the CUT output, the response analysis is still complex, and distortion still cannot be measured.

Mixed-signal BIST papers greatly increased in number and novelty in the late 1990's [5], [6], [7], [8], and they implied various definitions of BIST. Some BIST proposals, for example [8], required a low distortion sine wave from the tester, and others, for example [5] and [7], required the tester to perform computations before deciding pass or fail. Others relied upon the existence of a general compute engine or DSP on-chip. Some BIST circuits produced an output frequency or voltage "signature", or required pre-testing of the BIST's analog circuitry. An output frequency or voltage could be converted into a pass/fail bit by using a digital frequency counter and a digital magnitude comparator, or a voltage comparator and a DAC (perhaps constructed on-chip from a digital pulse density modulator and a low pass filter), however, these on-chip solutions have been rarely employed until recently, probably because a mixed-signal tester was used anyway and it could easily perform the comparison.

For a circuit to truly have BIST, the off-chip stimulus should comprise only an external clock, a power source, and a digital instruction (which might include parameter values) to start the test; the response should be a digital signature in which each bit must be correct. All mixed-signal testers can capture a multi-bit serial digital value, interpret it as a binary-coded number, and perform computations with the number before deciding pass or fail. But many digital testers have no capture memory; the tester decides pass or fail on the basis of each bit in isolation. For this reason, for mixed-signal BIST to permit testing a circuit on a digital tester instead of a (25% to 50% more expensive) mixed-signal tester, the BIST circuitry must generate a digital signature that permits a bit-by-bit comparison to the fault-free signature.

The techniques that relied upon a mixed-signal tester revealed the reason for their lack of adoption: if the tester must provide some analog stimulus or computations then it must be a mixed-signal tester, and if the test time and accuracy are similar to the non-BIST approach, then there is likely no real economic gain. Accelerating the test time for one of the tests but not eliminating all of the analog tests meant that a mixed-signal tester was still required. Claims of increased fault coverage were (and still are) hard to prove due to the lack of a widely accepted or corroborated analog fault model. Furthermore, the lack of a fault model that simulates quickly has hindered the development of any widely-used, systematic mixed-signal DfT method.

9.1.4 Fault Modeling

The availability of a simulation-efficient fault model (the stuck-at fault) that is widely accepted as being representative of typical IC faults is arguably the single greatest contributor to the development of digital DfT and BIST in the 1970's and 1980's. The model permitted clear demonstration that automatically generated, pseudo-random patterns, applied via a scan path, achieved much higher fault coverage than a painstakingly hand-written functional test pattern. Analog and mixed-signal DfT and test generation have not had this advantage.

The digital stuck-at fault is equivalent to a logic gate input or output shorted to ground or the power supply. In analog circuits, most faults do not cause the circuit output to become equal to a power rail voltage, and so most researchers judge the quality of their proposed DfT or test by measuring its coverage of all possible short and open circuits (not just those to ground or power).

Clearly, shorts and opens do not adequately address the spectrum of analog faults, however, simulating even this limited set has proven computation-intensive. Typically, simulation times for an analog circuit can range from several seconds to several hours (PLLs and sigma-delta modulators can take much longer). Each short or open circuit to be simulated requires an additional simulation, and a simple differential amplifier, like the ten node circuit in Figure 9-3, has a few dozens possible shorts and opens.

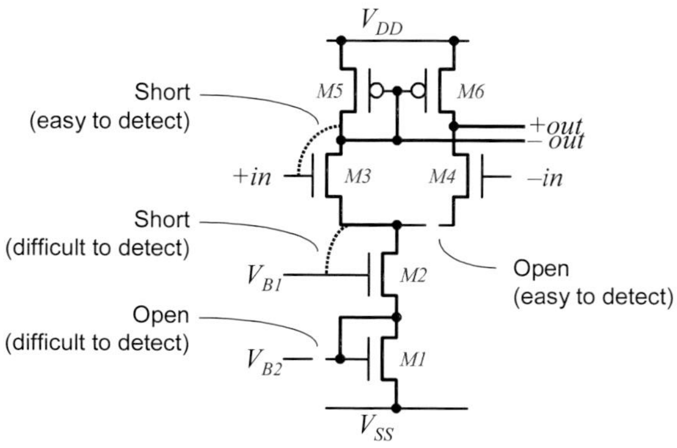

Figure 9-3: Analog circuit with example short and open faults.

Open circuits on the gates of CMOS transistors have proven very difficult to simulate for analog circuits: a true open circuit leaves the transistor gate floating, but whether it floats to an 'on', 'off', or partially-on voltage greatly affects whether the fault can be detected. A few papers [9] have analyzed real circuits to ascertain which model is best, and concluded that *all three* can occur and should be included in fault simulation. Similar analysis has found that "open" circuits can have impedances ranging from a few hundred kilohms to a few hundred megohms – typically 1 megohm is used in fault simulation (higher values are more difficult to detect).

Short circuits are much simpler to simulate and, based on analysis of real circuits, their resistance can range from milliohms to tens of ohms – typically 1 ohm is used in analog fault simulation. The key question has been how many short circuits are possible for a given circuit. For N nodes, in theory, there are N(N-1)/2 possible two-node shorts. To determine the number of realistically possible shorts, the circuit's layout can be analyzed so that only likely short circuits are included. For example, if nodes A and B can only be short circuited by a pathway that passes through a node C, then the AB path does not need to be simulated – it will already be accounted for by simulations for short circuits AC and CB.

Parametric faults have proven more difficult to define beyond saying that they are excess variation in the value of a parameter. The parameter could be a specified performance for which faulty behavior is caused by an element's variation or a spot defect, or the parameter could be an element's value.

One common fault model [10] defines a parametric fault to be any single circuit element value (or element matching) that exceeds a specified process maximum variation, for example, ±10%. The typical elements considered are resistances, capacitances, and transistor aspect ratios[3], and, in some cases, element matching. This model assumes that a circuit is designed to tolerate precisely this amount of variation, and that the circuit will fail some functional specification if any element's value varies beyond this range.

Another fault model [11] defines a parametric fault as any element (or element matching) deviation that causes a circuit performance parameter to vary beyond test limits. This model accounts for different tolerances for each circuit element, as well as multiple defects. For example, a design may require most transistors to vary by less than the process nominal range but tolerate some transistors to vary by significantly more.

A third fault model [12] defines analog faults as element deviations that cause a design parameter (that is not a specified performance) to vary beyond test limits, for example, *closed-loop damping*[4], *Q*[5], or *open-loop gain*[6]. This approach sometimes simplifies the testing, and relies upon the inherent theoretical relationship between the specified performances and the design parameters.

Analog fault coverage, as opposed to modeling, is not so well-defined. The purpose of calculating fault coverage is to estimate the number of defective devices that will pass a test. For a set of equally probable faults, and a test that detects a proportion F of the possible faults, the proportion D of defective ICs in the population of ICs that pass the test is estimated by the equation [13]:

$$D = 1 - Y^{(1-F)}$$

where Y is the yield (proportion of defect-free ICs of all ICs tested).

Although a few papers [14], [15] have provided interesting evidence that weighting faults according to their likelihood of occurrence will more accurately predict actual defect levels, the difficulty of assessing probabilities for every fault in any large digital IC has discouraged this refinement. One way to weight digital faults is to base the weight on the extracted interconnect capacitance. For (small) analog

[3] The aspect ratio of a transistor is its channel width divided by channel length.

[4] The damping ratio ζ of a system (e.g., PLL) is a generic indicator of its response to transients: $\zeta<1$ (under-damped) indicates there will be oscillatory overshoot; $\zeta>1$ (over-damped) indicates an unnecessarily slow response. $\zeta=1$ is denoted critically-damped. Percent overshoot for a second-order system equals $100\, e^{-\zeta\pi/\sqrt{1-\zeta^2}}$.

[5] The Q of a resonant circuit (e.g., filter) is a generic indicator of its quality and is equal to its centre frequency divided by its –3dB bandwidth.

[6] The open loop gain of a circuit having feedback (e.g., amplifier) is the total gain in the forward and feedback paths, without the gain-reducing effects of feedback.

circuits, weighting is similarly difficult for the shorts and opens, but even more complex for parametric variation faults [16]. One way to obtain weights is to simulate applying many spot defects randomly to a circuit's layout and detect how many times each wiring net is affected. In either case, digital or analog, no actual silicon validation of fault weighting has been reported.

Software for analog fault modeling has been developed primarily by universities. DRAFTS (discretized analog circuit fault simulator), developed by the University of Texas, was reported by a few papers [17] from that university. AnaFAULT, based on the simulator Eldo, was developed in Germany and reported in a few papers [18]. Commercially developed ANALOG fault simulators are mentioned briefly in Section 9.5.3.

Analog fault modeling has certainly evolved in the last 15 years, but it has not reached the level of use that digital fault modeling has, for various reasons. It is reasonable to expect that a digital stuck-at-one fault must be detected otherwise the circuit will almost certainly fail some functional requirement. However, it is not clear that a capacitor value 3% larger than the process maximum will cause a circuit's performance to be inadequate, because other element values might have compensating values, or the defect might not cause any performance specification to fail. It is also unclear whether all defects that don't cause a specification failure really matter. Many test engineers don't trust analog fault models enough to fail a circuit that passes all functional tests but fails a fault model-based test. Lastly, almost any analog test set that includes the typical tests for gain, noise, distortion, and bandwidth will detect all shorts and opens, thus, effort spent on fault modeling often appears to be unnecessary.

9.2 THE STATE OF THE ART

Present day drivers of development of mixed-signal ICs are computer video, wireless telecommunications, and home entertainment. The related test challenges are multi-megahertz 8~10-bit DACs for color video, Digital Versatile Disc (DVD) interfaces, and 44 kHz, 16~24 bit ADCs for high fidelity audio. The advent of software radio that performs radio frequency (RF) to baseband conversion directly via an ADC offers a new test challenge for high-volume applications: embedded converters that have both high resolution (>10 bits) and high sampling rates (>50 MHz). The performance of high-speed converters is often limited more by sampling jitter than voltage noise. The testing of PLL jitter is a related test challenge, shared by digital and mixed-signal test engineers. PLLs are often regarded as digital circuit blocks because their inputs and outputs are digital, but their function and usually their internal circuitry are clearly mixed-signal in nature. The testing of op-amps and switched capacitor filters has been the subject of extensive study. Generally, the huge variety of miscellaneous or "random" analog functions has been largely ignored by researchers, mostly due to the diversity of such circuits and the unlikelihood of finding a general solution.

9.2.1 Testing

Many mixed-signal DfT and BIST papers justify the need for more on-chip test circuitry by citing the high cost of mixed-signal testers. This is consistent with often-expressed corporate interest in any DfT that permits an IC to be tested on a digital tester instead of a mixed signal tester. It is worth briefly reviewing why a mixed-signal tester costs more than a digital tester.

At the very least, a mixed-signal tester requires a digitizer comprising an ADC and memory, an Arbitrary Waveform Generator (AWG) comprising a DAC and memory, a digital signal processor (DSP) for computing Fourier transforms, and clock generation that has very fine frequency resolution and relatively low jitter to permit coherent, low noise sampling. The digitizer and AWG require a variety of filtering, gains, differential/single-ended conversion, force/sense connections, and calibration facilities. A robustly grounded and electro-magnetically isolated device interface board is needed to minimize the contamination of low level analog signals by digital signals. Higher resolution and sample rates require greater electrical isolation. Last but not least, specialized software is needed to link and control all these facilities without making the task of test programming too complex.

Many digital testers are only able to compare a single bit to the expected value so that it can instantly decide pass or fail (though recently, so called DfT or structural testers have come equipped with many megabits of capture memory per pin). All mixed-signal testers require enough capture memory depth to be able to interpret the bits as sequences of serial binary-coded numbers which must be further processed before deciding pass or fail.

The additional cost for mixed-signal capability certainly depends on the number of high performance analog channels that can be tested simultaneously, and on the resolution and sample rate of the analog channels. Some large mixed-signal testers have only 2 or 4 analog channels, however, some "low cost" testers have medium performance analog capability on every channel. A small number of analog channels can limit the extent to which parallel or multi-site testing can be used.

Some test engineers have noted that the most expensive part of a mixed-signal tester is all the digital channels, but, in any case, the combination of digital and analog is more expensive.

We now elaborate on various test methods used in mixed-signal testing.

Force/sense and *force/measure* (also known as *Kelvin probe* or *four-probe* testing) are important and powerful techniques used to ensure that unknown resistance in a signal delivery path does not affect analog measurement accuracy, and is illustrated in Figure 9-4. The techniques are employed by most high resolution and high current, AC and DC signal delivery systems, such as power supplies, low frequency AWGs, and parametric (or precision) measurement units (PMUs).

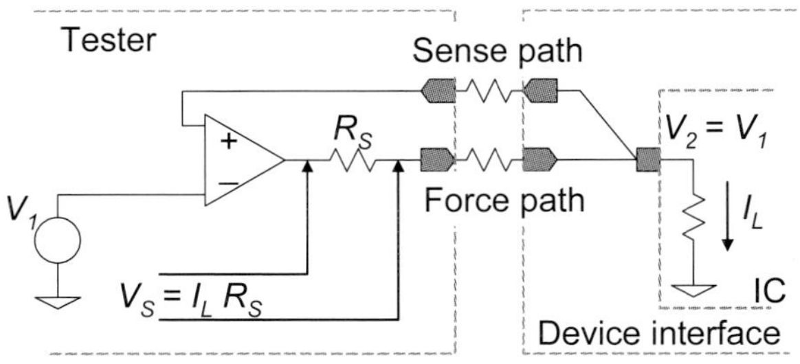

Figure 9-4: A circuit that uses force/sense and force/measure.

A driving voltage or current I_L is delivered to an IC via a low-impedance path and returned via a similar path. Any resistance in the drive path will cause the voltage delivered to the CUT to differ from the driver's voltage. The voltage V_2 that is actually delivered to the CUT is sensed or measured via a high-impedance path. The delivered current is measured by sensing the voltage V_S across a precision low-value resistance R_S in the driving path. The driving voltage adjusts to maintain the sensed voltage equal to the intended voltage (typically by the negative feedback loop of an op-amp) and/or to maintain the sensed current equal to the intended current. The technique is not suitable for megahertz testing because the delay around the feedback loop (due to resistance and capacitance) causes instability.

Thus, there are three useful modes: deliver a specified voltage; deliver a specified current and sense the delivered voltage; deliver a specified voltage and sense the delivered current. Different testers support this capability to different extents. All three modes are needed for analog testing, and four separate paths are needed in the device interface board to support these modes: force-high, sense-high, force-low, sense-low. "High" indicates the driving positive current path into the CUT; "low" indicates the return path. For mixed-signal testers, these paths usually extend into the interface board all the way to the CUT pins (or even further within the IC) so that any resistances in the path do not affect measurement accuracy. The current sensing resistor is typically within the current driver.

In wafer testing of ICs, the resistance of a probe-to-bond pad contact is typically a half ohm but may increase to a few ohms as a probe tests more ICs due to accumulated impurities on the probes (hence the need for periodic cleaning of probes). This resistance can be kept significantly lower for socket-to-package contacts due to the larger contact area and increased contact wiping (scraping) during connection.

The design of a device interface board is typically a major part of designing a test for an IC. The board typically contains a lot of analog circuitry, including for example, power supply and bias voltage decoupling capacitors, output load resistors, op-amps, discrete transistors, and feedback capacitors. Grounding is a crucial consideration for both high resolution analog circuits of any frequency and equally important for high frequency digital circuits [19]. Grounding includes how signals are referenced to a common node and how signals are electrically isolated from one another, and is the subject of whole books.

The first general reference book for mixed-signal testing which particularly focuses on DSP-Based Testing is [20]. It addresses coherent sampling and analysis for ADC and DAC testing, especially for telecommunications. The text, adapted from a tutorial, has been republished several times due to its continued relevance.

The current state of mixed-signal test engineering is best described in the textbook [21]. It includes chapters on DC parametric measurements, measurement accuracy, tester hardware, sampling theory and DSP-based testing, analog and sampled channel testing, focused calibrations, ADC and DAC testing, device interface board design, DfT, data analysis, and test economics. The well-written book covers almost all of the material in [20] plus much more, but is nevertheless only an introduction to this broad subject, especially for DfT and BIST.

A large part of mixed-signal testing relates to sampling theory. Sampling is most obviously performed by an ADC – periodic capturing of the instantaneous signal voltage – but is also central to DAC and switched capacitor filter operation.

Coherent sampling, as illustrated in Figure 9-5 refers to sampling at a rate that ensures that an integer number of cycles of the signal-under-test is captured, and ensures that every sample has a unique phase of the signal's period to ensure maximum information is captured. Coherent sampling is essential to achieving maximum signal-to-noise ratio in a minimum test time. If coherent sampling is not used, then windowing is typically required. Windowing is a mathematical operation that diminishes the relative weight of samples at the beginning and end of the sampling interval to minimize the impact of the discontinuity at those points. For coherent sampling, the sample rate is prescribed by the equation:

$f_T/f_S = M/N$, or $N/f_S = M/f_T$,

where f_S is the sample rate, f_T is the test sinewave frequency, N is the number of samples collected, M is the number of cycles of the sinewave sampled, and N and M are relatively prime (they have no common factors other than 1).

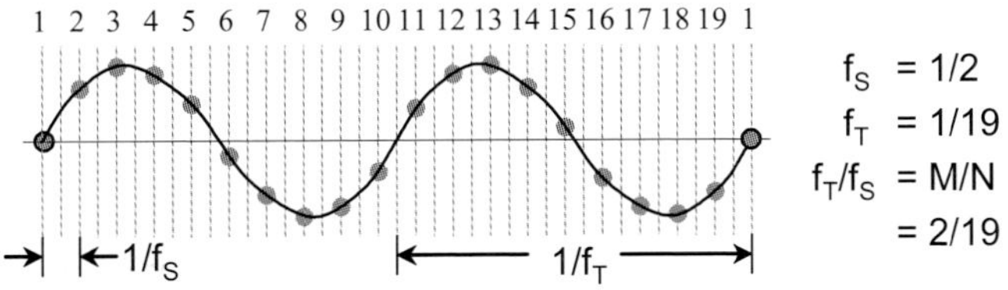

Figure 9-5: Coherent sampling.

Under-sampling occurs when the sample rate is intentionally much lower than the Nyquist rate[7]. This is used, for example, when testing very high frequency signals with a low sample-rate digitizer.

A signal is over-sampled when the sample rate is much higher than twice the signal's highest frequency of interest so that samples can be combined to achieve higher resolution, as most notably done in sigma-delta converters. This principle

[7] According to the Nyquist criterion, to capture all information about a signal, the sampling rate should be greater than twice the highest frequency of interest.

permits, for example, a medium-resolution (8~12 bits) digitizer or AWG to be included on-chip for BIST – the circuit comprises mostly scan-tested logic gates and only a few, relatively process-insensitive analog elements.

Testing the input threshold voltages of an ADC is typically done using a linear ramp, the DC servo method, the histogram method, or the fast Fourier transform (FFT) method. For 6- to 8-bit converters, a simple linear ramp, synchronously started, can be applied and digital output values can be compared to ideal values. Wider than 8-bit requires averaging so that noise can be tolerated.

In the DC servo method shown in Figure 9-6, remarkably like the first-order sigma-delta converter in Figure 9-2, a digital code is applied and if the ADC's output code is greater than this, an analog voltage applied to the ADC input is decreased; otherwise the voltage is increased. The voltage eventually settles to the threshold voltage of the upper limit of the applied digital code and is measured using a high accuracy DC voltmeter. The applied digital code is then increased to the next code of interest.

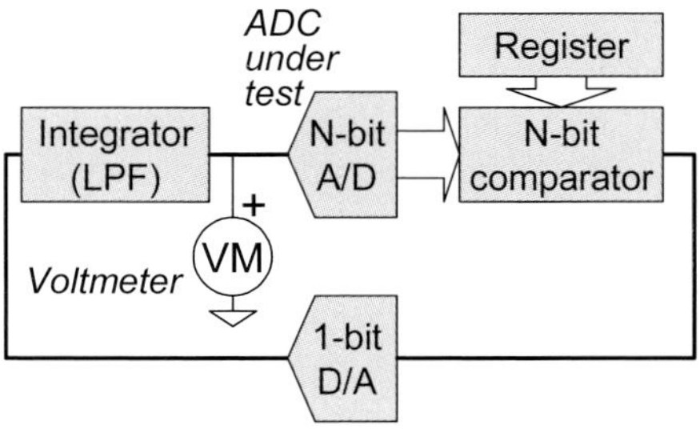

Figure 9-6: DC servo ADC test method.

In the histogram method, one or more cycles of a full-range linear ramp or sine wave voltage is applied to the ADC input. The number of times each output digital code appears is counted; the test should be long enough for many (e.g., 32) samples of each code. For a linear ramp, all codes should appear an equal number of times, thus, a plot of output count versus code should be a horizontal line, as shown in Figure 9-7. For a sine wave, the plot will be like a bathtub (curved up at both ends). For higher frequency converters, a low distortion sine wave is easier to generate than a low distortion linear ramp because any low pass filtering only reduces the amplitude of a sine wave but introduces 3rd harmonic distortion into a linear ramp.

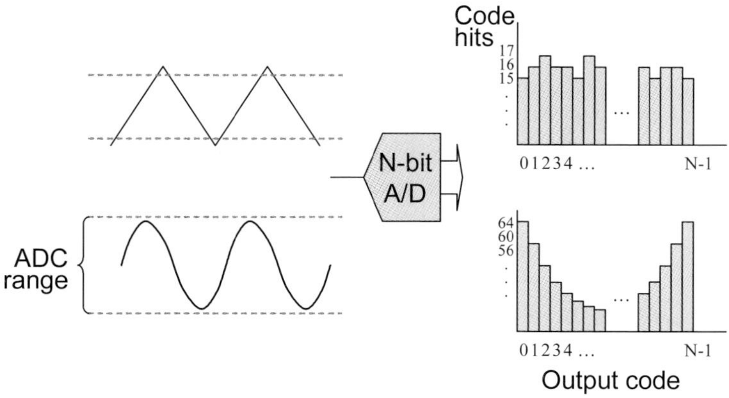

Figure 9-7: Histogram ADC test method.

The DC servo and histogram methods can be equally accurate, but which test is faster depends on noise of the ADC and the latency of the digital output [22]. The most important specifications that these two tests verify are *differential non-linearity* (DNL) and *integral non-linearity* (INL). *Differential linearity error* (DLE), briefly stated, is the difference between the actual code transition point step size and the ideal 1 LSB step size, and DNL is the maximum DLE for the converter across all codes. Integral linearity error (ILE) is the difference between the actual input code transition points and the ideal transition points, and INL is the maximum ILE across all codes. INL is the integral of DNL, but it cannot always be derived by simply measuring DLE and summing. If DLE is measured with coarsely quantized resolution (e.g., ¼ LSB), then a systematic 0.01 LSB error, for example, which accumulates across one thousand LSB intervals to become a ten LSB error, will be undetected. The linearity of the applied ramp must exceed that of the ADC under test for the INL test to be accurate, whereas it will have much less effect on the DNL test.

IC capacitors or resistors can be matched typically with 0.1% accuracy which permits the straightforward design of converters with 10 or fewer bits of resolution. And for these converters, thermal noise is typically less than the noise caused by quantization, referred to as quantization distortion (QD). Above 10 bits, more complex design techniques are required, such as *averaging* or *self-calibration*[8]. At sampling frequencies above a few megahertz, the signal transition between output levels also becomes more significant. Neither the DC servo method nor the histogram method is sensitive to transitions, noise, or crosstalk (which is both an advantage and a disadvantage). For these parameters, FFT-based tests are used to measure:

[8] For example, when implementing a resistive divide-by-two with two unequal resistors to obtain a voltage mid-way between two voltages, the resistors can be interchanged every sampling cycle so that their time-averaged values will be exactly equal. This can be done continuously or only during a self-calibration mode.

- *signal-to-noise ratio* (*SNR*): the ratio of the full-swing signal's fundamental frequency power to the power of all in-band non-harmonic noise;
- *signal-to-noise-and-distortion ratio* (*SINAD*): similar to SNR, except it includes harmonics in the noise power;
- *total harmonic distortion* (*THD*): the ratio of the signal's fundamental frequency power to the total power of the harmonic frequencies;
- *spurious-free dynamic range* (*SFDR*): the ratio of the signal's fundamental frequency power to the power of the highest amplitude harmonic or noise frequency.

These frequency and noise-focused tests reveal the impact that manufacturing defects may have on sinusoid signals, and can be performed in less test time than DNL, but they are also more complex mathematically and less diagnostic (for example, they don't indicate which code transition is faulty). Typically, a test suite need comprise only tests for SNR, SFDR, DNL, and INL, though it is theoretically possible to calculate DNL and INL from the Fourier transform results.

Measuring harmonic distortion for fundamental frequencies that are out-of-band (e.g., above half of the sampling frequency) is accomplished by measuring inter-modulation distortion (IMD). According to IMD, two fundamental frequencies are applied simultaneously to the CUT, the frequencies differing by a few percent. CUT non-linearity will cause the output signal to contain harmonics at frequencies equal to each of the fundamentals plus and minus integer multiples of the difference between the two frequencies.

Testing a DAC is similar in some respects to testing an ADC, but in other ways it is very different. Whereas code boundaries are tested indirectly for an ADC, the output voltage for each digital code can be measured directly for a DAC: histogram and DC servo methods are not applicable. However, FFT-based testing for a DAC is similar to testing an ADC. For more details, the reader is referred to textbooks [20], [21].

Test methods for ADCs and DACs may be extended to comparators, filters, amplifiers, and many random analog functions: for example, SNR is applicable to filters and amplifiers, and a comparator is similar to a 1-bit ADC. However, PLL test methods are completely different.

PLL Test

The most common application for a PLL is probably clock synthesis: generating an on-chip high frequency clock for the core logic of the IC, phase-locked to an off-chip low frequency reference clock. The most common PLL test is for maximum lock time. Lock time is the time it takes for a PLL to acquire phase lock after an abrupt change in the phase of its reference signal, usually based on a Lock Detector output as shown in Figure 9-8. This time can range from less than a microsecond to milliseconds. In some test programs for large digital ICs, the IC is powered-up, the reference clock is applied, the program pauses to allow phase lock, and then core logic test patterns are applied. Any incorrect output bits will be due to defects in the logic or due to the PLL not achieving phase lock. A more diagnostic but more complex test of the PLL applies a reference clock phase change and then checks the state of the PLL's lock indicator, or, if it does not have one, continuously measures the PLL's output frequency until it is correct. The lock time has surprisingly high fault coverage – it verifies mid-range performance of most of the PLL's circuitry and is sensitive to the PLL's damping and loop gain.

The lock range of a PLL is the minimum and maximum frequencies that a PLL can lock to (within its lock time). Sometimes a PLL has two significant lock times: one for small frequency deviations and another for maximum deviations. This test may be performed by applying the required minimum input frequency, verifying phase lock, then applying the maximum input frequency and verifying lock. The test is less commonly performed because it requires the application of multiple reference clock frequencies, the IC's core logic may need to be disconnected from the PLL (because it would overheat at the PLL's maximum frequency), and there may be many combinations of PLL feedback division, charge-pump current, and loop filtering to test.

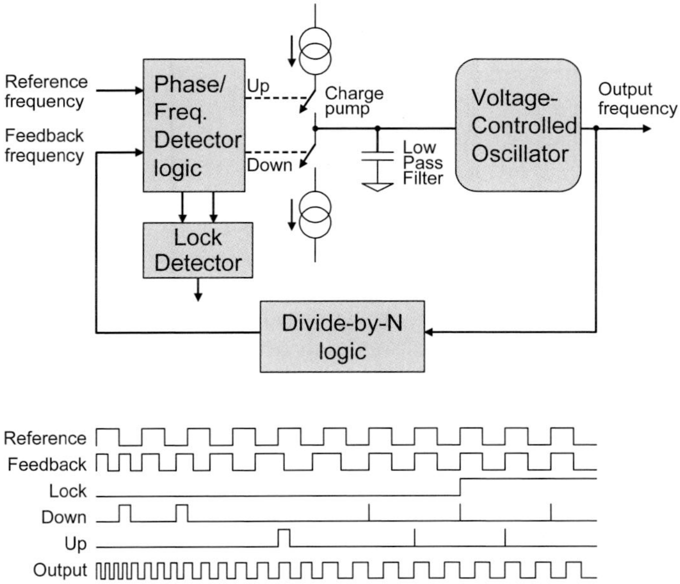

Figure 9-8: PLL and waveforms.

The jitter of a PLL is sometimes its most important performance parameter but the one that is most difficult to test. Specially-equipped testers test jitter by sampling the duration of random periods (time interval analyzers), or by phase-locking to the PLL output and sampling the signal around the transition time, at a constant reference voltage, until a histogram can be compiled (oscilloscopes, and parametric bit error rate testers). Measuring jitter requires sampling circuitry that has less sampling jitter than the signal under test, which can be less than 1 ps RMS for gigahertz clock signals. The path that conveys a clock-under-test to a tester can easily add significant jitter due to capacitive-coupling to other signals on-chip and transmission-line effects off-chip.

There are many ways to specify and measure jitter, which makes the task a specialty for some test engineers (and some companies) – entire careers can be devoted to this topic.

There are three commonly specified types of jitter [23]. *Timing jitter* is edge time variation relative to the ideal edge times, and is most important where synchronization of many data signals is required. *Period jitter* is variation in the period relative to the ideal (or average) period, and is the derivative of timing jitter (like DNL is the derivative of INL) – it is most important when circuit signal propagation delays must be less than a clock period for correct operation. *Cycle-to-cycle jitter* is the variation in each cycle's duration relative to the preceding cycle, and is the derivative of period jitter – it is most important when a clock cycle's duration is used to predict when the next signal edge will occur. For each of these types, the jitter can be random or deterministic (and there are many sub-types), and the maximum value for each is usually specified separately. It is also possible to measure the spectrum or phase noise for jitter using FFT-based methods or purely analog methods, especially for signals employed in radio transmission, where minimizing channel interference is very important.

For digital or mixed-signal applications, jitter is measured in the time domain and its magnitude is typically reported in units of picoseconds or percent of the clock period (unit interval). While it is the peak-to-peak value that affects a circuit's function, testing the RMS value produces a more repeatable result because the peak-to-peak value increases with the number of signal edges sampled, whereas the RMS value converges rapidly to a constant value. The frequency of the jitter is important: for crystal oscillators, jitter power per hertz below 100 Hz is orders of magnitude larger than jitter above 100 kHz. For analog or radio applications, jitter is measured in the frequency domain using a spectrum analyzer, and its magnitude is reported as spectral density in a specific frequency band.

9.2.2 DfT

The simplest summary of the state-of-the-art in mixed-signal DfT is that there are no standard, systematic approaches, unlike what is seen in digital DfT. Most of the mixed-signal DfT and BIST techniques proposed in the last dozen years are not used in industry for various reasons (which are explored in section 9.3.2).

Loop-around or loopback, shown earlier in Figure 9-2, continues to be one of the most common DfT techniques. It is simple, reasonably diagnostic when multiple loopback paths are provided, closely related to normal function operation, and allows simpler stimulus generation and analysis, for example, digital in/out. The disadvantages of loopback are that it may not detect compensatory faults (e.g., insufficient output drive compensated by excessive input gain), and it may not detect crosstalk between elements of the loop. Solutions to these problems include monitoring signals in the loop via a second output, and injecting a DC offset at a second input, but research in these areas is minimal.

The use of analog buses seems to be common in industry but not in published papers, likely because the approach is regarded as mature. ICs may contain one or many analog buses to accommodate frequency and capacitance considerations[9]. They are general purpose and suitable for conveying DC-to-multi-megahertz voltages and currents into and out of an IC, though frequencies above a few megahertz require significant design effort to avoid interference and distortion. The force/sense and force/measure technique used in interface boards can be extended into an IC via analog buses, but the added delay can reduce usable bandwidth to kilohertz frequencies. The practical fan-out of analog buses is quite limited, to less than 30 for 0.18 μm CMOS, by the capacitance and leakage current of each connected transistor's diffusion; finer geometry technologies may have even lower fan-out limits. The limit can be extended by using hierarchy, current mode signals, and/or high threshold access transistors that have lower leakage.

Ad hoc analog DfT, including analog bus access, appears to be the rule for random analog circuitry. Even when analog buses are used, the tests are typically not reusable.

The 1149.1 boundary scan standard [24] requires monitoring digital outputs to analog circuitry, and controlling analog inputs to digital circuitry. This systematic technique can improve diagnosis, test time, and fault coverage, regardless of whether the circuit's application requires boundary scan.

Digital scan access to the digital output of an ADC and input of a DAC (see Figure 9-2) has been exploited beyond testing the converters themselves: the tested DAC generates an analog stimulus for core analog functions in the IC, and the tested ADC monitors core analog signals. This approach conveys the core analog signals to the tester or on-chip DSP via a digital scan path, and hence band-limiting, distortion, and noise may be reduced compared to using analog bus access. The approach does, however, require that an appropriate speed/resolution ADC, DAC, or DSP exists in the IC's function.

DfT for PLLs ranges from analog access to the loop filter output, to digital frequency counters and dividers. Analog access to the voltage-controlled oscillator (VCO) input permits measuring its DC level after phase lock has been achieved for a known frequency, and some designers also permit driving the VCO input directly to measure its frequency versus voltage transfer function. The PLL's output frequency can be simply divided down to a low enough frequency that the tester can easily measure it, or a frequency counter might be included on-chip[10].

Presently, BIST is rarely used for commercial mixed-signal functions, for the reasons mentioned in section 9.1.3, except when it is a required function, for example in serializer-deserializer (SerDes) ICs where a built-in PRPG and bit error rate (BER) detector is required by some communication standards. One exception is self-calibration, which is used in many ADCs, but this is required to achieve the functional performance and not to reduce test costs or improve fault coverage.

---

[9] Audio and radio frequencies are always conveyed on separate analog buses to avoid interference and to accommodate the typically different VDDs of these circuit types.

[10] A frequency counter counts the number of cycles of the unknown frequency that occur in a known number of cycles of a known frequency.

One way to discover BIST developed by companies for their own use is to search the European and U.S. patent office databases. Detailed study of the approaches described in patents and patent applications often reveals the same problems seen in published papers: for example, inadequate fault coverage, sensitivity to noise, or lack of diagnosability. The harsh economic reality is that a DfT approach must *greatly* reduce total test time, complexity, or tester cost before it will be adopted by companies other than the one that developed the approach.

9.2.3 Fault Modeling

Other than a dwindling number of published papers on fault modeling, it is difficult to find evidence that industrial designers use fault modeling to guide their analog test plan. It appears that verification of specifications is still the key methodology, with feedback from the manufacturing Quality Control (QC) or Quality Assurance (QA) departments based on actual device failures, and on historical data for the company's design style, manufacturing process, and customer applications. Typically, when a failing device is found in the application, it is returned to the QA department, who retests the device, and if it passes, more analysis is performed until the root cause fault is found (in 25~50% of cases, *no fault is found*, and this is itself a problem). If a fault is eventually found, a new test is added to the production test program to catch that fault. In effect, this procedure weights the analog faults based on present reality rather than any model – potential faults that actually occur are weighted highly, and potential faults that haven't caused failures are weighted lowly. This method does not rely on modeling accuracy, but can obviously result in defects escaping to the end-user of an IC. The exception to this practical but non-preventative method is circuits that are intended for human life-critical applications, for which analog fault modeling and estimating fault coverage may be a contractual requirement.

A few companies have offered analog fault simulators in the last 10 years, but all seem to have withered (faultMaxx based on LIMsoft [10], and AnaFAULT) or are aimed at only life-critical applications, such as automotive circuits (Intusoft's Test Designer).

The most common fault simulation method is Monte Carlo simulation, partly because simulating random combinations of process parameters and conditions is a standard analog design verification technique (without consideration of a test's fault coverage). Many random variations of a circuit, typically with a Gaussian probability distribution, are tested in simulation to determine the proportion of circuits that pass a test and yet fail some specification.

Noise greatly affects fault simulation results and is often neglected to make the computation reasonably efficient, but microvolt differences between faulty and fault-free circuit responses cannot be detected with a comparator, so some form of averaging is essential.

9.3 ADVANCES IN THE LAST 10 YEARS

Mixed-signal test methodologies seem to be approaching the top of the S-curve[11] in terms of new developments, whereas DfT seems to be in the middle of the S-curve – there are lots of new techniques being proposed each year.

Here are criteria for evaluating any new mixed-signal test or DfT technique:

- *Yield impact* – The increase in silicon area due to DfT is the most common indicator of yield impact because increasing the area of an IC typically reduces its yield. Area for digital DfT circuitry can be assumed proportional to gate count (equivalent 2-input NAND gates) only if the gate delays are not critical, and area for analog DfT circuitry may have little relationship to its yield.
- *Fault coverage vs. yield coverage* – Generally, tightening test limits will increase fault coverage and reduce yield: a better technique must improve both. A ten-transistor DfT circuit with 99% yield coverage (fails 1% of defect-free circuits under test), has about the same yield impact as a DfT circuit that occupies 1% of the whole IC.
- *Test time vs. accuracy and precision* – Generally, test time can be reduced if accuracy or precision can be degraded. A new technique that reduces test time is only useful if test accuracy and precision are not degraded. At lower supply voltages and higher speeds, test precision and accuracy is more challenging than test time.
- *Test precision* – All analog signals include noise so some form of averaging or summing is essential to increase precision. Test repeatability is proportional to precision and measurement variance.
- *Test accuracy* – Systematic errors, such as offset, cause inaccuracy. Some form of subtraction of errors, for example focused calibration, is essential.
- *Cost of tester vs. completeness of test* – More test types may increase fault coverage, but require more tester capability. Fewer test types can reduce tester cost but may reduce test coverage: a better technique should improve both.
- *Process tolerance* – Manufacturing variation in the DfT circuitry and the CUT may be similar, but the DfT circuit variation should not affect the result. Typically, it must be accounted for with focused calibration or self-test.
- *Design impact* – The performance of analog circuits (and high-speed digital circuits) is affected by adding monitoring circuitry or multiplexers. The effects can be simulated but may increase complexity and risk of design error or reduce performance unacceptably, especially if DfT is not automated.
- *Diagnosis* – When analog circuits fail, or pass marginally, the designer or test engineer needs to know why, otherwise yield could continue to decrease. A better test or DfT technique should diagnose what performance is failing, including whether it is the CUT or the DfT circuitry.
- *Sampling* – Many test techniques involve sampling, for example, periodic under-sampling or random sampling of a signal's value or of a binary

[11] In some areas of learning, a graph of progress versus time, , looks like a stylized "S".

comparison between a signal and a reference signal. The significance of the inevitable loss of information must be assessed.

- *Reuse* – As test generation becomes a greater portion of product development time and cost, the ability to reuse a circuit's test suite along with the circuit becomes more important. Systematic approaches are more reusable than ad hoc approaches.

9.3.1 Testing

The advent of DSP for mixed-signal testers in the late 1980's led to rapid change in mixed-signal test methods. As single-IC implementations of DSPs have become faster and cheaper, it has become practical to have more DSP capability in testers – some testers have a DSP on every channel, and can perform a multi-kilosample FFT in less than a millisecond.

The introduction of sigma-delta ADCs and DACs required new tests. These converters soon migrated into testers, to permit higher resolution testing, and testers whose analog signal resolution decreases predictably with speed – for example, a tester's sigma-delta-based digitizer might be capable of 24-bit resolution at 1 kHz, 16-bit resolution at 1 MHz, and 12-bit resolution at 10 MHz.

The impact of clock jitter on the SNR of high speed, high resolution ADC, DAC, and SerDes ICs has led to testers having optional, very low jitter clocks. Jitter below 1 ps RMS is sometimes necessary[12]. Also, a growing number of testers have the ability to supply multiple asynchronous clocks, to be more representative of the applications that ICs are designed for.

One technique [25] for measuring very low jitter exploits the very low aperture jitter specified for many commercial high speed, high resolution ADCs – aperture jitter is typically less than one or two picoseconds. The clock signal to be tested is connected to the clock input of a high performance ADC whose analog input is a low jitter sine wave. The sine wave is relatively easy to generate using a standard signal generator and a high-Q passive band-pass filter. The digital output of the ADC is compared to an ideal jitter-free response, with computational complexity equivalent to an FFT. Jitter measurement accuracy was reported to be about one picosecond – the best reported for any technique that does not use gigahertz bandwidth hardware. The lack of subsequent reports after [25] suggests that the technique might not be suitable for testing a large number of signals or embedded signals.

The early success of I_{DDQ} testing for digital CMOS circuitry has enticed many researchers to investigate analogous approaches for analog circuitry. Both the DC value of the quiescent current and the spectrum have been measured for general analog circuits and PLLs, but the fault coverage has been disappointing. Milliamps of quiescent current, with perhaps 10% variation due to processing, make most DC

---

[12] The output noise caused by an ADC clock's period jitter is proportional to the derivative of the input signal voltage; the noise in units of LSB is also inversely proportional to the LSB voltage step. Sampling a 1 V, 20 MHz sine wave, with a 14-bit ADC having a 2 V range and 1 ps RMS sampling period jitter, generates 1 LSB RMS output noise.

current faults undetectable, and minimizing the supply current's dependence on instantaneous output signal voltage is a typical analog circuit design objective.

A very detailed attempt to systematically develop a test program based on fault simulation was described in [26]. The close relationship between yield and parametric fault coverage analysis was demonstrated, and the test simplification and test time reduction achieved was consistent with conventional product engineering methods. The approach involved many simulations, but the circuits were relatively simple (op-amp and variable gain amplifier).

The complex relationship between analog circuit defects and the voltage/current variations that are detectable at a circuit's pins (input, output, references, and power) led another research group to a general, and similarly computationally complex approach [27]. First, in simulation, a variety of pseudo-random analog voltage sequences were applied to the CUT. The resultant output responses for various faulty and defect-free circuits were correlated to the functional specification-based responses, using *multi-variate adaptive regression splines* (MARS). An optimal stimulus was found by a genetic algorithm that compared fault coverages for the various pseudo-random stimuli. Lastly, the stimulus was further optimized by repeating the experiment using a representative sample of actual ICs. The results for production testing of op-amps indicated an order of magnitude reduction in test time compared to conventional specification-based tests, with no significant yield impact (high yield coverage). The authors have extended the technique to address RF circuits by mixing the stimulus with a constant-frequency RF signal.

The advantages of this test technique are its applicability to many analog circuits, the simple stimulus generation and response capture, and its high correlation with specification-based tests. The disadvantages are the apparent difficulty in using the technique for testing analog functions embedded in an SOC, and the relatively small reduction in absolute test time (for example, 300 ms test times were reduced to 30 ms: a 10X relative reduction but only 250 ms absolute reduction).

Aside from these general hardware advances, the methodology of testing mixed-signal circuits does not seem to have advanced dramatically in the last 10 years. Most of the advances have been in DfT techniques.

9.3.2 DfT

As sampling rates and resolutions of mixed-signal functions increase, and as the functions become more embedded in a system on a chip (SoC), the need for DfT increases. Evidence from the last dozen years indicates that DfT for discrete mixed-signal components, such as converters, PLLs, and gain functions, is both unnecessary and ineffective – the potential reduction in test time (if any) or tester cost does not justify the increased IC area or uncertainty in accuracy. The intriguing theoretical challenge of implementing BIST for mixed-signal functions has led a number of academic researchers into this field after a long career in digital DfT. While experience in another field ought to bring fresh ideas into mixed-signal testing, results thus far indicate that merely applying modified digital BIST techniques to analog circuits is fruitless.

Digital circuits are suitable candidates for reducing test time by using parallelism in testing – scan-based testing applies digital stimulus in parallel to the many logic gates that comprise a function block. Random patterns can quickly test digital functions more thoroughly than manually written, specification-focused patterns because the stimulus is easy to generate and the results are easy to analyze. However, when testing mixed-signal functions, test accuracy is always proportional to test time, and generally, the complexity of the output signal analysis is inversely proportional to the complexity of generating the stimulus.

Mixed analog-digital BIST (MADBIST) [7] is an example of how the mostly-digital implementation of the sigma-delta principle makes it suitable for DfT circuitry; it is illustrated in Figure 9-9. The heart of the approach is an all-digital oscillator that generates a sine wave encoded in a sigma-delta bit stream. Fairly crude low pass filtering of the bit stream, which might be accomplished by the anti-alias filter of an ADC-under-test, is sufficient to form a very high linearity, analog waveform. The digital output of the ADC is analyzed using a DSP (which is assumed to be part of the on-chip function – it is too costly to include a DSP only for test purposes). After successfully testing the ADC, it can be used to test a DAC, and the two combined can test other analog circuitry on the IC.

The primary advantages of MADBIST are its mostly digital construction, which is process insensitive and can be scan tested, its accuracy, and its general-purpose nature and diagnostic capabilities. The primary disadvantages are the tens of thousands of logic gates needed (unless the CUT's circuitry can be reused – but this adds complexity), the 4th-order analog filtering needed, and the ratio of the clock frequency to the sine wave frequency (typically greater than 100). Other authors from the same university later improved the method by simplifying the sigma-delta oscillator (instead using only an 8-Kbit shift register), and by eliminating the need for an ADC (a high speed comparator and mostly-digital DAC could be used instead) [28].

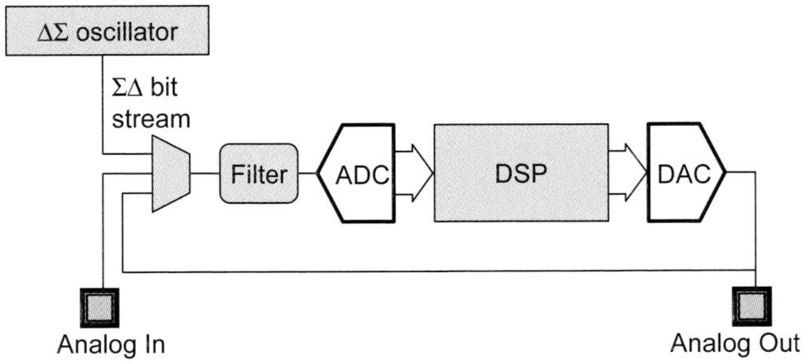

Figure 9-9: MADBIST.

Oscillation BIST (OBIST) [6] is an example of a radically different DfT technique, and is a function-based structural test. Briefly stated, the technique comprises connecting the output of a CUT to its input in such a way that oscillation is induced, as shown in Figure 9-10 (and often the result of inadvertent positive feedback). For above-unity-gain analog circuits, the feedback can be as simple as a transmission gate switch; for below-unity-gain circuits, a digital comparator or inverter is needed in the feedback path; for purely low-pass circuits, such as op-amps, the feedback must accomplish high-pass filtering. In essence, the oscillation frequency is at the convergence of a function's low-pass transfer function and the feedback's high-pass transfer function. The authors later extended the principle to include tests for ADCs and DACs [29] though test times and dependence on DfT circuit accuracy make the approach impractical in some cases.

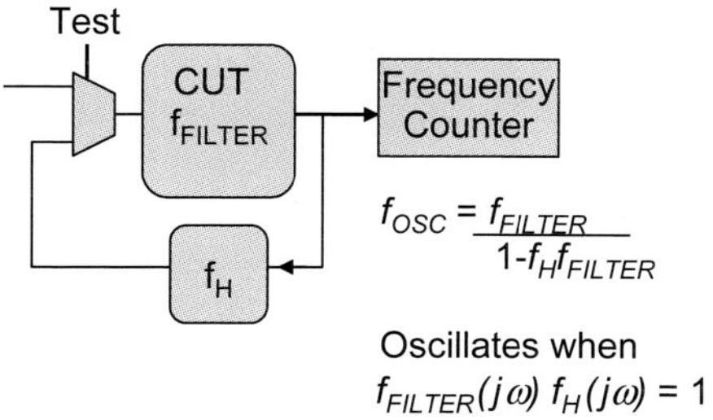

Figure 9-10: Oscillation BIST.

The primary advantages of OBIST are that no analog stimulus is needed (hence signal inaccuracy cannot contribute to measurement inaccuracy), and the measurement can be as simple as measuring the oscillation frequency with a digital frequency counter. The primary disadvantages are that OBIST is not diagnostic (especially if the fault prevents oscillation), and it does not test parameters closely related to specifications (especially noise) or design parameters, instead relying on fault model accuracy to assess the fault coverage. When testing many embedded op-amps, it is hard for any other method to beat the simplicity of OBIST.

The first commercially offered mixed-signal BIST, dubbed adcBIST, was another radically different approach [30]. The technique applied a unique staircase ramp, comprising four digitally generated RC-exponential curved steps, to an ADC, and digitally accumulated the ADC's output throughout each of the four steps to obtain four sums that were arithmetically combined to create four signatures. The signatures corresponded to the offset, gain, sum of even harmonics (mostly 2nd harmonic), and the sum of odd harmonics (mostly 3rd harmonic).

The primary advantages of adcBIST are its accuracy in the presence of noise (16-bits linearity was demonstrated) and its mostly digital construction. The disadvantages are its need for a continuous-time low-pass filter (which must typically be off-chip), its inability to measure noise-related parameters, and its lack of generality (it is only for ADCs and ADC/DAC pairs).

As power rail voltages decreases, most analog circuits are becoming differential to improve tolerance to power rail noise and to increase the effective signal amplitude. A general DfT technique proposed for differential signals is to monitor the sum of the two signals of a differential output: it should be a constant voltage, for any signal. It has also been reported that the DC bias of differential signals (that do not swing rail-to-rail) is highly correlated to the presence of defects [31]. The DC bias or average voltage can be obtained easily using a first-order RC low pass filter.

A variety of signals of interest are very low frequency, so very simple single-order sigma-delta ADCs have been proposed as a way to measure these signals with very high resolution and accuracy. As discussed previously, most of a signal-delta converter is digital circuitry that can be tested via scan, so the sensitivity to process variations can be reduced to a comparator and an integrator. A few authors have proposed using these converters for on-chip measurements. For DC measurements, the digital filtering can be reduced to a single digital counter, with a (sin x)/x low-pass response.

Key requirements of any DfT method are to increase the observability and controllability of the CUT. For digital circuits, scan path access meets both of these requirements. For analog circuits, it seems natural to attempt an analogous concept: an analog shift register. Various techniques have been used in the "early" days of IC design, such as *bucket brigade devices* (BBDs) and *charge-coupled devices* (CCDs). BBDs used MOS switches to connect one capacitor to another, in sequence, in a long chain, but because the switch divides two quantities of charge when it turns off, each transfer adds noise that is inversely proportional to the capacitances. CCDs convey analog samples as complete quantities of charge within the substrate of a semiconductor – the charge is not split in two – so that the charge transfer is almost lossless and does not add noise, which is one reason why CCDs are still used for digital camera imagers and high speed delay lines. A few researchers have recently proposed using capacitors and op-amps to construct an analog shift register, much like a BBD. The achieved speed and accuracy, even in simulation, is not good enough for most mixed-signal testing. Analog buses can appear to function very similarly to analog shift registers, without involving multiple charge transfers, and are therefore preferable.

A multi-company effort began in 1992 to develop a formal mixed-signal DfT approach, initially focused on analog boundary scan testing for circuit boards containing mixed-signal ICs – discrete components on boards were getting too small to probe mechanically. The outcome of this effort was the 1149.4 Standard for a Mixed Signal Test Bus [32]. The standard specifies a two-wire analog bus (or four-wire if differential) and a multitude of analog access switches controlled via a digital scan register and an 1149.1-compliant Test Access Port (TAP). The circuit infrastructure [33], shown in Figure 9-11, builds on the force/measure principle, the 1149.1 boundary scan standard, and the accuracy of low frequency measurements in the presence of digital switching noise. In its simplest implementation, the switches are CMOS transmission gates with 0.5~5 kΩ "on" resistance and on-chip analog buses that connect directly to off-chip analog buses, totaling 50~500 pF load capacitance – resulting in a –3 dB low-pass bandwidth of less than a megahertz. One intended measurement method is to apply a known AC+DC current to a circuit node via one of the analog buses, and to measure the resultant voltage at the node via the second bus.

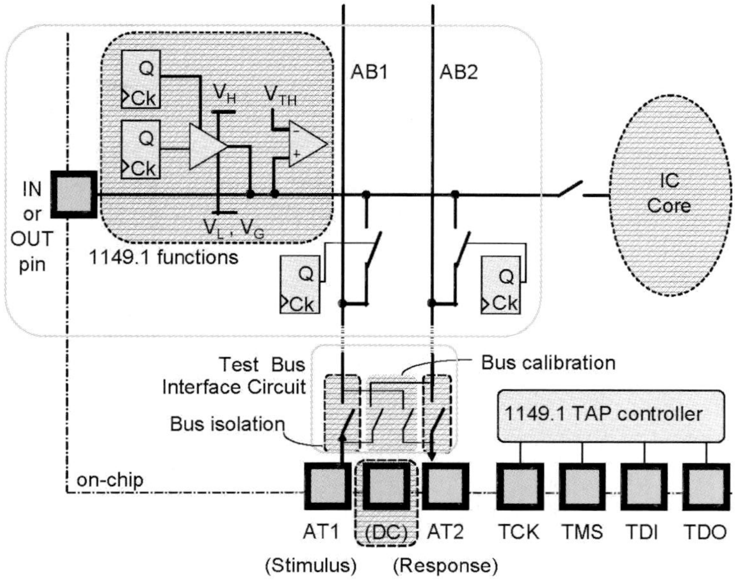

Figure 9-11: 1149.4 Mixed-signal Test Bus.

Referring to Figure 9-11, to measure a parallel resistor and capacitor connected (not shown) between the IN/OUT pin and ground, 100 mV DC combined with 100 mV sine wave at 1 kHz is applied via the AT1 pin while its two series-switches are enabled. The voltage at the IN/OUT pin is simultaneously monitored at the AT2 pin while its two series switches are enabled. The current is monitored via a small resistance at the stimulus generator (see Figure 9-4). Examples, including the mathematics and hardware results, are available at the 1149.4 Working Group's web-site[13].

A few test ICs have been constructed by companies, and one design was commercially distributed [34]. A small but growing number of academic and industrial researchers have proposed DfT techniques that are compatible with or exploit the standard. These techniques range from measuring pin DC parameters and testing power rail quality [35] to testing high speed converters and RF circuits.

The primary advantages of the 1149.4 mixed-signal test bus are its general applicability, its support for robust measurement techniques [36], and the fact that it is a standard – tests can be reused from design to design, at the IC and board level, and between companies.

The primary disadvantages of the mixed-signal test bus are its IC area (about 1% for a 0.18 μm CMOS IC), its low frequency range, and its impact on accessed circuitry. The area can only be justified if the bus is used for many tests whose greater accuracy, reduced test access costs, or faster test times decrease test costs by more than 1% of the IC's total cost. As a result of these disadvantages, the standard

[13] http://grouper.ieee.org/groups/1149/4/

is not yet used as widely as expected, though there is a growing trickle of corporate interest.

A few researchers have focused on on-chip linear ramp generation as the key to testing ADC linearity. The mostly analog techniques use a constant current to charge a linear capacitor [37], and adjust the current until the charging time is a specific number of clock cycles [38]. Up to 14-bit linearity has been reported, however, there is no separate test of the stimulus linearity before testing an ADC.

High speed, high resolution ADCs can be constructed using pipelined low resolution ADCs. Each stage of the pipeline compares its input signal to a reference voltage to generate an output bit, and outputs the difference between the reference and the input signal as a residual voltage, multiplied by two, to the next pipeline stage. DfT techniques have been reported [39] that specifically address this converter type, by measuring differences between the stages (which are all identical in design).

The concept of a check sum, as used to verify the integrity of a software file or data transmission, has also been explored in analog DfT [40]. The sum of selected node voltages of a CUT is generated by connecting high-value resistors between the nodes and a single virtual ground of an op-amp. Specific stimuli are used, though this type of sum could also be used for on-line testing (meaning continuously while mission mode signals are present). Generally, the technique is too sensitive to offset voltages and too insensitive to AC faults.

The variety of analog DfT, BIST, and fault modeling techniques that has been proposed in the last decade is described in significant detail in [43]: chapters include defect-oriented testing, fault simulation, ATG algorithms, DfT, 1149.4, spectrum-based BIST, and tests for switched-current circuits. The book covers most of the theory relevant to these sub-topics, and benefits from the experiences of multiple authors, but does not have sufficient silicon results needed for the methods to be considered for immediate use in industry. However, this is true of most published mixed-signal DfT techniques: they do not demonstrate sufficient resolution (>16 bits DNL), accuracy (>16 bits INL), noise tolerance, process tolerance (30% variation), or speed (hundreds of MHz). Test engineers continue to search for methods that achieve more resolution to diagnose low-level noise sources and high frequencies (wide noise-bandwidths).

PLL DfT

DfT for PLLs has proven to be unique to that function, however, analogies with ADC testing can be found. The most commonly important requirement for a PLL is low jitter, which is comparable to ADC noise, and this is the most difficult parameter to test with on- or off-chip techniques. On-chip jitter (or noise) measurement is hindered by the need to have measurement circuitry that has less jitter (or noise) than the PLL (or ADC). Multi-megahertz PLLs typically have 5~500 ps RMS jitter, while multi-gigahertz PLLs may have 0.5~5 ps RMS jitter. Most published on-chip jitter or delay measurement circuits use digital delay lines of some type, and these delays can have 5~15 ps RMS jitter. If the delay line is used in an oscillator, the jitter effects are cumulative. Off-chip (tester-based) test techniques for measuring jitter are based on sampling, for which the sampling clock and the signal path can each contribute 1~10 ps RMS jitter.

The other PLL parameters of importance are the lock range (compare to voltage range of an ADC), the lock time (compare to ADC minimum sampling period), and loop gain (compare to sigma-delta integrator gain).

The first PLL-specific BIST that measured sub-100 ps jitter [41] used a delay line comprising many logic gates in series, and compared the PLL output edge timing to the input reference clock edges, for a digitally-controlled incrementing delay, to deduce a cumulative histogram of the output jitter relative to the reference clock edges. The BIST also measured the loop gain by modifying the reference clock phase for a small number of clock cycles and then measuring the resultant change in the PLL's output frequency. Other techniques modified the apparent input phase by directly controlling the outputs of the PLL's phase frequency detector.

Other promising approaches for measuring jitter used Vernier oscillators[14] and are summarized in [42]. One of the oscillators is initialized synchronously with a first event (a first rising edge), and the other is initialized synchronously with a second event (a subsequent rising edge). When the two output oscillation signals are detected to be exactly in phase, the number of periods that has occurred in each oscillator since each event, multiplied by the respective duration of each period, is computed and the difference between the two products is the time difference between the two events. Variations on this scheme have been attempted, but in each case, jitter in the oscillators, plus their tendency to synchronize to each other, has limited the accuracy achieved in silicon to 10~20 ps RMS, the same that has been achieved with delay lines.

9.3.3 Fault Modeling

Other than the approaches mentioned elsewhere in this chapter, there have been no significant new techniques developed in the last 10 years, though papers on the subject continue to be published. The subject is interesting, but very subtle, complex, and difficult/expensive to prove any theory experimentally. One recent paper proved, arguably, that for some analog faults, it is impossible to prove that they can be detected [44]. Perhaps this should be considered progress, if there is further corroboration.

9.4 EMERGING TECHNIQUES AND DIRECTIONS

Mixed-signal ICs have been rapidly increasing in terms of frequency and resolution in the last few years – a challenging combination for test. The increase in frequencies increases the noise bandwidth, and the increase in resolution increases the apparent noise level in units of LSB. The only test aspect that is helped by faster sample rates is test time.

[14] Two oscillators tuned so that their periods differ by a constant amount between 1 and 50 picoseconds, for example.

9.4.1 Testing

The emergence of so-called "low cost" or structural testers has had a significant impact on new digital testers, partly because it motivated the makers of "big iron" testers to develop lower-cost products. These testers may have fewer or lower performance capabilities but are intended to benefit from DfT and embedded test. These testers can also be distinguished by their mixed-signal capabilities or lack thereof – most of the lowest-cost testers (less than US$100K) have no mixed-signal facilities, but their manufacturers are striving to change this because of the large percentage of new IC designs that require analog testing.

Simple 12-bit linearity analog I/O test capabilities, with DSP, are quite cheap (less than US$1K) and available as plug-in modules for small PXI-based mainframes. Costs increase significantly when better than 16-bit resolution is needed. A recently published ADC test technique [45] demonstrated 18-bit resolution using a (real) tester with only 12-bit linearity by accumulating samples twice; the second set was accumulated while adding a small, stable offset to the applied voltage ramp signal. A reasonably complex mathematical algorithm was used, but it is less complex than an FFT.

A trend that is increasing tester cost is the rapidly growing number of digital SerDes ICs with differential pins operating at serial data rates above 2 Gb/s. At these speeds, analog parameters dominate performance, such as jitter, frequency response, crosstalk, and phase delay. The number of pins with these signals, per IC, is also increasing, into the hundreds. Conveying the small-swing signals (<0.5 Vp-p) to the tester, without significantly affecting the signal's jitter (<2 ps RMS) and wave shape, is becoming extremely difficult – all wires must be considered as transmission lines and the effect of even a corner in a wire is measurable. One solution is to move the tester's gigahertz sampling and generation circuitry onto the interface board to be closer to the CUT: in [46] a board-mounted multiplexer and demultiplexer permit multiple channels of the tester to be combined to generate and sample 5 Gb/s signals. Another new solution is to exploit a CUT's loopback self-test: the tester simply adds jitter and voltage to looped-back signals.

9.4.2 DfT

Researchers are beginning to publish papers that explore extensions to OBIST [6] in which the amplitude and other parameters of the oscillation are also considered. As stated before, this technique entirely eliminates the need for stimulus accuracy.

The accuracy of sigma-delta analog signal generation that uses a re-circulating bit pattern in a shift register [28] has been proven in hardware, and other researchers are beginning to explore its capabilities. The approach is suitable for generating any analog waveform and can reuse scan shift registers that already exist in logic elsewhere in the IC (though this prevents simultaneously testing that logic). The primary limitation is the maximum output frequency, but mixers can be used to translate any waveform to RF frequencies.

Two well-proven principles, force/sense and frequency mixing, have been combined to achieve accuracy at high frequencies [47]. Although 16-bit linearity at

100 M samples/s was claimed based on simulations, the technique will likely need to be combined with other techniques, such as [45], to further improve linearity.

In the area of PLL jitter testing, DfT techniques focusing on SerDes are where the most progress is likely to occur because of the test challenges and growing production volumes. One digital, under-sampling technique [48] achieved sub-picosecond resolution by using the sampler of the CUT itself, and a median-centered histogram accumulation algorithm that excludes the low frequency jitter contributed by reference frequencies that does not affect SerDes bit error rate.

In general, tester and DfT solutions are beginning to overlap, now referred to as *Test Resource Partitioning* (TRP)– putting each part of a test resource where it is most effectively implemented. For example, the highest-speed switching and sampling can be performed on-chip, but it might be relatively inaccurate; the highest accuracy can be achieved in a tester, but it is often slower than the CUT and conveying the signal to/from the tester degrades accuracy and speed. Partitioning the hardware helps by putting, for example, crude high-speed samplers, mixers or comparators on the chip or interface board to produce a low-frequency analog or digital output. This signal is easily conveyed to highly accurate, low-speed ADCs and DSPs in the tester. A calibration step may be needed.

9.5 EDA TOOLS FOR MIXED-SIGNAL TESTING

There is a wealth of Electronic Design Automation (EDA) tools for digital, mixed-signal, and analog design, and for digital DfT, but for mixed-signal DfT there are almost no commercially available tools. The tools that do exist are created by university researchers, or by corporations for their internal use, or by start-up companies that have disappeared.

9.5.1 Testing

Almost all tools for automating the generation of tests for ADCs, DACs, and general analog paths, are tester-specific. They are provided with the tester, and "know" about the tester's signal generation and signal capture resources, and clock frequency limitations. The test patterns generated by this software are only transferable to other testers of the same model family, and sometimes tester-specific translation software must be applied.

These tools automatically ensure that sampling is coherent, and choose from the finite number of discrete frequencies available for a specific tester – without this automation, coherent sampling can be a tedious exploration of frequency requirements versus frequency capabilities.

Other mixed-signal test tools are part of specific DfT capabilities in the IC and enable selection of tests and run-time parameters, or set test limits for on-chip comparisons.

9.5.2 DfT

There are no mixed-signal tools analogous to scan path insertion for digital circuits. That is because there is no widely used, systematic mixed-signal DfT method, and because there is no commercially-available layout synthesis for general analog circuitry. A few layout synthesis products targeted at some common functions, like op-amps and PLLs, have been announced in the last few of years, but none for DfT.

A few BIST products require software to choose the synthesis options for the on-chip circuitry, and, for a manufactured IC, to choose the test settings.

HABIST (histogram-based analog BIST) [49] applied a periodic analog waveform to an ADC, gathered the output digital samples in the form of a histogram, then subtracted the histogram from the expected histogram (after appropriate scaling), and lastly analyzed this delta histogram to deduce offset, gain, and non-linearity. The computations could be performed on-chip with a small microprocessor or off-chip (though the latter is not BIST).

PLLBIST [41] requires the user to supply PLL parameters so that appropriate BIST circuit register-transfer-level (RTL) code can be generated; the RTL is then synthesized into logic for automatic layout. Additional software is used to generate test patterns that select the BIST test to be performed, apply test limits, and read the pass/fail bit for each test limit.

9.5.3 Fault Modeling

A variety of analog fault modeling software has been described in published papers, and a few have been available within universities or were commercially available, though only one or two are still available. This is more an indication of the low value that industry presently attributes to analog fault modeling than proof that modeling is impractical.

FaultMaxx, developed by Opmax (later Fluence), was the first parallel analog fault simulator. It created a sensitivity matrix for all of a circuit's nodes based on only two Spice simulations, and, from the node sensitivities, the relationship between any element variation and a circuit performance could be calculated (for the circuit operating in its linear region). By applying the various fault models described earlier in this chapter, the fault coverage of tests could be estimated in an extremely computation-efficient manner.

One analog fault simulator that is still commercially available is Test Designer, developed by Intusoft. It uses Spice simulations, and a graphical user interface, to guide the user in developing analog tests that detect targeted faults.

9.6 FUTURE DIRECTIONS

The development of mixed-signal DfT has completed its childhood and is now entering adolescence. New techniques can no longer simply be creative: they must build on proven principles, and prove that they can solve significant problems in the real world. This is especially true for BIST. Most techniques published thus far claim to offer an alternative to expensive testers but are incomplete and unproven.

As in so many fields, faster progress is made with small solid steps towards an end goal rather than with a single leap. Mixed-signal DfT that solves a portion of a problem well, combined with other such solutions, will eventually produce embedded tests that offer a true alternative to ever-higher performance testers.

The growing dominance of electronics in consumer products means that designers and test engineers will become even more risk-averse than those in telephony in the 1980's because the cost of major defects in products can be prohibitive after the products have been shipped to millions of customers. The barriers to new test and DfT techniques will be higher. On the other hand, very low-cost consumer-electronics products are notorious for having minor defects. This points to an increasing trend of grading defects on the basis of their impact rather than simply declaring an IC defective or non-defective. The growing number of ICs with field-programmable analog functions may further contribute to this defect categorization.

New ICs continue to improve the resolution of A/D and D/A converters significantly beyond 16 bits. The 24-bit resolution typically reserved for professional audio is now being seen in consumer applications, mostly because of the home theatre sound quality and the DSP needed to provide sound effects. Thus, new test and DfT methods will need to have unlimited resolution. Techniques such as force/sense extended into the CUT, sigma-delta conversion, and DSP, only offer unlimited resolution at low frequencies, so frequency conversion techniques (modulation, demodulation) will also be required.

Noise and jitter testing is proving to be the most challenging field for DfT. The testing of linearity and distortion can exploit the ability of averaging to cancel noise effects, but testing noise phenomena requires the DfT circuitry to have less self-noise and a greater bandwidth than the CUT, and averaging diminishes the level of the signal of interest (noise). For example, to measure jitter, DfT that adds on-chip delay lines and oscillators will not work, especially at gigahertz frequencies – they have too much self-jitter and contribute jitter to the CUT via the power rails and substrate. Some emerging techniques focus solely on noise test, and some use noise to improve resolution – noise becomes a friend instead of the enemy.

The number of publicized start-up companies focusing on mixed-signal DfT appears to be increasing. Recent examples include Q-Star, which provides circuitry for measuring I_{DDQ}, Ardext, which has commercialized the analog test technique described in [27], and DFT Microsystems, which is commercializing the BIST techniques described in [28]. Prior to the last few years, the only publicized start-ups were Opmax, that provided BIST described in [6], [47] (though no longer offered), and LogicVision, which continues to provide digital parametric DfT and BIST.

The complexity for mixed-signal DfT must shift to digital analysis algorithms and away from analog circuitry. It will be necessary to minimize, but not eliminate, analog portions in the test circuitry. New techniques must continue to embody the essential characteristics: summing for precision, subtracting for accuracy, self-testing, accuracy/time trade-off, and diagnosis for yield improvement.

REFERENCES

[1] M. Sachdev, "A Realistic Defect Oriented Testability Methodology for Analog Circuits", *Journal of Electronic Testing: Theory and Applications*, vol. 6, no. 3, pp. 265-276, June 1995.

[2] A. Grochowski, D. Bhattacharya, T. Viswanathan, K. Laker, "Integrated Circuit Testing for Quality Assurance in Manufacturing: History, Current Status, and Future Trends", *IEEE Trans. on Circuits and Systems-II*, vol. 44, no. 8, August 1977.

[3] P. Fasang, D. Mullins, T. Wong, "Design for Testability for Mixed Analog/Digital ASICs," *Proc. of IEEE Custom Integrated Circuits Conf.*, pp. 16.5.1-5.4, May 1988.

[4] M. Ohletz, "Hybrid Built-In Self-Test (HBIST) for Mixed Analogue/Digital Integrated Circuits", *Proc. of European Test Conf.*, 1991.

[5] C. Pan, K. Cheng, "Test generation for linear time-invariant analog circuits," *IEEE Trans. on Circuits and Systems-II*, vol. 46, no. 5, pp. 554-564, May 1999.

[6] K. Arabi, B. Kamaniska, "Testing Analog and Mixed-Signal Integrated Circuits Using Oscillation Test Method", *IEEE Trans. on Computer Aided Design*, vol. 16, no. 7, pp. 745-753, July 1997.

[7] M. Toner, G. Roberts, "A BIST Scheme for SNR, Gain Tracking, and Frequency Response Test of a Sigma-Delta ADC", *IEEE Trans. on Circuits and Systems-II,* vol. 42, no. 1, pp. 1-15, January 1995.

[8] A. Chatterjee, B. Kim, N. Nagi, "DC Built-In Self-Test for Linear Analog Circuits", *IEEE Design & Test of Computers*, pp. 26-33, vol.13, no. 2, 1996.

[9] A. Brosa, J. Figueras, "Characterization of Floating Gate Defects in Analog Cells", *Journal of Electronic Testing: Theory and Applications*, vol. 14, nos. 1/2, pp. 23-31, February/April, 1999.

[10] S. Khaled, N. Hamida, D. Marche, B. Kaminska, "LIMSoft: Automated Tool for Sensitivity Analysis and Test Vector Generation", *IEEE Proc. on Circuits, Devices and Systems*, pp. 386-392, December 1996.

[11] W. Lindermeir, H. Graeb, K. Antreich, "Design Based Analog Testing by Characteristic Observation Inference", *Proc. of Int'l Conf. on Computer Aided Design*, pp. 620-26, 1995.

[12] A. Balivada, J. Chen, J. Abraham, "Analog Testing with Time Response Parameters," *IEEE Design & Test of Computers*, pp. 18-25, vol.13, no. 2, 1996.

[13] T. Williams, N. Brown, "Defect Level as a Function of Fault Coverage", *IEEE Trans. on Computers*, pp. 987-88, December 1981.

[14] J. Teixeira de Sousa, F. Goncalves, J. Teixeira, C. Marzocca, F. Corsi, T. Williams, "Defect Level Evaluation in an IC Design Environment", *IEEE Trans. on Computer Aided Design of Integrated Circuits and Systems*, vol. 15, no. 10, pp. 1286-1293, October 1996.

[15] P. Maxwell, R. Aitken, V. Johansen, I. Chiang, "The Effect of Different Test Sets on Quality Level Prediction: When is 80% Better Than 90%?", *Proc. of Int'l Test Conf.*, October 1991.

[16] S. Sunter, N. Nagi, "Test Metrics for Analog Parametric Faults", *Proc. of VLSI Test Symp.*, pp. 226-34, April 1999.

[17] N. Nagi, A. Chatterjee, J. Abraham, "Fault Simulation of Linear Analog Circuits", *Journal of Electronic Testing: Theory and Applications*, vol. 4, no. 4, pp. 345-360, November, 1993.

[18] C. Sebeke, J. Teixeira, M. Ohletz, "Analog fault extraction and simulation of layout realistic faults for integrated analogue circuits", *Proc. of European Design and Test Conf.*, pg. 464, 1995.

[19] H. Johnson, G. Martin, *High Speed Digital design: A Handbook of Black Magic*, Prentice Hall, Englewood Cliffs, NJ, 1993.

[20] M. Mahoney, *DSP-Based Testing of Analog and Mixed-Signal Circuits*, IEEE Computer Society Press, 1987.

[21] M. Burns, G. Roberts, *An Introduction to Mixed-Signal IC Test and Measurement*, Oxford University Press, USA, 2001.

[22] S. Max, "Testing High Speed High Accuracy Analog to Digital Converters Embedded in Systems on a Chip", *Proc. of Int'l Test Conf.*, pp. 763-771, September 1999.

[23] T. Yamaguchi, M. Soma, D. Halter, R. Raina, J. Nissen, M. Ishida, "A Method for Measuring the Cycle-to-Cycle Period Jitter of High-Frequency Clock Signals", *Proc. VLSI Test Symp.*, pp. 102-110, April 2001.

[24] Std. 1149.1a-1993, *IEEE Standard Test Access Port and Boundary-Scan Architecture*, IEEE, New York, 1993.

[25] S. Cherubal, A. Chatterjee, "A High-Resolution Jitter Measurement Technique Using ADC Sampling", *Proc. of Int'l Test Conf.*, pp. 838-847, October 2001.

[26] L. Milor, A. Sangiovanni-Vincentelli, "Minimizing Production Test Time to Detect Faults in Analog Circuits", *IEEE Trans. on Computer-Aided Design of Integrated Circuits*, vol 13, no. 6, pp. 796-813, June 1994.

[27] P. Variyam, S. Cherubal, A. Chatterjee, "Prediction of Analog Performance Parameters Using Fast Transient Testing", *IEEE Trans. on Computer-Aided Design of Integrated Circuits*, vol. 21, issue 3, pp. 349-361, March 2002.

[28] M. Hafed, G. Roberts, "Test and Evaluation of Multiple Embedded Mixed-Signal Test Cores", *Proc. of Int'l Test Conf.*, pp. 1022-30, October 2002.

[29] K. Arabi, B. Kaminska, J. Rzeszut, "BIST for D/A and A/D Converters", *IEEE Design & Test of Computers*, pp. 40-49, vol. 13, no. 4, 1996.

[30] A. Roy, S. Sunter, D. Appello, A. Fudoli, "High-Accuracy Stimulus Generation for A/D Converter BIST", *Proc. of Int'l Test Conf.*, pp. 1031-1039, October 2002.

[31] G. Monté, B. Antaki, S. Patenaude, Y. Savaria, C. Thibeault, P. Trouborst, "Tools for the Characterization of Bipolar CML Testability", *Proc. of VLSI Test Symp.*, pp. 388-395, April 2001.

[32] Std. 1149.4-1999, *IEEE Standard for a Mixed Signal Test Bus*, IEEE, New York.

[33] *Analog and Mixed-Signal Boundary Scan*, ed. A.Osseiran, Kluwer Academic Publishers, 1999.

[34] S. Sunter, K. Filliter, J. Woo, P. McHugh, "A General Purpose 1149.4 IC with HF Analog Test Capabilities", *Proc. of Int'l Test Conf.*, pp. 38-45, October 2001.

[35] S. Sunter, B. Nadeau-Dostie, "Complete, Contactless I/O Testing – Reaching the Boundary in Minimizing Digital IC Testing Cost", *Proc. of Int'l Test Conf.*, pp. 446-455, October 2002.

[36] K. Parker, J. McDermid, S. Oresjo, "Structure and Metrology for an Analog Testability Bus", *Proc. of Int'l Test Conf.*, pp. 309-22, October 1993.

[37] B. Provost, E. Sanchez-Sinencio, "Auto-Calibrating Analog Timer for On-Chip Testing", *Proc. of Int'l Test Conf.*, pp. 541-548, September 1999.

[38] F. Azaïs, S. Bernard, Y. Bertrand, X. Michel, M. Renovell, "A Low-Cost Adaptive Ramp Generator for Analog BIST Applications", *Proc. of VLSI Test Symp.*, pp. 266-71, April 2001.

[39] E. Peralias, G. Huertas, A. Rueda, J. Huertas, "Self-Testable Pipelined ADC with Low Hardware Overhead", *Proc. of VLSI Test Symp.*, pp. 272-277, April 2001.

[40] A. Chatterjee, B. Kim, N. Nagi, DC Built-In Self-Test for Linear Analog Circuits. *IEEE Design & Test of Computers*, vol. 13, no. 2, 1996.

[41] S. Sunter, A. Roy, "BIST for Phase-Locked Loops in Digital Applications", *Proc. of Int'l Test Conf.*, pp. 532-540, September 1999.

[42] S. Tabatabaei, A. Ivanov, "Embedded Timing Analysis: A SoC Infrastructure," *IEEE Design & Test of Computers*, vol. 19, no. 3, pp. 22-34, May-June 2002.

[43] *Analog and Mixed-Signal Testing*, ed. B.Vinnakota, Prentice Hall, USA, 2000.

[44] J. Savir, Z. Guo, "Test Limitations of Parametric Faults in Analog Circuits", *IEEE Trans. on Instrumentation and Measurement*, vol. 52, no. 5, pp. 1444-1454, October 2003.

[45] L. Jin, K. Parthasarathy, T. Kuyel, D. Chen, R. Geiger, "Linearity Testing of Precision Analog-to-Digital Converters Using Stationary Nonlinear Inputs", *Proc. of Int'l Test Conf.*, pp. 218-227, September 2003.

[46] D. Keezer, D. Minier, M. Paradis, M. Caron, "Modular Extension of ATE to 5 Gbps", *Proc. of Int'l Test Conf.*, pp. 748-757, October 2004.

[47] S. Sunter, "Testing High Frequency ADCs and DACs with a Low Frequency Analog Bus", *Proc. of Int'l Test Conf.*, pp. 228-235, September 2003.

[48] S. Sunter, A. Roy, J-F. Côté, "An Automated, Complete, Structural Test Solution for SerDes", *Proc. of Int'l Test Conf.*, pp. 95-104, October 2004.

[49] A. Frisch, T. Almy, "HABIST: Histogram-based Built In Self Test", *Proc. of Int'l Test Conf.*, pp. 760-767, November 1997.

Chapter 10

RF Testing

Randy Wolf, Mustapha Slamani,
John Ferrario and Jayendra Bhagat

10.1 INTRODUCTION

Today's wireless communication products are increasingly complex and more integrated than ever before. The low prices that consumers pay for wireless phones in a competitive market demand low-cost Radio Frequency Integrated Circuits (RF ICs). The test cost has become an important factor in determining the profit margin. To economically test high volumes of RF ICs at a fraction of the IC cost, we must adjust our existing test methods and define new test strategies. As pointed out in International Technology Roadmap for Semiconductors (ITRS) 2003 [1], "*Customer requirements for form factor and power consumption are driving a significant increase in design integration levels. . . Test complexity will increase dramatically with the combination of different classes of circuits on a single die or within a single package. In particular, for System-in-Package (SIP) increased focus on known good die and sub-assembly test will be driven by the cost issue*". The commercial wireless industry has driven a need for very low cost RF ICs built with very low cost packages and manufacturing processes. A key contributor to the cost of manufacturing an RF IC packaged part is the module final test. Up until that step in the manufacturing process, the components can be handled in a batch mode with standard high volume wafer fabrication and package part assembly equipment. Once the part hits RF test, it must be individually placed in a precision socket with precision pressure, and electromagnetic isolation, and tested at a very narrow band,

D. Gizopoulos (ed.), Advances in Electronic Testing: Challenges and Methodologies, 337-369.

high frequency and low signal level. The ability to mechanically handle individual components and place them in a precision socket quickly and repeatable has been addressed by the commercial handler manufacturers with a range of efficiencies.

The actual RF ICs are electrically tested with either of a rack and stack bench top equipment connected to a PC, or with commercially available Automatic Test Equipment (ATE). Usually, the most costly and complex component in these systems are the RF receiver, or spectrum analyzer/digitizer, and the RF source(s). Figure 10-1, gives an idea of how basic ATE test cost increase when incorporating mixed signal and RF options to it. Most systems are configured to handle only one receiver per system and up to four optional sources. Receivers must handle a frequency range between 100MHz and 6GHz and have a very high dynamic range capable of measuring stringent two tone signals such as Adjacent Channel Power (ACPR) or Third Order Intercept Point (IP3). These signals are difficult to measure because they consist of a primary high power frequency or tone at 900MHz to 6GHz which is adjacent to a very low level noise tone 1MHz away which must be measured repeatable to 0.1dB accuracy. The high susceptibility of the Device Under Test (DUT) to electromagnetic noise from its immediate surroundings and the need for an extremely sensitive, precision RF receiver to be able to make these types of measurements prohibit parallel site testing. The sources must be capable of providing up to 6GHz with low phase noise and a power output between –120dBm to 13dBm in 0.1dB steps.

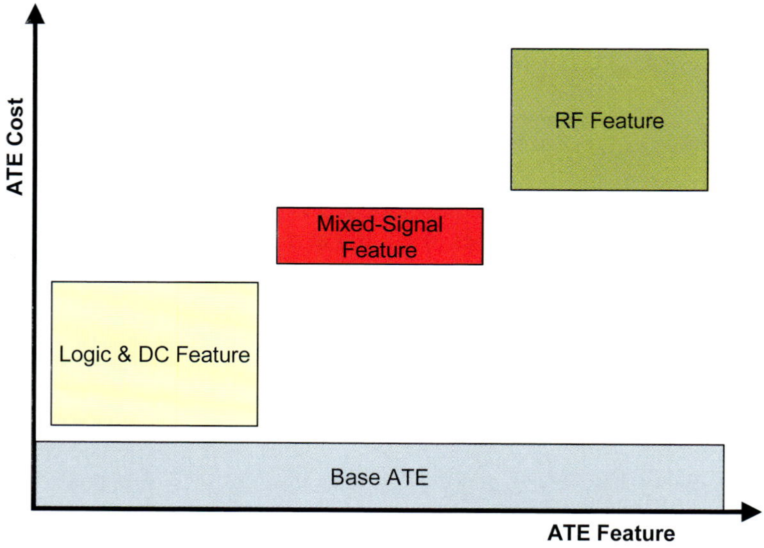

Figure 10-1: ATE cost increase with additional mixed signal and RF features.

This chapter describes methods to address the constraints of RF testing. It provides a discussion of the characteristics of an RF test system developed in IBM that incorporates sub-circuits that can be included to the RF test board to convert the RF signal to a DC signal. This critical step has a major impact to the cost of test for an RF device by converting the test system from a complex RF single site tester to an extremely fast, inexpensive multi-site DC tester. The result of this approach drives the cost of test of these systems to that of a high throughput DC parametric tester. The sources are designed with high precision components while the key components, such as a low noise Voltage Control Oscillator (VCO), are designed for the frequency band of interest for the Device Under Test (DUT). Each successive frequency band utilizes the same circuit board with the same VCO package.

Several RF testing approaches have been recently proposed [2], [3], [4], [5], [6]. In this chapter we keep our focus on the description of the development process of a specific low-cost oriented RF test system.

Before proceeding to the details of the test architecture, we discuss the types of RF ICs and the tests required for them along with the challenges in developing circuits to perform the tests: to ensure they are capable of making accurate measurements required by the test, and they are fast enough and repeatable to keep the cost down for high volume manufacturing testing.

10.2 TESTING RF ICS

10.2.1 RF IC Categories

There are three basic categories of RF ICs:

1. Pure RF ICs; e.g. Low Noise Amplifier (LNA), Power Amplifier (PA), Voltage Control Oscillator (VCO), Mixer, etc.
2. Combined RF/Mixed Signal ICs; e.g. Wireless LAN (WLAN), Global System for Mobile Communication (GSM) or Digital Audio Broadcasting (DAB) Transceiver.
3. Combined RF/Mixed Signal/digital Base Band ICs; e.g. WLAN, GSM or Global Positioning System (GPS) System-on-Chips (SOC).

The first category performs a single RF function and has a low pin count requiring power, an RF receiver, possibly an RF source and a few, if any, digital controls. Typical RF tests required for the first category are:

1. Gain for the LNA, PA, Mixer.
2. Noise Figure (NF) for the LNA, Mixer.
3. 1dB Compression Point for the LNA, PA, Mixer.
4. Third Order Intercept Point (IP3) for the LNA, PA, Mixer.
5. Standing Wave Ration (SWR) for the LNA, PA, Mixer.
6. Adjacent Channel Power (ACPR) for the PA.
7. Phase Noise, Tuning Voltage (VTune), Frequency Range for the VCO.

The second category performs several Radio Frequency (RF) and Intermediate Frequency (IF) functions. In addition to the requirements of the first category, these usually require more digital pins and complex programming such as automatic gain control. Additional sources and receivers might be required to handle the dual bands that are characteristic of transceivers. Tests such as selectivity and sensitivity of the receiver require more than one source, and testing the transmitter's ACPR and harmonics/spurs requires more stringent receivers.

The third category is the most complex combining RF/IF functions with digital such as quadrature I and Q baseband signals, Analog to Digital Conversion (ADC), Digital to Analog Conversion (DAC) and Digital Signal Processing (DSP) functions, often called *System-on-chip* (SOC). These are high pin count devices requiring the most complex programming of the IC to thoroughly test all its functions. Additional tests include error vector magnitude (EVM), bit error rate (BER), phase locking, jitter, response times and digital test operations such as scan-based test, Automatic Test Pattern Generation (ATPG), and memory test.

10.2.2 RF Test Challenges

RF test challenges are summarized in signal integrity, de-embedding, modeling, correlation and DUT specification [7].

Signal Integrity

Many factors contribute to the complexity of RF testing. Signal integrity, the requirement for a clean DUT socket-to-board-to-tester path, and a 50-Ω environment are key elements for obtaining an accurate measurement. Minimum discontinuities in the signal path can disturb the measurement accuracy. A stable RF measurement requires high-performance contactors and a precise contact pressure. Because the RF signal levels are very low, good electromagnetic isolation and external noise immunity are required during testing, and the surrounding environment should emulate end use. The lack of good signal grounds near the DUT pins affects signal integrity. The situation becomes adverse in a SOC environment where digital blocks with 1.2V to 3V switching signals are physically near RF blocks with low-level RF signals, for example, –100 dBm for a receiver. In this case, isolation between the RF and high-speed digital signals in the DUT and test board becomes a requirement.

De-embedding

The objective of de-embedding is to find the losses between the DUT and the ATE system either in a vector or scalar form. The losses are used to offset the value of an RF measurement obtained by the ATE system. Usually, the procedure uses the measurements obtained by using short, open and load calibration standards and performs some mathematical operations to solve the equations.

During testing, the main concern is the uniformity of off-chip interconnects and the interface with the ATE, probes, sockets, load board, and so forth. This has direct influence on measured behavior, and an accurate measurement can be only obtained by a painful de-embedding procedure.

Modeling

Modeling the impact of the contactor and the test board is important for design stability and increasing test margins. Much work must be done to accurately model the RF signal path between the DUT and the tester before a successful first pass of silicon and the test board can be achieved. Parasitic effects on DUT performance (not just on measurement accuracy) are typically not well understood. Poor RF test hardware design methodology can lead to substantial delays and higher costs. Robust test hardware design methodology, a rich library, two- and three-dimensional simulation, parasitic simulation, and powerful simulation tools reduce the risk of an unsuccessful manufacturing test solution.

Correlation

Another major issue with RF testing today that can impact time-to-market (TTM) is correlation from tester to tester, customer to production, fixture to fixture, and offsets versus absolute accuracy. RF devices are typically very sensitive to their environment, a characteristic that manifests itself through board-to-board and tester-to-tester variation. The fudge factor (or offset) between the "golden"/expected and tester data saves time, helps solve hardware variation problems, and does not impact the go-no-go test decision.

DUT Specification

In recent years, RF circuits have performed better than the test equipment; the complexity of RF test specifications represents a technology bottleneck. Parameters such as noise, jitter, nonlinearity, bit error rate (BER), error vector magnitude (EVM), and modulation require state-of-the-art instrumentation. Digital modulation schemes, such as QAM16, QPSK, and W-CDMA, drive aggressive ATE hardware and software requirements.

Given all of these serious issues that need attention in testing RF components, how one tackles these items while improving quality and reduces test development time, and thus cost, is discussed in the next section.

10.3 RF TEST COST REDUCTION FACTORS

There are several factors that cause delays, and thus increase cost, in developing a successful RF test solution. Ways of decreasing delays and cost of RF test are:

<u>*At the test development stage:*</u>

1. Performance:

   Robust test hardware design methodology: library, 2D and 3D simulation, parasitic simulation tools.

2. Reduce Tester Cost:

   Utilization of Design-for-Test (DFT) on the test board such as:

   - Conversion of RF signals to low frequencies in the test board.
   - Assistance to the tester's digitizer to capture signals with high dynamic range.
   - Selection of the appropriate devices in the test board (like military spec devices).
   - Availability of an on-board RF source.

At the manufacturing stage:

1. Easy transition from test development stage to manufacturing stage:
   - A good de-embedding methodology: improve accuracy and variation from board-to-board and tester-to-tester.
   - Fudge factor (offset) between golden and tester data: save time and solve variation problems; no impact on go/no-go test.
2. Decrease the Device Interface Board (DIB) test circuitry utilizing the DUT itself:
   - Chain test, loop-back test.
3. Improve throughput:
   - Parallel test, Ping Pong test.

It is important that the test engineer has a thorough understanding of the above items because there are other non-ideal test environment complications that are not always avoidable such as:

- Parasitics added by the socket/wafer probes causing decrease in isolation, stability and power/gain. This results in deviation from the "soldered down" test.
- RF ICs not designed with testability in mind – some argue this is an IC designer issue.
- Tests added during development stage.

10.3.1 Resources and Test Time Cost

The costs associated with test equipment for the time spent testing each device are the primary factors that impact the overall test cost as shown in Table 10-1 [8].

Contributor Test Time	Contributor Test Costs
Handler Capability	Test System Capital
Index Time of Handler	Handler Capital
Tester Capability/Speed (Electrical Test Time)	Operations Overhead (Operator, Building, Maintenance)
Handler/Tester/Controller Communications Time	Test Hardware & Software Development Engineering

Table 10-1: Test time and test cost contributors.

To better understand the impact these factors have on the test cost for one device, we generated a simple model from the following approximate cost assumptions:

- A handler costs $300K and depreciates at a fixed rate for five years.
- An ATE RF tester costs $1M and depreciates at a fixed rate for five years.
- Operation and maintenance of a tester and handler on the manufacturing floor costs $50 per hour. This estimate is a fixed cost that contains everything except capital equipment or test development engineering.
- Hardware and software engineering time to develop the test solution costs $150K. This estimate assumes the engineer spends three to six months on the test solution over a two-year program life.

Figure 10-2 emphasizes the relative importance of the test time on the overall test cost per device. The cost factors associated with testing a device at various test times are shown. The lowest limit in the test is set to 280 ms (a 200-ms handler index time[1] plus an 80-ms test time).

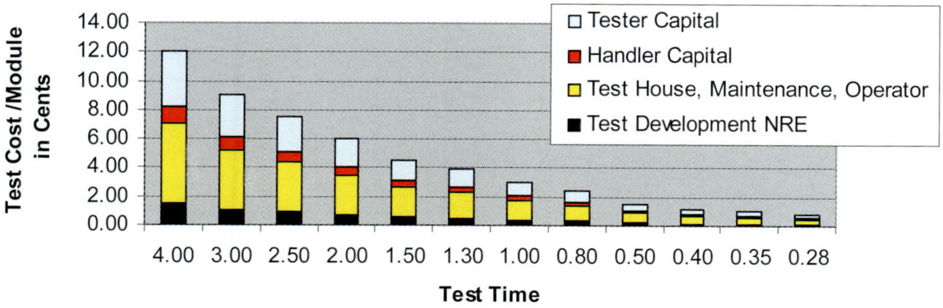

Figure 10-2: Test Cost per Module versus Test Time on a $1 Million Tester [8].

Note that the primary contributor to the device cost is the operations overhead cost followed by the test system capital cost.

According to this model, a 330 ms total test time using a $1 million tester leads to a one-cent-per-module total test cost adder. At 330 ms, each tester is capable of testing 50 to 90 million modules per year. For example, if the manufacturing tests floor runs 17 hours per day, 7 days a week, and 48 weeks per year (or 5712 hours/year) and the total capacity at a 330 ms total test time is 62 million modules. At this capacity, one or two testers can normally accommodate all the product requirements, reducing the additional cost of maintaining multiple systems correlated across a test floor.

Based on the above assumptions, an RF IC test system is needed that could achieve a 300 ms (or less) per device test time using a $100K tester to reach the targeted one-cent-per-device test cost adder.

How to accomplish these two factors will be discussed starting with test time reduction and then the hardware cost.

Two things influence test times the most: the time for the test program to run and how fast can the handler move the parts between the bins and the socket or move the wafer. One way to reduce the program test time is to simplify the test measurement required such as converting a test signal to a DC parameter instead of digitizing it. This is covered in Section 10.4, but before moving to it we discuss about the handlers.

---

[1] An index time of 200 ms is given for a rotary-mode handler. Index time is the time from the end of testing a device to the start of testing the next device.

10.3.2 Handler

With test times in the milliseconds and on the order of the handler index time, test system designers recognize the importance of not decoupling the handler from the tester. In the sub-second regime, the interaction of the handler, tester, and controller play a critical role in the total test time. To develop a millisecond tester, the following steps are recommended:

1. Find the handler with the fastest index time that can serially place a component into the socket.
2. Develop the RF hardware to optimize the handler throughput and minimize capital investment.
3. Fine-tune the composite tester/handler combination for optimum throughput.
4. Run all general-purpose interface bus (GPIB) communications during the idle handler load time.

In item 1 note that the handler determines the tester used. It is not recommended developing a test on the best RF tester available and then moving it to a handler. Both the handler index time and the electrical test time must be understood to achieve test times below 1 second per device.

Handler Types Considered

Three types of handler operation modes have been considered and analyzed.

Rotary handler—The rotary handler has multiple arms and queues devices in an assembly-line fashion. Rotary handler tasks can be divided into shorter, parallel tasks. This feature enables the handler to reduce the effective index time to that of the longest individual task. The index time of the rotary handler used herein was 150 ms.

Pick-and-place handler—The pick-and-place handler can operate serially or in parallel to pick up the devices and place them in a test socket. The index time of the pick-and-place handlers used herein was approximately 1.2 s per device.

Pick-and-place handler in ping-pong mode—A pick-and-place handler in ping-pong mode has two sockets on the test board. While the tester is testing the device in one socket, the handler removes the device from the other socket and queues up a new device over the empty socket. When testing is finished on the first device, the handler quickly places the queued device into the second socket. If the test time is greater then 1.2 s, the index time is 0.3 s. If the test time is less then 1.2 s, the total test time plus the index time remains at 1.2 s.

In Figure 10-3, throughput for each of the three types of handlers is plotted against test time. Only serial pick-and-place or rotary handling was considered in the described RF testing system. Parallel testing of wireless devices was not pursued because of electromagnetic field coupling effects, increased noise level, and limited RF test receiver resources.

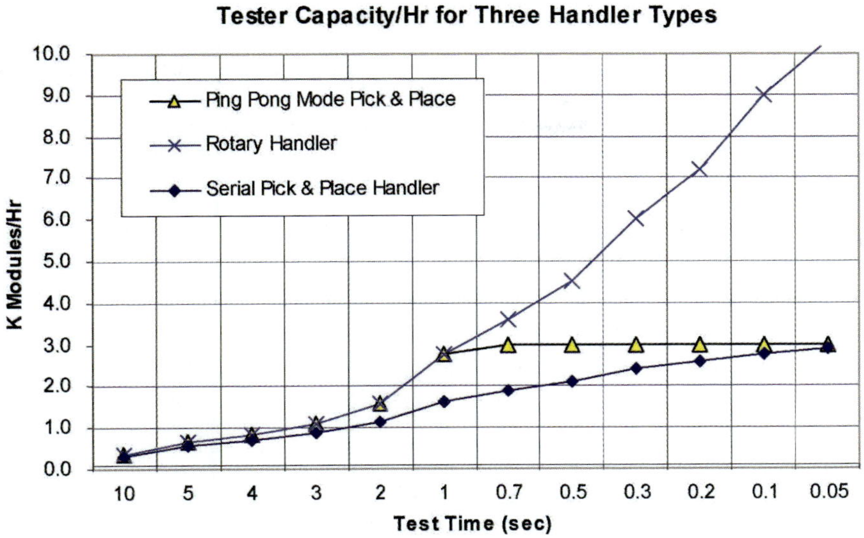

Figure 10-3: Tester capacity per hour for three types of handlers [8].

In Figure 10-3, note that ping-pong mode testing outperformed serial testing on a pick-and-place tool. For test times below 1 s, the rotary handler had a sizable throughput advantage. For test times below 0.1 s, the rotary handler throughput can be three times that of the pick-and-place handler. Although the rotary handler enabled dividing the test into eight subtasks, division was not necessary for this test. In this case, parallel testing may have been possible due to the longer distance between fixtures. Parallel testing would have enabled shielding, but would not have saved as much time as the parallel pick-and-place handler.

Handler Types Not Considered

Other handlers exist that were not considered in the comparison that took place during the development of the described RF testing system. Examples of these handlers are:

Strip handlers[2]—Strip handlers were not considered because of the high tester and handler capital cost. Strip handlers require major changes in the packaging lines and cause engineering problems, such as those associated with trying to probe multiple adjacent RF components. To touch the lead frame of a device in a strip handler, we use a probe similar to the one used for wafer test instead of a socket. When using a probe instead of a socket, it is more difficult to get the matching circuit of a lead as close to the device. This added distance increases the probe inductance and makes matching more difficult and sensitive to device variations. In addition, because the ground plane is shared in a strip handler configuration, the number of parallel devices can amplify the noise induced in the ground plane. If the device is tested serially on a strip handler to avoid potential crosstalk issues, the individual index

[2] For more information see http://www.amkor.com/services/test/striptest.cfm

time is removed, but the RF tester receiver must be routed to each of the test sites. For package manufacturers currently using strip handlers, further investigation into the business case for developing RF tests on a strip handler may be worthwhile as parallel RF testers become more economical.

Multisite pick-and-place or gravity-feed handlers—we have not considered these type of handlers because of the RF crosstalk effects between adjacent sites on the test board and the complexity of parallel RF measurement circuits. These handlers can be used to serially test the RF portion and parallel test the direct current. When used this way, we found that the RF tests consumed 60–80% of the test time. For example, given a 1 s test time with parallel testing on 30% of the tests, then:

(1 s for 2 devices + 0.7 s serial test overlap) $\geq$ 0.85 s per device

(1 s for 4 devices + 2.1 s serial test overlap) $\geq$ (1 + 2.1)/4 = 0.78 s per device

(1 s for 8 devices + 4.9 s overlap) $\geq$ 5.9/8 = 0.74 s per device

Therefore, as the number of sockets increases:

- the test time per device decreases,
- the jam rate associated with getting four or eight small RF devices correctly aligned in the socket increases (we struggled with two devices in parallel), and
- the complexity of the tester and a test board capable of switching the RF signal to multiple sites increases.

Multi-site parallel testing may be economical on devices which have only a short test time for RF section and a long test time for mixed signal section. In this case RF tests can run serially while the mixed signal tests are run in parallel. This avoids the RF coupling problem between sites. Given the above-stated problems with parallel RF test, we concluded that only serial handlers either pick-and-place or rotary can be used for RF testing.

10.4 TEST HARDWARE

The cost of getting an RF product onto a tester is primarily driven by the test fixture hardware design, build and debug. The test code development time is a known, predictable quantity that normally takes less time than the design of the hardware. A poor hardware methodology on the other hand can easily lead to substantial delays and cost over runs. Having robust RF board design and simulation skills and tools is absolutely critical to containing schedules, development costs and surviving in the RF test business. Regardless of the tester selected, an RF test engineer is in the RF board design business. Because of this, the subject of RF test hardware development methodology is extensively covered in this chapter. Another necessary topic that is covered is the development of sub-circuits providing the RF test functions and the cost reduction methods associated with them since $100K testers do not come with many options, especially RF test capability. First however, a "universal" board design is discussed because it provides several advantages in reducing costs associated with test. These advantages are:

1. Speed in verification of the hardware design since its ATE platform independence allows verification with established calibrated bench equipment.

2. Flexibility to the RF testing company in using different ATE manufacturers.
3. Lower building cost than the full ATE platform boards that interface with the ATEs test head.

10.4.1 Universal Test Board

What makes an RF test board "universal" is that it allows one to port the RF test solution from one ATE platform to another or to a test bench setup. To accomplish this, we have developed a system of interface boards that bridge the uniqueness of various testers: an inner common design RF board surrounded by an outer unique "patch" board is employed (see Figure 10-4 and Figure 10-5). The patch board contains the connections to the ATE using the form factor dictated by the ATE. The RF DUT test board (inner board) contains the test socket and test circuits. Its shape is not dictated by the ATE and so it is designed to mate only with the outer boards. Its circuits are designed specifically for each RF product and independently of the tester platform. The outer board and inner board are connected together by several high-density, controlled impedance ribbon cables. The key benefit of doing this is that it allows RF development on one specific model of tester with manufacturing production occurring on another model without redesign. This opens up the number of test houses available to perform testing for this product. Since the outside interface board usually requires being thick with hundreds of connectors, or pads with vias to route internally, the cost is high. This is another cost reduction because it requires building only one of these boards per tester, and swapping in and out the less costly, project specific, inner board.

Using a third-party tester as an example, Figure 10-4 shows the "Outer-board/ Octo Inner board" arrangement.

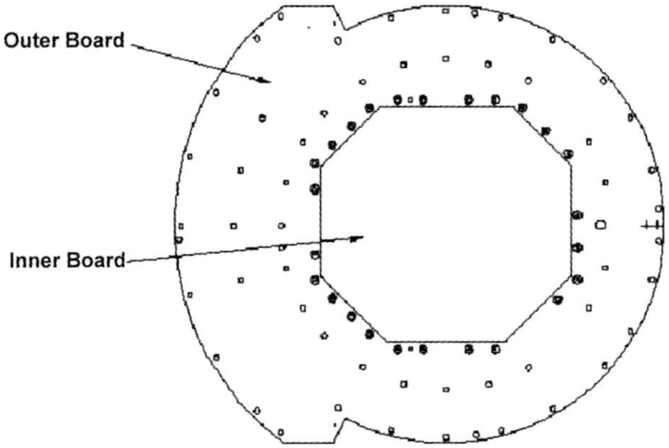

Figure 10-4: Outer /Inner test board.

A photograph of such a test board is shown in Figure 10-5. The product is currently in volume production within IBM.

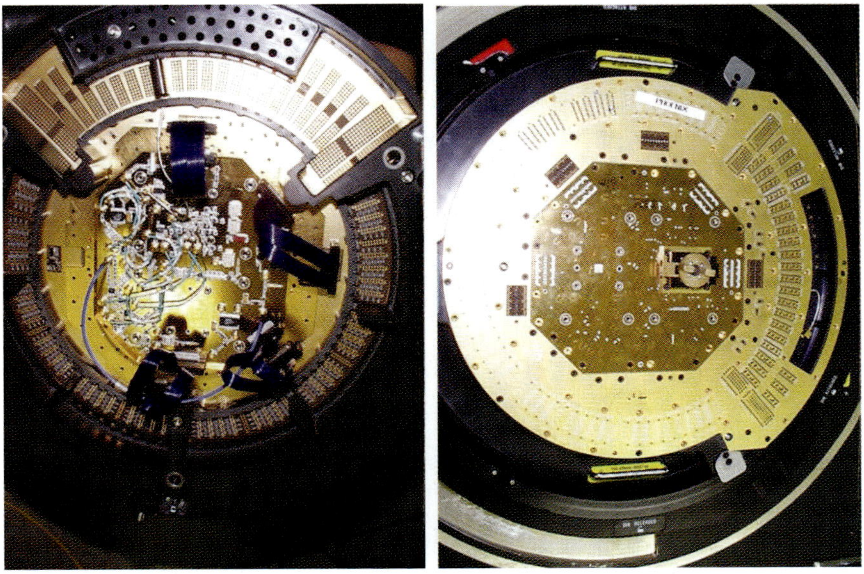

Figure 10-5: Bottom and top view of ATE support board and Octo board.

10.4.2 RF Test Function Sub-Circuit Design

Commercial RF ATE systems consist of microwave sources and receivers. Both the sources and the receivers are designed to cover a wide range of frequencies and power levels. In the initial stages of characterization, this flexibility is a necessary feature, which enables very rapid program changes. But once the part reaches the final stage of production, only a few worst-case frequencies or power levels are actually required to guarantee the performance of the device. The degree of purity and dynamic range of these few critical frequencies can directly affect the test time. Noise in the signal requires an increase in measurement averaging which results in increase test time. When the measurement requirement becomes very limited, such as the measurement of one tone that is in proximity to another tone that is higher in power, then a focused narrowband receiver can often do a better job than a broadband receiver. The ability of a receiver to select a signal without interference from an adjacent signal is called *selectivity* [9]. Another receiver parameter is *sensitivity*. This is the receiver's ability to measure a given signal relative to the receiver's noise floor [9]. A circuit that is specifically designed with passive filters for one frequency, and mounted on the test board as close as possible to the DUT, has a higher signal to noise ratio and requires less averaging. The reason for this is that the narrowband filter has less noise bandwidth going to the detector/digitizer. Other factors such as the dynamic range of the digitizer or detector diode can have a role in this parameter as well. The cost of having a fully loaded ATE system and the cost of test time becomes a large factor of the total test cost when the part goes into mass production test. To eliminate the need of purchasing fully loaded ATE systems and to reduce the test time, five key circuits can be developed that are easily added to the test board directly or connected to it as a "daughter" board. These circuits are discussed in the following subsections and have been implemented in the studied system.

RF Power Gain/Noise Figure Detector

Noise Figure is a sensitive test that requires good calibration and involves measuring very low level signals that are easily corrupted through external RF interference. Figure 10-6 is a circuit used to measure the noise figure and gain of a device under test (DUT). The noise diode generates a broad band RF signal which is filtered in the narrow band filter. This RF signal is referred to as *"white" noise* since it ideally has a Gaussian power distribution equal across a broad frequency spectrum. The power can be determined by the *Excess Noise Ratio* (ENR) given for that diode. As the name applies, the ratio is the difference between the diode's equivalent temperature (T_e) and 290°K, over 290°K. The relationship between T_e and the diode's power is determined with Boltzmann's constant. Example: at room temperature standard of 290°K, the power density is –174dBm/Hz. A good noise diode has an ENR between 20 and 30dB. The ENR is calculated by measuring its output power, but care must be taken since amplification, which adds its own noise, is often required to measure the low power output from the diode. A good reference for understanding noise generation and Noise Figure measurement can be found in [10].

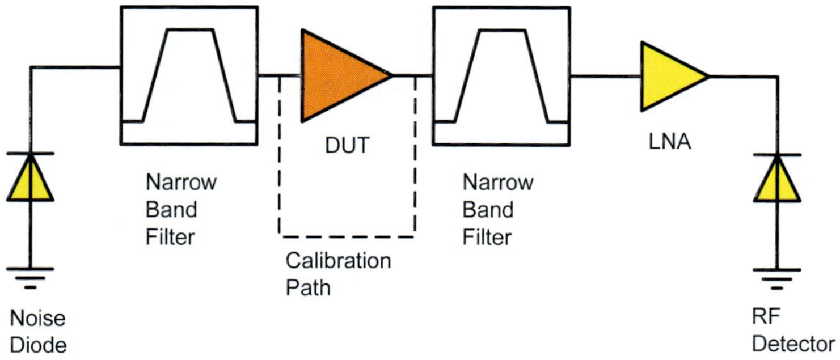

Figure 10-6: Noise figure measurement block diagram.

What is not shown in this circuit is the switching between "hot" and "cold" paths that take place between the noise diode and the filter. The "hot" path allows the full power from the diode to pass into the filter, which is typically –155dBm/Hz to –145dBm/Hz depending on the construction of the diode. The "cold" path attenuates the power by several magnitudes so the output power is just above -174dBm/Hz so it is more stable with fluctuating temperature. On the receiver side, a narrow band filter is used to filter out any out-of-band noise. The Low Noise Amplifier (LNA) mounted within an inch of the part is used to amplify the signal into the RF detector. Use of temperature compensating biasing results in a circuit that is not sensitive to temperature. The total uncompensated error in the circuit between –20° C and 80° C was 1dB. We did mount a thermistor on the board in our initial design for temperature compensation, but so far, we have found it unnecessary. Alternating "hot" and "cold" noise through the DUT, results in two power measurements. The ratio of these powers with hot over cold allows the Y Factor Method[3] to be used to determine the gain and noise figure of the DUT and the receiver. The effects of the receiver can be subtracted by determining the noise figure of the receiver using the same method, but without the DUT. In practice, the receiver's noise and gain contribution is previously determined and kept in a look up table. Calibration is done by switching to a bypass circuit of the DUT to the receiver where power measurement is compared to the look up table. This circuit has been used on a dual band LNA receiver chip to reduce the test time from 1 sec/part to 150msec/part. Both the source and the detection of the circuit have been done with an HP4142 DC source monitor channel.

Phase Noise and the FM Discriminator

Figure 10-7 is a Phase noise circuit that was used to measure the phase noise of free running RF voltage control oscillators (VCO). In this circuit the RF signal is split with a 90 degree phase shift so that the signals that meet at the mixer will cancel if the phase noise is zero. If the phase noise is not zero then only the low frequency phase error comes out of the mixer. The signal is then passed through a low pass filter to eliminate the high frequency term from the mixer. A DC sense line is used to adjust "VTune" on the VCO so the frequency through the delay line is quadrature to the other path at the mixer's inputs. The delay line is to de-correlate the paths; the more the delay, the higher the sensitivity of the discriminator, but a compromise has to be made with the loss in the line and frequency offset limitation due to the sinc function [11]. A narrow band filter is used to place a window around the phase noise offset of importance which is consistent with the test requirements. The quality factor Q of this filter is extremely important but achievable with off the shelf components. The RF detector then measures the power in that window without the need for averaging or stepping through the window in discrete frequency steps. A complex RF phase noise measurement is now reduced to a simple DC voltage measurement.

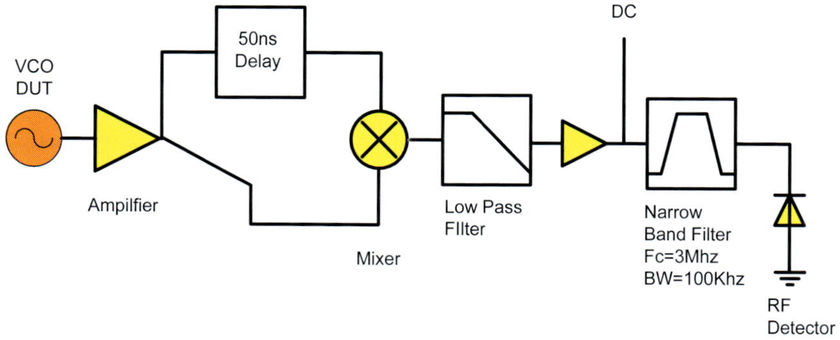

Figure 10-7: Block diagram of FM discriminator.

[3] The Y Factor Method consists of a calculation of the ratio (Y) of the "hot" power measurement over the "cold" power measurement and then a calculation of the noise factor f given by the equation: $f = [(T_h/T_o - 1) - Y (T_c/T_o - 1)] / (Y - 1)$. T_h is the equivalent temperature of the hot source, T_c is the equivalent temperature of the cold source and T_o=290K.

ACPR and IP3 Measurement

IP3 or ACPR measurements both involve measuring a small tone adjacent to a larger tone like those shown in Figure 10-8.

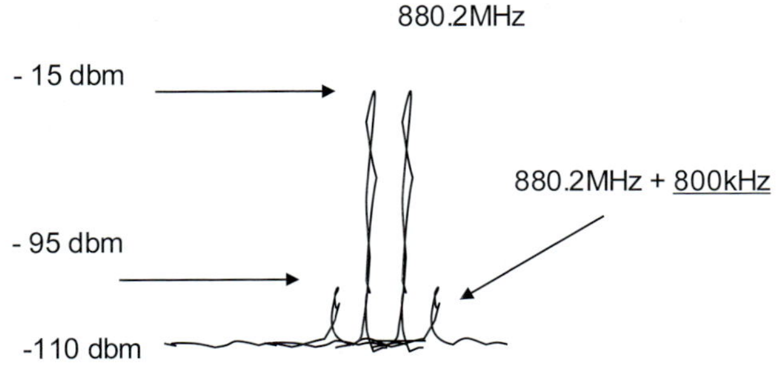

Figure 10-8: Two tone test for IP3 measurement.

The problem is that the larger tone will saturate the receiver due to limitations in the receiver dynamic range. Filtering out the primary signal is not an option due to how close the adjacent signal is to the primary relative to the high primary signal frequency. So the signal is down converted with a mixer to a lower frequency where it is possible to filter out the primary signal (see Figure 10-9) because of the availability of higher Q filters at these lower frequencies.

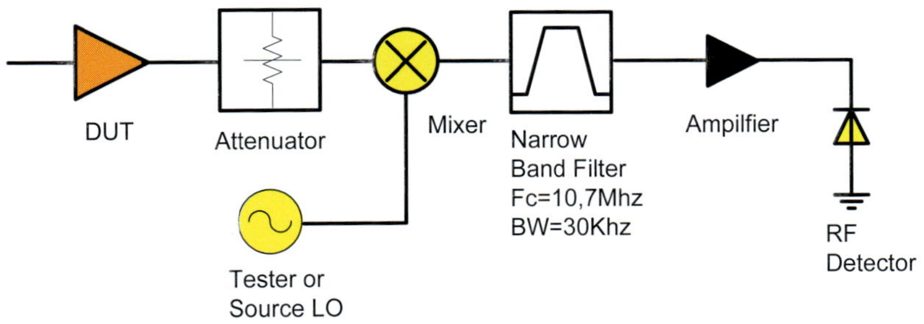

Figure 10-9: Block diagram of high selective receiver.

An attenuator is placed in between the DUT and the mixer to prevent reflections from the mixer getting back into the DUT. The narrow band filter is lined up to include just the adjacent tone. The RF detector is used to convert the power measurement into a DC voltage. A complex time-consuming IP3 measurement is reduced to a DC voltage measurement. Prior to using this circuit the IP3 test had been the longest test for LNA and mixer components. The ATEs take longer time and are less precise for IP3 measurement due to the limitation of their receiver dynamic range. After implementation of the circuit the test time was reduced by 100 times for both the overall test as well as this specific test.

RF and Local Oscillator Source

Figure 10-10 is a circuit used when a specific tone is required for either a local oscillator (LO) or a mixer input and replaces the tester's RF source. The frequency of the circuit is limited to a narrow band of the VCO. The PLL has digital control lines which are used to adjust the frequency. The RF sense detector is used to monitor the variable attenuator, VCO and amplifier power. An external frequency counter is used to periodically check the VCO frequency. The PLL was an off-the-shelf, self contained part. There are commercially available modules that include the entire circuit and are controlled via several digital pins. We preferred this approach so we would have more control over the VCO source and filtering. The spectral purity of this source was compared with the company's commercial RF manufacturing test systems. Not only did the source have better phase noise but it also leads to a lower test time when used as an LO source for an IP3 measurement. The drawback of using these circuits as a source is the limited frequency range. The wider the frequency range of the VCO, the worse the phase noise, so we tended to use very limited frequency range VCOs.

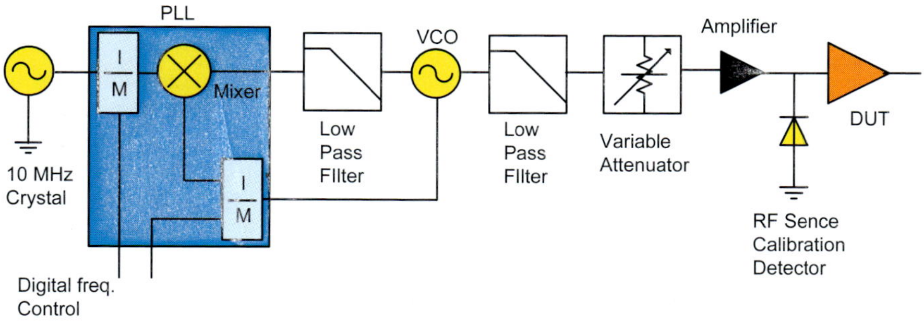

Figure 10-10: Block diagram of frequency and amplitude adjustable source.

S-Parameter Measurement

Testing Scattering (S)-parameters[4] requires the accurate measurement of two power levels. For S11 or S22 measurement, the powers to be measured are the incident and reflected powers. Figure 10-11 shows a block diagram of the key components in accomplishing this. A directional coupler is often used in place of the circulator because it has a wider bandwidth, but circulators have the advantage of no coupling loss, which means it can measure lower return loss. In both cases, the main limitation in accurate S11 and S22 measurement is the isolation of the Circulator, or in the case of using a directional coupler, the directivity. The reason is that the "forward" power going to the DUT is often much higher than the reflected power so any forward power that leeks into the return path to the detector can mask the reflected power. Knowing the isolation/directivity will determine the error in measurement for a given return loss measurement [9].

---

[4] S parameters are reflection and transmission coefficients of a network.

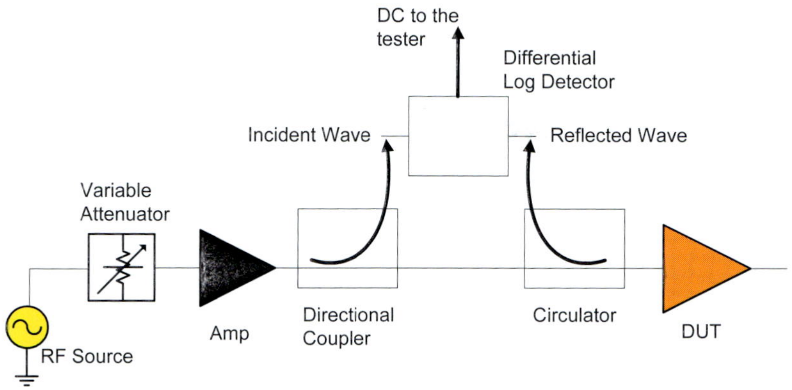

Figure 10-11: Block diagram showing S11 measurement.

S21 measurement can be made by implementing the circuit shown in Figure 10-12. Directivity is not as important for this measurement as it is for the previous one. All S-parameter measurements can be made using couplers and differential log detectors. Several configurations can be done to allow measurement of S11, S22, S21 and S12 such as the use of switches or multiple couplers in the path of the DUT. Differential log detectors that are available on the market have large dynamic ranges in power measurement to allow measurements to be made up through the cellular bands. The DC output vs. RF power input is characterized with a sine wave. Some detectors also characterize other inputs available like Gaussian noise or standard WCDMA, but sine wave is often used for these measurements.

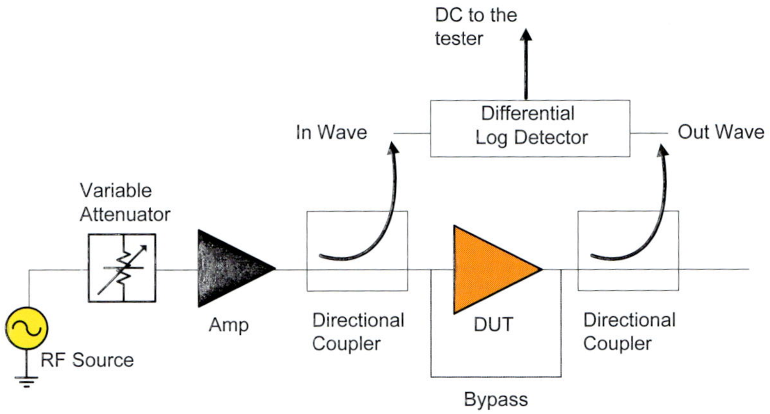

Figure 10-12: Block diagram showing S21 measurement.

10.4.3 Complete Test Architecture

The individual sub-circuits discussed in the previous sections are assembled in an RF test architecture that allows the use of low cost testers in order to reduce the overall RF test cost. The universal test structure utilizing the RF building blocks is shown in Fig 10-13. The on-board circuitry is used to interface with the low cost tester. The

goal is to make the architecture composed of RF building blocks more portable across multiple IC projects and test platforms. As an example, the main signal path for testing the LNA input (S11 for example, see Figure 10-11) is illustrated in Figure 10-13 with (highlighted) heavy black lines.

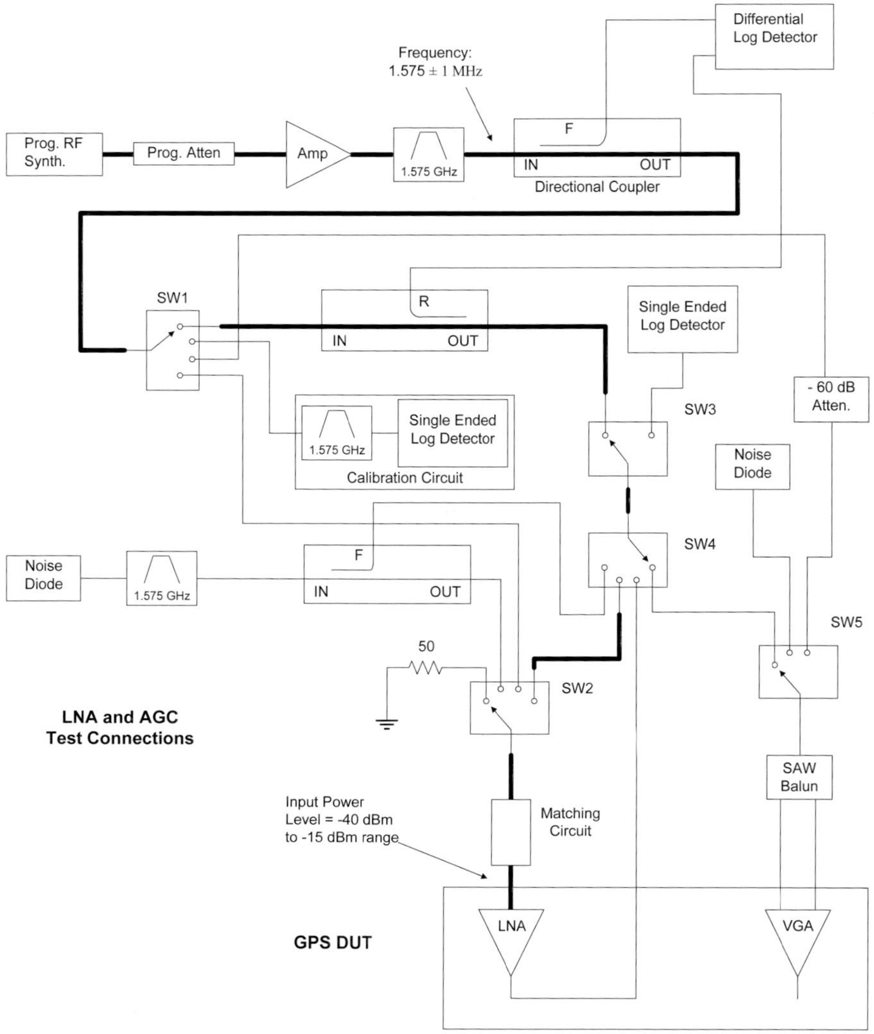

Figure 10-13: Block diagram of on-board design for test circuitry.

In this GPS (Global Positioning System) DUT testing example, all of the RF testing involves the testing of the front end low noise amplifier (LNA) and the following receiver variable gain amplifier (VGA) input circuits. Beyond the VGA, in this example, the signal is down converted internally by the GPS device to frequencies below RF so that testing reduces to digital and mixed signal testing. For the LNA, however, the test specification requires that S11, S22, S21, gain compression and noise figure to be measured at RF frequencies. For the VGA and

receiver circuits that follow, the tested measurements are VGA S11 (RF test), automatic gain control (AGC), loop sensitivity (lower frequency digital test), AGC noise sensitivity (lower frequency digital test), AGC image rejection (lower frequency Fast Fourier Transform - FFT), AGC gain (lower frequency FFT), and AGC dynamic range (lower frequency digital test). The stimuli for these tests are either the RF frequency synthesizer outputs or noise signals that have been band-pass filtered. The frequencies of interest in both cases are in the vicinity of 1.575GHz and the signal levels vary from –30dBm to –100dBm.

The test structure of Figure 10-13 is composed of a programmable RF synthesizer, Logarithmic power detectors (Log Detectors), noise figure measurement circuitry, directional couplers, band pass filters, fixed and variable attenuators, calibration circuitry and switching matrix. Calibration circuitry is an important feature in this test architecture. Calibration of the signal sources and detectors must be performed initially. It is always a good practice to calibrate periodically, in order to make sure that both the calibration monitoring circuits are functioning properly and that the test measurement circuitry is still within calibration limits.

There are two RF sources types that need calibration in Figure 10-13. One is the programmable RF synthesizer and the other is the noise diode. There are several RF test points located on the test board so that RF bench equipment can be plugged into the board to monitor signal power levels and spectra during the calibration process. There are also two RF logarithmic detectors that need calibration. They have RF test points available for connecting external equipment. Software has also been written to determine the slope and offset of the output DC voltage from these detectors with varying input power levels at the frequency band of interest.

The RF switching matrix allows us to configure the connection of these sources to the RF detectors during calibration or the connection to the DUT during testing.

Test Improvements

Reuse of circuits allows new projects to pick up the design with reduced development time and implementation cost. In addition, for those tests where RF measurements are implemented with software algorithms, such as Discrete Fourier Transform (DFT) algorithms, there have been improvement factors of 7 times in test time and factors of 2 times in standard deviation [7]. Focusing specifically on test time reduction, the use of these building block circuits over the "as-delivered" solution on RF ATE is summarized in Table 10-2 for typical RF tests.

Parameter	Test time using ATE	Test time using On-board DFT	Test time improvement
Gain	40 ms	10 ms	300%
S11	60 ms	10 ms	500%
S22	60 ms	10 ms	500%
NF	80 ms	20 ms	300%
IIP3 with 85dB dynamic range and 800KHz spacing	384 ms	120 ms	220%

Table 10-2: Test time improvement for different parameters.

Table 10-3 shows the benefits of this methodology for a complete suite of tests applied to a particular device type. Data on the improved repeatability of test when utilizing these methods has been obtained and is presented in Table 10-4 for various RF measurements. The ATE standard deviation is the measured standard deviation of testing a single part repetitively using a conventional ATE outfitted with RF measurement capability. The on-board DFT column shows the standard deviation obtained on the same part utilizing a similar ATE without the RF measurement capability but employing the on-board design for test methods presented in the previous sections. In all cases, major improvements were observed.

Product	Test time reduction
Si Bipolar Preamp	70%
SiGe Integrated Transmitter	25%
SiGe Integrated Receiver	25%
802.11a,b WLAN	40%

Table 10-3: Test time improvements by product.

Parameter	ATE standard deviation	On-board DFT standard deviation
Gain	0.010 db	0.0038 db
S11	0.031 db	0.0066 db
S22	0.031 db	0.0062 db
NF	0.084 db	0.0160 db
IIP3	0.280 db	0.090 db

Table 10-4: Test repeatability data.

10.5 HARDWARE DEVELOPMENT PROCESS

The topic of hardware development process is discussed in this section. In order for the hardware development to be cost effective, a methodology must be devised that keeps development or debugging effort below the effort required to test the IC on a tester that has full parallel RF test capability. The engineering cost must be lower than the cost to either upgrade or buy a parallel RF tester. We have taken several actions to minimize the development cost associated with this approach:

1. Adopted a robust RF board design strategy.
2. Built measurement circuit schematics as separate layout libraries, using common components and common packages as much as possible, and adding a limited self-test and calibration capability in every circuit.
3. Built the individual measurement circuits on individual "sub-circuits," and attach the sub-circuits to the main test board, or universal board, using high-density connectors for the low-frequency lines and small coaxial connectors for the RF lines.
4. Maximized schematic, layout, and sub-circuit reuse.

The first item is required regardless of whether the engineer is using a $2M tester or a $100K tester, but it is more important for the latter because more complex circuitry is required. In either case, the RF IC test fixtures consist of a test board and a socket to contact the package, or a probe to contact a wafer. The test board maps the DC, analog, digital and RF measurements, and stimulus resources of the tester into the DUT. The test board or Device Interface Board (DIB) contains all of the RF matching and load circuit components required to drive the DUT under the appropriate application conditions. It also holds the buffers, relays, and the measurement circuits, which bridge the test circuit to the limited capability of the used ATE. For a low pin count RF IC, like a Low Noise Amplifier (LNA) or Power Amplifier (PA), the test board may only contain the matching circuit and the socket. Initially, these boards were designed in IBM by using mechanical CAD tools. The test board designer would layout a number of potential matching circuits on separate boards. When the first RF IC modules became available, the modules would be soldered to multiple boards, and multiple matching circuits would be built until the optimal matching configuration and components could be determined. Once the matching was determined, the process was repeated for a board with socket. This iterative approach to RF test board development is still seen in the RF test industry due to inadequate RF layout tools, skills and the necessary support models for simulating the RF board, socket and RF IC package; it is a hard-to-control process, often referred to as "black magic" engineering. The steps of the standard test board development process are shown in Figure 10-14. This process, though simple, is prone to manual errors in RF performance, and is unmanageable for building larger, multi-layer, integrated test boards following an aggressive schedule.

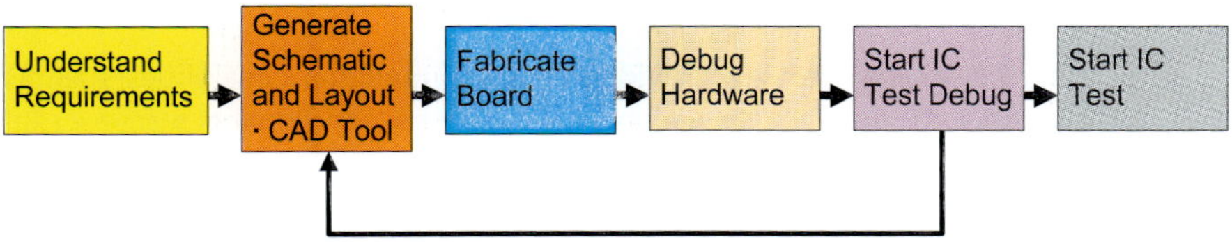

Figure 10-14: RF test board design flow.

To deal with this problem, a new process was developed that separated the process into smaller steps, identifying the risks at each stage, and then correcting each of the steps. To address the deficiencies of the current process, a new RF board design methodology has been implemented. In this implementation, the test board designer starts with the necessary information to design a schematic, and calculates the appropriate matching circuit components and layout for the test board. Ideally, inputs into the design process consist of: a test specification identifying the operation of the die and the list of required tests, information about the input and output impedance of the device, optimally in the form of S-parameters generated by the RF IC simulator, the package RF model and the socket RF model. This information enables the test engineer to generate a test strategy and design a test board. The test board design then goes through three design reviews.

In the first review, the type of tester and required tester functions are reviewed. It is critical that engineers with diverse skills review the test requirements, so that the test solution selected is based on the part requirements, not based on one test engineer's proficiency on a particular tester platform. In general, if the part is a low volume part, a test solution is selected that minimizes the time invested in engineering development; If the part is a high volume part, it is more important to reduce the time of manufacturing test, which is described by the following equation:

Total Cost = Engineering Development Time + Manufacturing Test Time

For high volume, cost sensitive, wireless phone components, it is often worth the added engineering time to develop a custom test solution that is focused on taking the exact measurements of the component as fast and as inexpensively as possible. For wireless phone components, the manufacturing cost contributes more to the overall test cost of the product than the development cost.

The next two reviews in the test board design process consist of a schematic and layout review. The focus of the schematic review is to ensure that the device under test has the correct bias and load conditions, and that the appropriate measurement circuitry has been designed. A fully detailed application board schematic, from the design team, is used to save time and reduce potential errors in communicating the load conditions. After the board layout, if the appropriate tools have been used, the layout can be electronically verified against a set of fabrication design rules and a set of in-house RF design rules, which ensures that all boards are built consistently and will meet the board fabricator's capability.

Once the test board requirements are identified, the board designer must be able to layout a complex multi-layer board and simulate complex electromagnetic effects. Next, two design tools are described, which were thought to be the best available tools, at the time, to accomplish both of these objectives; the Agilent ADS[5] CAD tool and the Cadence PCB[6] CAD tool.

The Agilent ADS tool provides a wide range of simulation capabilities such as DC, transient response, and linear and non-linear frequency domain responses; it also allows the designer to build a library of components, which includes the schematic symbol of each component, along with the model and physical footprint. Then, these components can be used to build the critical portion of the schematic, simulate it, and then place it in the "complete" schematic, which contains all of the components and connections required by the tester. The designer can then "push" the components onto a layout where placement, routing and generating Gerber files takes place. ADS also provides the designer the capability to design critical transmission lines in the layout and simulate them with great accuracy. This feature, called momentum, is explained in greater detail in section 10.6.2. The drawbacks of the Agilent ADS tool are its editing features and the fact that its built-in design rule checking requires custom programming if the designer wants to verify that the fabrication rules are not violated.

The Cadence PCB tool has been chosen in this project over ADS for designing the complex multi-layer PCBs because it addresses the layout deficiencies of the Agilent ADS tool and provides a more sophisticated design rule checking function.

---

[5] http://eesof.tm.agilent.com/products/adsoview.html

[6] http://www.cadence.com/products/si_pk_bd/index.aspx

The Cadence tools allow the designer to place restrictions in both the schematic (Cadence Concept) and layout tool (Cadence Allegro) to ensure that the critical components and layout are completed according to design guidelines. With the Cadence PCB tool, it is easy to query components for the schematic, generate a complete "Bill of Materials", place components in layout and route multi-layer boards. The Librarian maintains and enters new components into a central library, located on a network, so that designers have access to them for their schematics. Maintaining the central library involves building the schematic symbol and properties of the component, organizing the library structure. The engineer then builds the schematic from this library, but to utilize the library to its fullest capability, the engineer must become familiar with the tool's ability to query specific parameters to select components, and to place layout restrictions, such as controlled impedance, line lengths and component placement. When the design is ready for layout, the Cadence Allegro tool is tied to the schematic, so that the correct footprint of each component is ready, along with the nets and restrictions input by the engineer. A technician skilled in the Cadence tool and the capabilities of the PCB manufacturer is required due to the complexity of the tool, and because the technician is required to set up the design rules. Disadvantages of the Cadence PCB tool are that it takes time to learn how to use this tool effectively because of the number of features it contains, the tool's lack of frequency domain simulation, and the tool requires several people to support it if it is to be practical for design purposes. The Agilent ADS libraries and the Cadence PCB libraries can be synchronized so the IFF format can be used to import ADS designs into the Cadence tool. This however requires personnel who understands both program structures to build their libraries to be compatible with each other.

10.6 HIGH FREQUENCY SIMULATION TOOLS

Independently of the RF tester used, the RF test board requires, at a minimum, controlled impedance lines with matching circuits, isolation issues and possible differential to single ended transformation, switching, specific electrical delays and splitting/combining of RF power. Lower cost testers will require more complex designs, or sub-circuits, on the board to provide the RF test functions. In either case, the board designer requires an accurate CAD tool that allows them to verify their designs to reduce the need for re-designs. As previously mentioned, re-designs cost money in project time, engineer's time and additional hardware costs. The CAD tool should provide schematic design verification and layout verification. In some cases, a full 3D simulation verification capability might be required for the interface between sensitive RF IC designs and the socket pins or wafer probes.

10.6.1 Schematic Simulation

Schematic simulation can be accomplished with different degrees of complexity. There is a high level simulation, such as function blocks specifying the general parameters, or a low level simulation that includes the circuitry contained in each of the function blocks. The simulation can also be a combination of high and low level blocks. In high bandwidth PCB designs, the parasitic portion of the components should be added for a more accurate simulation. However, the lower the level, the more time required building and running the simulation. The engineer must make a judgment on how much detail to include in the simulation, depending on the complexity of the design, frequency and time. Even if a perfect simulation result can be reached at a schematic level, it is not until the final board layout is complete that the full schematic can be realized. The parasitics of the board layout such as trace patterns, location of components and ground plane and the parasitics of the socket or probe can add effects that change the performance of the board at higher frequencies.

Often in test, there are DUTs with single ended ports and differential ports with unknown impedances that need to be matched. The schematic simulation tool assists the designer in matching the device, by incorporating the unmatched data from the tester into an N-port data block, where N equals the number of ports of the DUT, and optimizing the match according to the goals setup by the designer. The designer obtains the unmatched data by de-embedding the test board using calibrated terminations. In addition, the designer must setup the matching topology that is to be used in the circuit. An example of this for a two-port, single ended LNA is shown in Figure 10-15; the matching was optimized for return loss, gain and noise figure as shown in the goal sections.

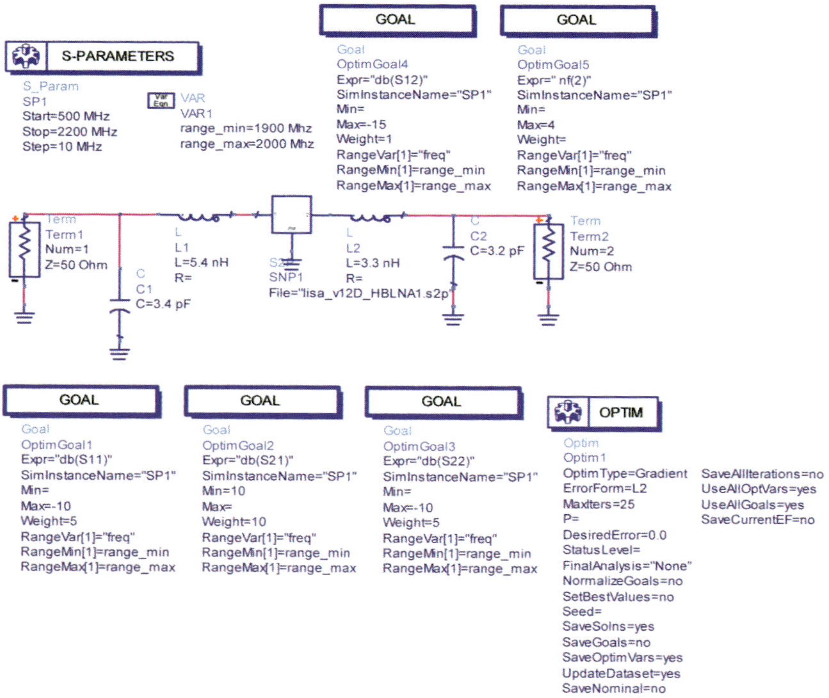

Figure 10-15: Design tool to optimize match of a two port network.

Differential ports increase the complexity because the testers are single ended and therefore, require a balun[7]. To match in this situation, the designer can capture the data from the tester, input it into the schematic tool (including the ideal or characterized data of the balun and line lengths), extract the differential load, and use this load to optimize a match similar to Figure 10-16.

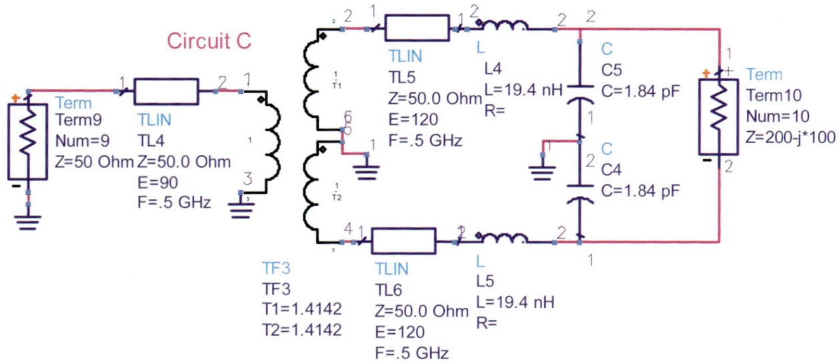

Figure 10-16: Design tool to match complex differential load to transformer.

Figure 10-17 shows the matching for a 200-j100 differential load using a balun with a 2:1 impedance transformation. Marker 5 shows the input match with the matching components, whereas marker 4 shows the return loss without the matching components.

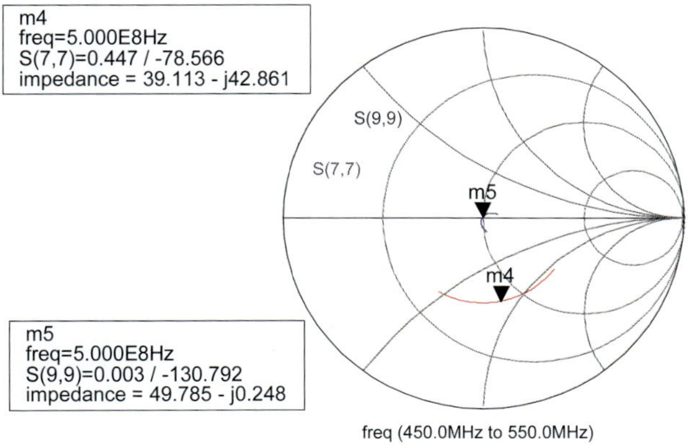

Figure 10-17: Smith chart showing transformer to load match.

[7] Balun: A circuit that transfers a Balanced circuit to an Unbalanced circuit [4].

10.6.2 2.5D RF Board Simulation

Schematic design may include some of the parasitics of the layout pattern, but not all of the parasitics - coupling/crosstalk. In order to address the effects of the layout pattern and to design the test board successfully with one pass, board layout pattern simulation and modeling become necessary for high frequency test board design. For an RF test board, 2.5D Electromagnetic simulation tools have proven effective in accuracy and time considerations in studying the parasitics of the board layout patterns, as well as some 3D components such as socket pogo pins and packages with all-rectangular structures. An example of a 2.5D EM simulation tool is the Agilent Momentum/Momentum RF. This tool can model multi-layer structures where microstrips, striplines, vias, slots and substrates are required. This tool can model pogo pins by stacking the vias and setting the dielectric constant of the substrate equal to the socket properties, or air, whichever is the case. All high frequency effects, such as skin effects, impedance line variation, substrate impact, couplings, radiations etc, will be accounted for in Momentum. Momentum RF does not consider the radiation and is designed for electrically small structure simulations, but it uses memory efficiently.

Normally, there is a very fine layout pattern for a socket, or probe card, on a test board, but in order to fit the test fixture of testers, the size of a test board is relatively large - the larger the size, the more memory and time required simulating the layout pattern. Fortunately, the components on the test board are concentrated in an area close to the socket or probe, and the impedance lines are easily and accurately modeled in a schematic simulation. These attributes make it possible to simulate/model only part of the layout pattern and provide accurate results/models.

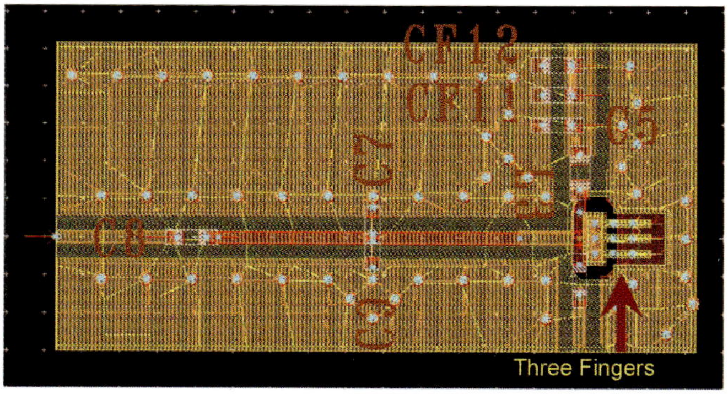

Figure 10-18: Simulation model for layout effects using 2.5D simulation tool.

Figure 10-18 is an example of a layout simulation for studying the effects of component placements and the effect of short traces for component connections. The three fingers (traces) on the right are the pads for contacting to the socket pins. A multi-port S-parameter matrix is obtained on this simulation, but the components are not taken into consideration. To consider the components, a simulation must be run in a schematic window, which incorporates the layout simulation results along with the components; this will provide a more meaningful two-port S-parameter matrix for analysis and comparison.

An example of using the Momentum tool to perform a "three dimensional" simulation is shown in Figure 10-19. Figure 10-19 provides a 3D view of a simulation structure for a socket utilizing pogo pins. An equivalent circuit model can be extracted from the simulation, as shown in Figure 10-20.

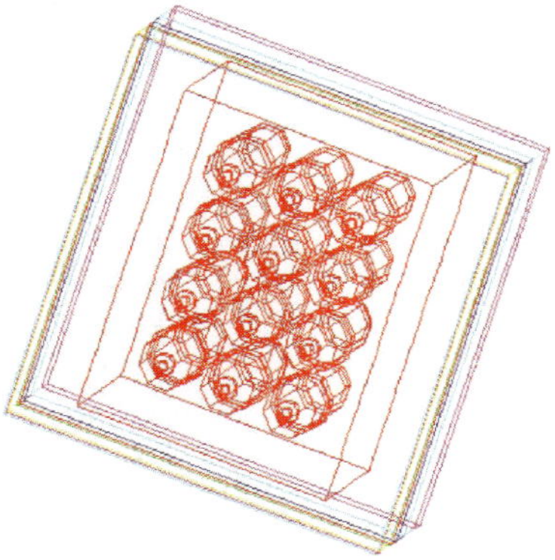

Figure 10-19: Simulation structure for a socket utilizing pogo pins.

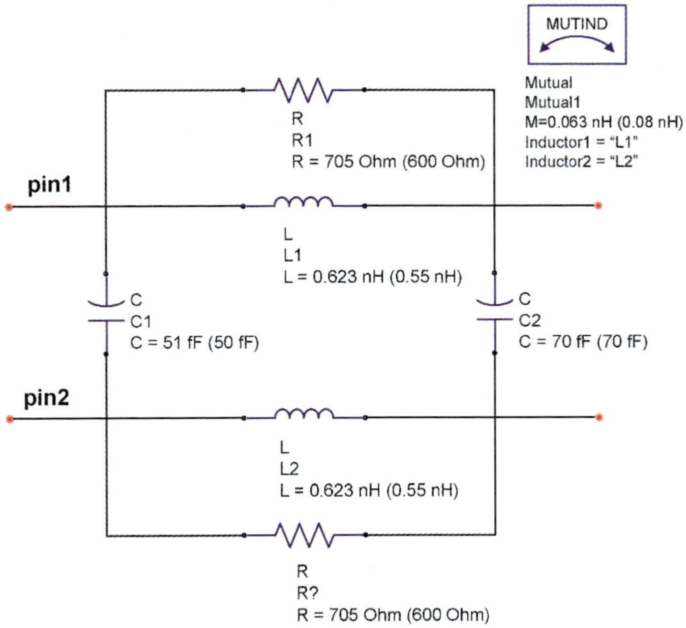

Figure 10-20: Equivalent circuit model (values in parentheses are based on a measured model).

The 2.5D simulation tool can be utilized in the design of microstrip or stripline structures, such as power splitters, couplers, phase shifters, matching stubs, etc. The designs can be saved and reused for other designs, which enables the designer to obtain a specific device quickly in-house and reduces the cost of new designs. An example of a 5GHz power splitter is shown in Figure 10-21. This layout was designed setting up the dielectric properties using 10mils Rogers 4350. Five ports are required—three ports are external and two are internal (for the 100Ohm matching resistor). This five-port dataset is incorporated into the schematic with the resistor attached as Figure 10-22 shows; simulation results are shown in Figure 10-23.

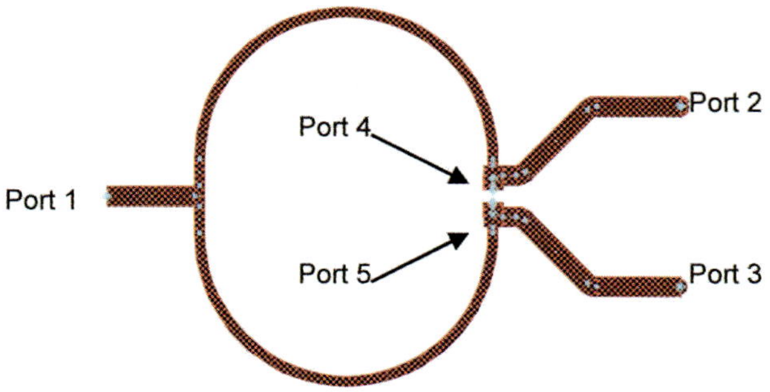

Figure 10-21: 5GHz splitter.

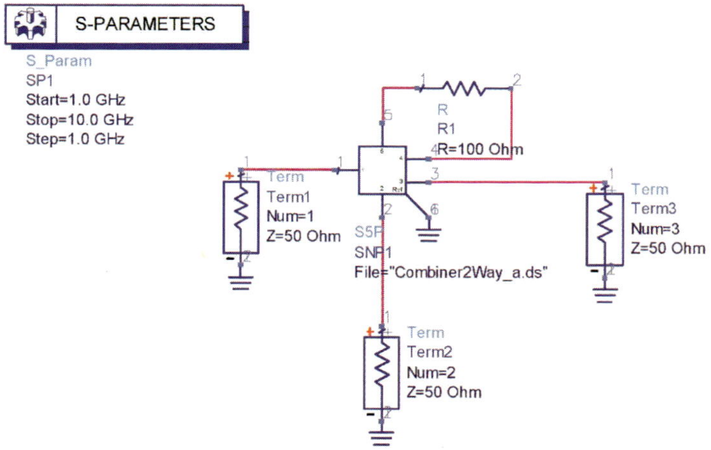

Figure 10-22: Design tool simulating the splitter.

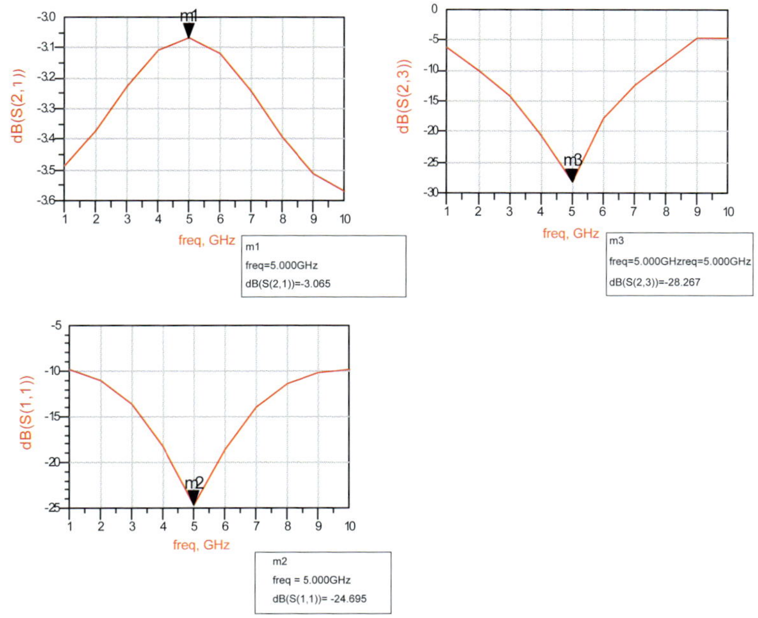

Figure 10-23: Results with 100 Ohms resistor added between Port 4 and 5 for matching. Top Left: Insertion Loss + 3dB Power Split. Top Right: Isolation between Port 2 and 3. Bottom: Port 1 S11.

Additionally, this 2.5D simulation tool can be used in the debug and correlation of the RF test board, which assists in tuning the following: matching component values, the effect of the balun, the insertion loss of the traces, connection pads and vias on the test board.

10.6.3 3D RF Socket and Package Modeling

EM simulation tools can be used for the layout pattern simulation and modeling, but it is more time consuming than the 2.5D tools, without a significant increase in accuracy. However, some examples of test structures that do benefit from 3D simulation tools are wafer probes, sockets with non-straight pins (S-shape or J-shape), packages with non-rectangular structures or parts of the packages (bond wires). The size of the structure to be simulated in 3D tools is limited by the memory of the computer and the ratio of the outline size to the minimum structure dimension.

An example of a 3D simulation tool is Ansoft HFSS[8]. This is a full wave 3D EM simulation tool, providing flexible structure generations for various structure shapes to be simulated and modeled. It is suitable for characterizing sockets and probes and the S-parameter data can be converted into an equivalent circuit model, but the engineer needs to understand which topology best fits the socket or probe configuration being modeled. Limitations in characterizing sockets, or probes,

---

[8] http://www.ansoft.com/products/hf/hfss/

depend on the number of pins/probes, and the boundary condition being simulated. Typically, the socket or probe is setup with two adjacent, or cattycorner, pins/probes as the ports with the surrounding pins/probes grounded.

A model of a single bond wire inside a QFN20 package is shown in Figure 10-24, and a model of a probe card is shown in Figure 10-25.

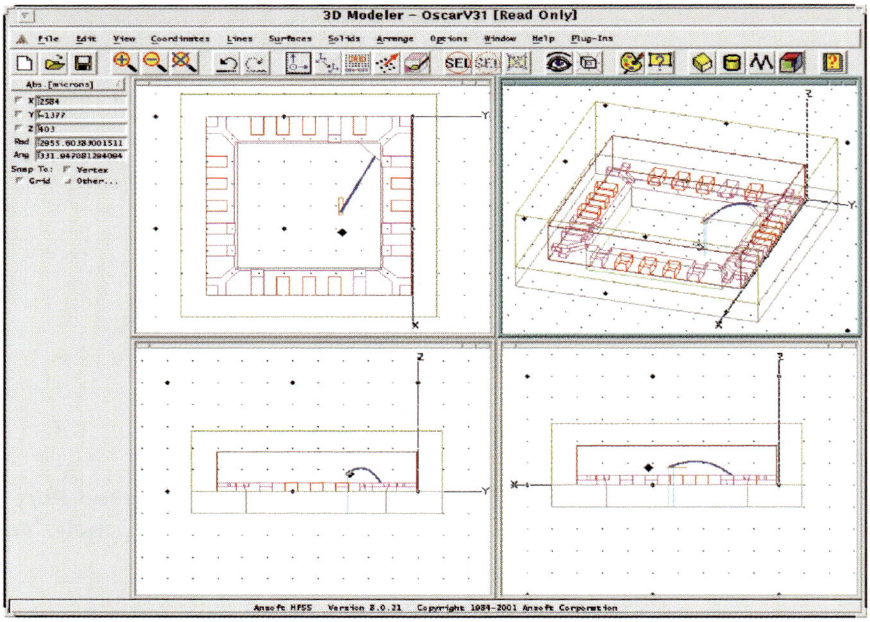

Figure 10-24: QFN20 package showing bond wire.

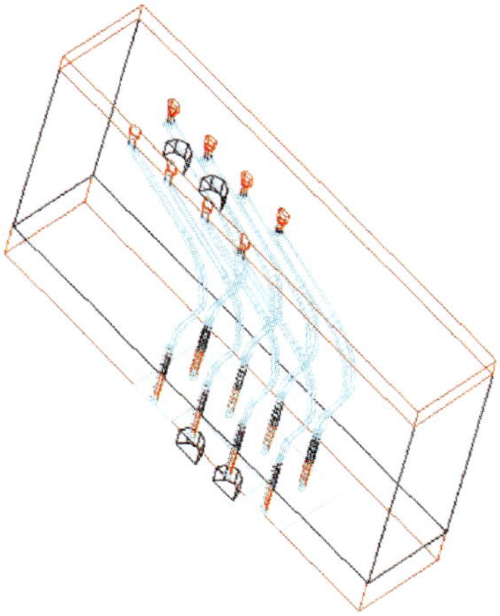

Figure 10-25: 3D model of IBM cobra probes.

10.7 DEVICE UNDER TEST INTERFACE

Sockets and probes have been only mentioned vaguely so far in this chapter. They are necessary in RF test (and any IC test in general) because they provide the interface between the DUT and the DIB and allow the DUT to be tested without soldering or wire bonding the DUT to the DIB. It is extremely critical to choose the appropriate socket or probe for the type of DUT to be tested. Not doing so, will cause the test to fail no matter how well the program is written and the test circuit on the DIB is designed.

10.7.1 Sockets

For RF module test, a good test socket has pins that provide decent travel to accommodate deviation in planarity between the board pads and between the module pads/leads. The pins need to provide enough force during compression along with their "scrubbing" capability to penetrate of any oxide buildup on the DUT's leads. This provides a low resistant contact. Last but not least is a pin that is short with no sharp edges along the length of the pin to provide low inductance. This keeps the reactance low at higher frequencies to provide less SWR loss and radiation loss. The latter keeps high isolation between pins. Having low loss between the DUT and DIB keeps ground bounce to a minimum and less potential for instability. Manufactures of high frequency sockets are capable of providing sockets for a wide variety of packages, such as SOIC, QFN and BGAs. The pin count for these sockets is exceeding 111 pins and provides less than 1dB loss beyond 10GHz. Much detail on their electrical performance is provided by these vendors on their web sites.

10.7.2 Wafer Probes

"Traditional"[9] wafer test probes are more problematic because of the length of the probes in use. Both probe styles are designed this way to allow flexibility in reaching the non-uniformity in I/O pad spacing often found on wafer die and to accommodate for deviations in the wafer's planarity. It also allows the probes to handle the small pitch required by the die. Some probes, like the Cobra probes, as shown in Figure 10-25, are designed to act as a spring, while cantilever probes are shape like a plane in the z-direction (up and down in travel). Both designs provide flexibility and strength in the probe to allow it to penetrate the oxide making a low resistant contact over 1 million times without damage to the probes – including solder balls.

Length is the problem since it equals inductance, and without proper ground returns, that means higher radiation loss and/or reflection loss. Radiation loss results in less through power and coupling between adjacent pins. Reflection loss results in lower signal integrity and higher probability of instability. There are high frequency wafer probes available such as coaxial design based and "membrane" based, but these have their limitations in either spacing or more frequent cleanings. Both cost several times more than traditional probes. To find the limitations of Cobra probes, several studies have been done by IBM in simulating Cobra probes and use them to

[9] Traditional being cantilever or Cobra style probes.

measure RF wafer die. Its primary candidate would be a die that was designed properly providing ground pads next to the critical RF pads to allow signal-ground (S-G) or ground-signal-ground (G-S-G).

10.8 CONCLUSIONS

The three major categories of RF ICs were discussed along with the concerns of how to provide an accurate, repeatable *low cost* test solution for these ICs through the description of an RF testing system at IBM. The major contributors that affect test cost such as the tester, the handler and time to measure were discussed. Several methodologies to reduce cost and increase accuracy were given starting with the required circuitry to provide the needed measurements and the design procedure to accomplish this was discussed. Implementing the methods described in this chapter has steadily increased turn around time, test coverage and accuracy because the design engineers are more involved with their design, thereby creating more rounded engineers, by exposing them to the complete design process and limitations. Faster turn around, better test coverage and accuracy have reduced cost and improved customer satisfaction by meeting their schedules and supplying them with the data they requested. As more components are implemented into the tools libraries, there is more time to focus on increasing the accuracy of the models. Increased accuracy and faster turn around will give the engineers more confidence and time to focus on combining these components to create complex, yet flexible, test functions, such as down converters, synthesizers, samplers and modulators to be used in a hierarchically structured design. Additional test functions being worked on are adding field programmable devices to sub-circuits to provide more sophisticated processing of modulated signals. The idea is to provide a low cost alternative to high cost options offered by ATE manufactures, a major effort towards effective RF testing solutions. High quality and accuracy RF testing at reduced test costs is the most important concern of future methodologies in this domain.

ACKNOWLEDGEMENTS

The authors wish to thank Dana Brown for editing assistance, Hanyi Ding for assisting in the 3D models and Jing Li for his contribution in design and test of the circuit shown in Figure 10-13.

REFERENCES

[1] International Technology Roadmap for Semiconductors 2003: Executive Summary.

[2] J. Ferrario, R. Wolf, H. Ding and M. Slamani "Moving from Mixed Signal to RF Test Hardware Development", Proceedings International Test Conference, 2001.

[3] J. Pineda de Gyvez, R Amine and G Gronthoud "VDD Ramp Testing for RF Circuits", Proceedings International Test Conference, 2003, pp. 651-658.

[4] S. Bhattacharya, A. Halder and A. Chatterjee, "Automatic Multitone Alternate Test Generation for RF Circuits Using Behavioral Model", Proceedings International Test Conference, 2003, pp. 659-664.

[5] J. Ferrario, R. Wolf, S. Moss, and M. Slamani "A Low-Cost Test Solution for Wireless Phone RFICs", IEEE Communications Magazine, Sept 2003, Vol. 41 No 9, pp. 82-88.

[6] D. Brown, J. Ferrario, R. Wolf and J. Li, "RF Testing on a Mixed Signal Tester", Proceedings International Test Conference, 2004.

[7] M. Slamani, J. Bhagat, J. Ferrario, and R. Wolf, "Challenges of incorporating an RF test system on a board", 3rd Workshop on Test of Wireless Circuits and Systems, 2004.

[8] J. Ferrario, R. Wolf and S. Moss, "Architecting Millisecond Test Solutions for Wireless Phone RF ICs", Proceedings International Test Conference, Washington D.C., Oct. 2002.

[9] P. Vizmuller, *RF Design Guide: Systems, Circuits, and Equations,* Artech House, Inc., Massachusetts, 1995.

[10] Hewlett Packard Application Note 57-1, *"Fundamentals of RF and Microwave Noise Figure Measurements".*

[11] Hewlett Packard Product Note 11729C-2, "Phase Noise Characterization of Microwave Oscillators: Frequency Discrimination Method".

Chapter 11

Loaded Board Testing

Kenneth P. Parker

Several chapters of this book focus on Integrated Circuit (IC) test topics. This chapter gives an overview of how loaded[1] printed circuit boards are tested. While testing a modern IC can be extremely challenging, testing a board populated with 200 – 500 such devices (digital, analog and mixed signal) along with a collection of 2,000 – 3,000 analog discrete components can also pose a challenge. Boards are virtually always mixed signal or even purely analog circuits and virtually never "pure digital" designs. This greatly complicates the process and forces us to abandon well-known tools that can be used at IC test, as many such tools will not handle mixed signal designs nor designs of such magnitude. A further complication is that at board test, it is very important to *locate* defects, not just perform a go/no-go test.

Before jumping into the topic of board test, we need to understand what it is we are testing for. This is notably different than what we test for at the IC level, so it is an important distinction to make.

[1] This does not include the topic of bare-board (unloaded board) testing. Here we assume the board itself has already been tested and that it is defect-free.

D. Gizopoulos (ed.), Advances in Electronic Testing: Challenges and Methodologies, 371-406.

11.1 THE DEFECT SPACE AT BOARD TEST

In the early 2000s, board test engineers began to rigorously define just what a defect was [1] and something called the PCOLA/SOQ model[2]. In fact, the words *defect* and *fault* are also given rigorous definition. These may be different than the corresponding definition for an IC tester.

Rigorous definitions and a model of the important board test defects allow engineers spread across the manufacturing spectrum[3] (not to mention the globe) to compare test coverage and produce metrics of test effectiveness. A designer at an Original Equipment Manufacturer (OEM) may create a circuit design specification and architecture. This may be sent to a design house where the detailed circuit design is finished. Then it could be farmed out to a layout service company that generates the actual board layout and manufacturing data. From there, it could be sent to a contract board manufacturer who will make thousands of them. But before that happens, a contract test development house may create the board test needed to assure defect-free boards are shipped. There could be a lot of independent engineers involved in this chain. If they cannot communicate about defects, this will guarantee confusion and an OEM that is at risk of missing schedules for product shipments, or the quality goals thereof.

11.1.1 What is a "Defect"?

A defect is an *unacceptable* deviation from a norm. A defect is therefore undesirable and cause for some remedial action, from discarding the board, or repairing it, or at the very least, fixing the process step[4] responsible for it. Some examples of defects are:

- An open solder joint.
- A solder joint with insufficient, excess, or malformed solder. There may be no electrical manifestation of this defect.
- A short caused by excess solder, bent pins, device misregistration.
- A dead device. For example, an ESD[5]-damaged IC or a cracked resistor.
- The placement of an incorrect device.
- A missing device.
- A polarized device rotated 180 degrees.
- A misaligned device (typically laterally displaced).

---

[2] The PCOLA/SOQ model is analyzed in detail in section 11.1.3.

[3] Before the 2000s and the rush to 'outsourcing', manufacturing and test was often all part of a vertically integrated process within a single company. Within that environment, each engineering group would work out with its neighboring groups some form of ad-hoc agreement on what was being tested and how test coverage could be measured.

[4] Some would argue that the defect is actually in the process step. This is true from a root-cause analysis point of view. We restrict our view here to the board itself, which is what you will ship to a customer. Thus the "norm" is a contract between the manufacturer and the customer.

[5] Electro-Static Discharge.

All of these defects can be enumerated by examining the structural information of the board, typically the bill of materials and XY position data. This enumeration is called the defect universe. Notice that no assumption of how testing will be conducted is used in developing this list of defects. This is at variance with past practice, where the capabilities of the target test system were considered in this enumeration, as if those untestable defects could not occur. In the past, certain defects were completely ignored because the popular board tester[6] technology of that time had no way to detect that defect. This was pointed out in [2], where it was noted that the proliferation of power and ground pins on larger ICs was distorting the measurement of test coverage, since open solder joints on these pins were not being counted.

The key word in the definition is "unacceptable". You as a manufacturer of products know what is unacceptable and what can be shipped. Depending on your product, this may be variable. For example, if you make cardiac pacemakers, your view of "unacceptable" may be driven by liability concerns. If you make inexpensive digital wristwatches, your standards may vary from this. But even if you do make cheap, high-volume, low-risk products, you may still want to know about defects so you can improve your process and eliminate them before they drift into a range where you can no longer ship product.

11.1.2 What is a "Fault"?

A fault is a physical manifestation of a defect. (The word "fault" is often used synonymously with "failure".) Thus a fault reveals the presence of a defect. A single defect may cause several faults (i.e., have several different manifestations) and a single fault could be the manifestation of several different defects.

A defect "shows up" as a fault. For example, a missing bond wire on an input to a logic gate may cause it to see a permanent logic 1 rather than a time varying signal. Several other defects can produce this same fault behavior. For example, missing solder between the input pin and the board, an electrostatically damaged input buffer within the IC or a broken printed circuit trace between the upstream driver and the input, will all exhibit the same faulty behavior. Similarly, one defect may cause several fault manifestations. For example, an open solder joint on an input pin (particularly a reset input) may cause an IC to show incorrect results, and these incorrect results vary in time[7].

An observed fault is not always a reliable pointer to a defect. For example, if an IC loaded onto a board has defective solder on one input pin causing an open circuit, this may appear to the IC to be a permanent, stuck-at-1 fault on that input. This faulty behavior may not be readily apparent since the effect of the erroneous logic 1

---

[6] There are several systems that test (or inspect) boards. See sections 11.2 and 11.3.

[7] Simple example: consider a counter circuit with a defect on its reset line that prevents it from being cleared. Any test that first asserts the reset and then issues some counting pulses will likely show different results on the counter outputs every time the test is run, because the defect prevents the reset of the count.

must propagate through the internal workings of the IC before its effects (improper output behavior) are seen. When this faulty output behavior is finally observed, it can be a highly challenging[8] task to relate this observed behavior to the input stuck-at fault caused by the defective solder. This is illustrated in Figure 11-1. Careful examination is necessary so that one does not replace IC2 on the board although the fault is in the IC1 surrounding area.

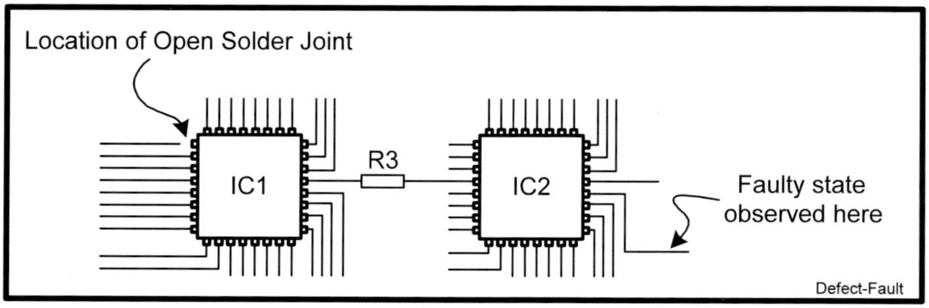

Figure 11-1: A defect may be far removed from where a fault is observed.

11.1.3 The "PCOLA/SOQ" Model

Using only a bill of materials for a board, and positional data (XY location data for all devices and their pins) one can develop the defect universe for that board by using the PCOLA/SOQ model documented by Hird, et al [1]. Note that netlist information (the logical connectivity of signals, power/ground distribution, etc.) is *not* used in this model. Nor is there any consideration given to what test technology (tester) might be used.

The bill of materials gives a complete list of all devices[9] that will be placed on a board. Each device has five properties that are important. They are presence, correctness, orientation, liveness and alignment. The first letters of each property, strung together, yield the acronym "PCOLA". If a device is missing from a board, that is a presence defect. If a device is present, but is the wrong device, that is a correctness defect. If a polarized device (e.g., a diode or electrolytic capacitor) is reversed (180 degrees) then that is an orientation defect. If a device is inoperable or dead in a gross sense[10] of the term, that is a liveness defect. Finally, if a device is mis-aligned, for example, offset by 2 millimeters or rotated by 8 degrees, this is an alignment defect. The first four properties are called "fundamental" in that they usually must be defect-free for the board to work. The last is called a qualitative

[8] Remember, the IC may not be a simple digital device, but may be analog or mixed signal. Familiar "stuck-at" models employed in modeling digital ICs rarely work at board test.

[9] This is a broad term including ICs, discrete analog components, bolts, screws, heat sinks, barcode labels, etc.

[10] Subtle defects, such as a delay fault deep inside a large digital IC are not considered here. We expect such defects to be eliminated earlier in the overall manufacturing process. Trying to eliminate such defects at board test is extraordinarily expensive.

property in that a board may still be functional when a qualitative defect occurs, but it is none-the-less, an unacceptable deviation. For example, a surface mounted resistor may flip up on its side (an alignment problem called "billboarding") but still be soldered to the two board pads and thus electrically functional. This may be deemed "unacceptable" and thus is an alignment defect. For some devices, the orientation property is a "don't care". A resistor is a case in point, so we need not test a resistor for proper orientation, and its orientation should not be counted in the defect universe. However, its alignment is important and will be counted[11].

Next, most devices have *connections* to the board that are established during manufacturing. Most of these are solder joints, but some may be press-fit connections, or even nut-and-bolt connections. Connections have the following properties: shorted, open and quality. These yield the SOQ acronym of the model. The first two are fundamental properties, which when they occur often affect board performance. The quality property of a connection may have no discernable electrical effect (during manufacturing) but may affect long-term reliability. An example is a solder joint where the volume of solder is too low, below a margin considered safe for long term reliability under thermal cycling stress. This joint will conduct current adequately now, but how about 15 months from now?

Connections may be shorted together, but in the past (since back then board test data often did not include XY location data) we tested board *nodes* against each other for shorts, even when there was no practical way they could ever be shorted. The SOQ model enumerates connections as the focal point for shorts. Connections that are closer together have the opportunity to be shorted, for example, by solder bridges or bent pins that touch each other. Connections that are beyond this "shorting radius" are not considered for shorts testing. Since the netlist data is *not* considered, two pins that are close to each other but that are both connected to (say) ground, are still candidates for a shorting defect. This short is included in the defect universe even though one might argue "it wouldn't matter". A process engineer would disagree because this short indicates something wrong with the manufacturing process. Another case where in the past we were sloppy was with nodes connected to very low impedance devices like jumpers or small inductors. It wasn't easy to determine if the two connections of these devices were shorted together, so, they simply weren't counted. The rigor of the PCOLA/SOQ model prevents this.

On a large printed circuit board there can easily be 50,000 connections. Each has a potential open and quality property. Then, depending on the proximity of other joints, there is a (possibly empty) list of possible shorts from a given connection. Two neighboring pins will have each other in their respective "shorted-to" lists, but their lists may be quite different.

Of course, the defect universe is always incomplete since as soon as one puts a bound on the membership of defects, someone will argue that some class of defect has been omitted. There is one mechanism in the PCOLA/SOQ model to handle this. An "intangible device" can be specified that can be added to the device list, beyond those already included in the bill of materials. A favorite example is the

---

[11] In Hird et al [1], this is handled by weighting properties for each device type. For a resistor, the orientation property is given zero weight.

"circuitware"[12] downloaded into a non-volatile programmable logic device (PLD). This data must be present and correct (orientation, live and alignment are not important and thus unweighted). A process will download those configuration bits and a test will verify they are indeed there and correct. Thus we want to include this circuitware in the test process. Since circuitware has no connections, the SOQ properties for this intangible device do not apply.

A summary of the PCOLA/SOQ model is given in Table 11-1 with some places in the manufacturing process where defects occur, along with some possible root causes for them.

Defect	When it occurs	Possible Cause
Missing component *(Presence)*	Placement, soldering	Shock, abrasion, too little glue
Wrong component *(Correct)*	Placement setup, inventory, handling	Handling error, mismarked packages, operator error, wrong specifications
Misoriented component *(Orientation)*	Placement setup	Handling error, operator error
Dead component *(Live)*	Handling, placement, soldering	Dead on arrival, handling damage, electrostatic damage
Device alignment *(Alignment)*	Placement, reflow	Improperly located device, solder surface tension
Shorts between pins *(Shorts)*	Wave/reflow soldering, through-hole insertion	Too much solder, solder stencil defect, pin misreg-istration, bent pins
Solder open *(Opens)*	Solder paste application, wave/reflow soldering	Too little solder, solder stencil defects, tombstoning, bent pins
Solder quality *(Quality)*	Solder paste application, reflow	Insufficient paste, too much paste, improper reflow, temperature/time

Table 11-1: Components of the defect universe for board test.

11.1.4 Test Coverage

Test coverage can be measured once we agree on the defect universe for a board. This is done by inspecting a given test and asking, "What does it mean when this test passes?" for each member of the set of defects. This question is posed about a passing test since a failing test can be very misleading.

A simple example is as follows. Say we have a way to measure the actual resistance of a resistor on a board, by contacting the two nodes[13] on either side of the

[12] The configuration information of the PLD.

[13] This simple example assumes we have access to some place on each node but not actually on the resistor itself. This means current must flow through the connections of the resistor. We

device. For each potential defect in the defect universe we would ask, "If this defect were present, could this test have passed?" For instance, if the resistor was not present, could the resistance measurement pass? If the resistor value was wrong by a significant amount, could this test pass? If this resistor was cracked across the middle (implying an internal open that kills it) could this test pass? If this resistor was rotated by 5 degrees (such that the solder was still intact) could this test pass? If either of the two connections of the resistor was open, could this test pass? If the two connections were shorted together, could this test pass? And if the quality of either joint was unacceptable (yet still conducting) could this test pass? The coverage of this simple resistor test is summarized in Table 11-2.

Resistor Defect	If the test passes...	Cover	Comments
Missing component ***(Presence)***	It must be present, this defect is covered	Tested	Possibly not tested if the resistance value is so high as to strain the upper limits of the measurement
Wrong component ***(Correct)***	It might be correct, defect is possibly covered	Partially tested	Not all wrong components can be detected. A ¼ watt resistor could measure the same as a 1 watt resistor
Misoriented component ***(Orientation)***	Not applicable	Not applicable	Orientation is irrelevant (zero weight)
Dead component ***(Live)***	It must be alive, this defect is covered	Tested	
Device alignment ***(Alignment)***	It could still be mis-aligned	Untested	This property is untested by measuring electrical resistance
Short between pins ***(Shorts)***	The pins cannot be shorted. This defect is covered	Tested	Possibly not tested if the resistance value is so low as to strain the lower limits of the measurement
Solder open ***(Opens)***	The pins cannot be open. These defects are covered	Tested	Possibly not tested if the resistance value is so high as to strain the upper limits of the measurement
Solder quality ***(Quality)***	It could have unacceptable connection quality	Untested	This property is untested by measuring electrical resistance

Table 11-2: Grading a simple resistor test for coverage.

---

further simplify by assuming there are no other components in parallel with the resistor that require guarding techniques.

The comments in Table 11-2 indicate cases where a property might remain untested, for example, if the resistor value is at some extreme (upper or lower) with respect to our measurement capability. Note also the concept of "partial" detection, as in the case of correctness. There, the value measurement indicates the right resistance value, but we cannot know if it is the right size (wattage) of resistor from this. The orientation property is a don't care, so we don't score it at all.

A complete board test is made up of thousands of measurements and other "subtests" that each examines some limited subset of the board's properties. These can be accumulated into an overall score. Note that several tests may each score some test coverage on a device or connection. The highest level of test is accepted (the Max function) but even several "partial" tests do not add up to a full test score for a given property. See Hird [1] for details on scoring and weighting coverage.

11.2 IN-CIRCUIT TEST (ICT)

In-Circuit test has been used for about 30 years to test loaded printed circuit boards. It is typically done right after soldering has been completed and before any functional[14] or system testing[15] Of the PCOLA/SOQ defect model, all except 'A' and 'Q' (device Alignment and solder Quality) are candidates for testing.

The name "In-Circuit" is derived from the fact that individual components are tested "In-Circuit", meaning, while they are connected to other components around them. This is a divide-and-conquer method that gives good diagnostic resolution that is critical for repairing defects and shipping good boards. As there may be thousands of components on a modern board, along with many tens of thousands of connections, a comprehensive In-Circuit test program will contain thousands of smaller subtests that focus on subsets of circuitry, from small groups of components down to individual components.

The key to In-Circuit testing is *access*, whereby the resources of the tester can be connected to important board nodes. This is done with a "bed-of-nails" fixture such as shown in Figure 11-2. Such a fixture contains thousands of sharp-pointed spring-loaded probes that are carefully arranged under (and sometimes over) the board. The board is then forced down onto the probe field (by vacuum pressure and/or mechanical actuators) where the individual probes[16] contact each node. The contact point may be a deliberately placed test point target, a conveniently placed trace via, a component connection pad or a through-hole pin.

In past years, virtually all the nodes of a board could be accessed, but this is not true today for two major reasons. The first is circuit density causing space to be at a

---

[14] In-Circuit testing is necessarily a "low-frequency", "divide and conquer" test technology where the board is not operated either at speed, nor performing its intended function. Functional tests are often performed "at-speed", with the board doing something from its native repertoire.

[15] System testing is where a board may be plugged into a complete system (where all other components are in known-good working order) and then exercised functionally, for example, by booting up the system and running diagnostic programs.

[16] Probes and "nails" will be used synonymously.

premium, even for small (for example, 28 mil diameter) probe targets.[17] The second is that node counts for larger boards may well exceed the nodal capacity of any available tester. Thus, today, In-Circuit testing is challenged to find ways to test boards when full nodal access is not possible. This is addressed by Design-for-Test (DFT) methodologies such as Boundary-Scan [3], [4] testing.

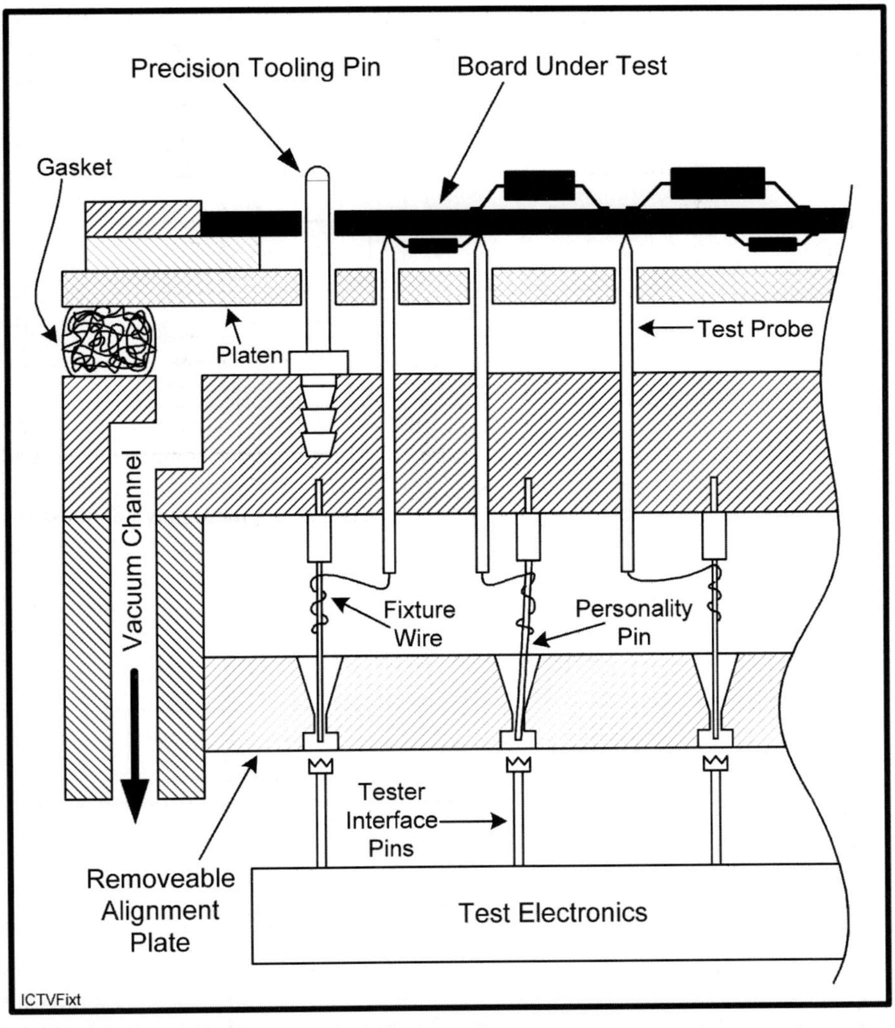

Figure 11-2: Cutaway drawing of a board resting on top of an In-Circuit vacuum-actuated fixture, the bed of nails. Vacuum under the board allows atmospheric pressure to press the board onto spring-loaded test probes.

---

[17] A 28 mil diameter circular target on a 5 mil wide trace is a significant aberration (electrically) and often necessitates moving nearby traces out of the way.

It often surprises people that a large amount of In-Circuit testing is done *without* applying power to the board. There are certain attractions to this; perhaps most important being finding shorts on a board before power[18] is applied (to avoid damaging the board). The next sections discuss powered and unpowered testing. The basic elements of a board test process are shown in Figure 11-3.

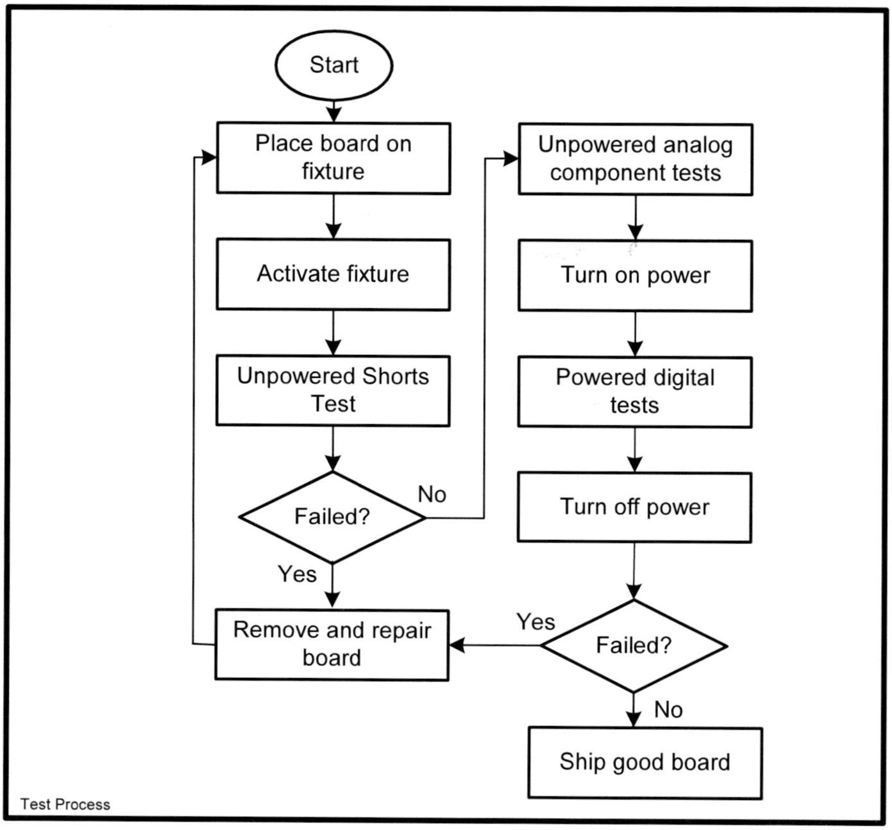

Figure 11-3: Board test process steps.

11.2.1 Unpowered Shorts Tests

For the purpose of this discussion, assume a board has full nodal access. Consider the (fictional) example shown in Figure 11-4. This circuit has 17 components connected to 11 nodes (labeled with circled numbers). We want to test this board for unwanted connectivity among nodes. This can be done by testing a selected node for unexpected current flow to all other nodes. Consider node 6 to be the selected node. We ground all but the selected node by closing appropriate relays in our tester connected to the bed of nails. Then node 6 is connected to a stimulus source. By

[18] Note that Boundary-Scan testing, used to fill in coverage lost when access limitations occur, must be conducted with power applied to the board.

using a small voltage source in series with a load resistor, we can both limit the voltage and current of this stimulus. We monitor the voltage across the load resistor. The voltage is chosen to be 0.1 volts. This means any silicon junctions on the board (see transistor N1 and zener diode Z1) between the grounded nodes and node 6 cannot turn on. If a resistor is stimulated (R2) some current will flow through it and the load resistor. We monitor the voltage across the load resistor and compare it to a threshold that would be observed if the board resistor was greater than a small value, for example, 8 ohms[19]. If the compared voltage indicates the board resistor is greater than this threshold, then we know the selected node cannot be shorted to any other node. By using this strategy of not turning on silicon junctions and comparing against a low resistance threshold, we make many of the devices on the board "disappear", as pictured in Figure 11-5.

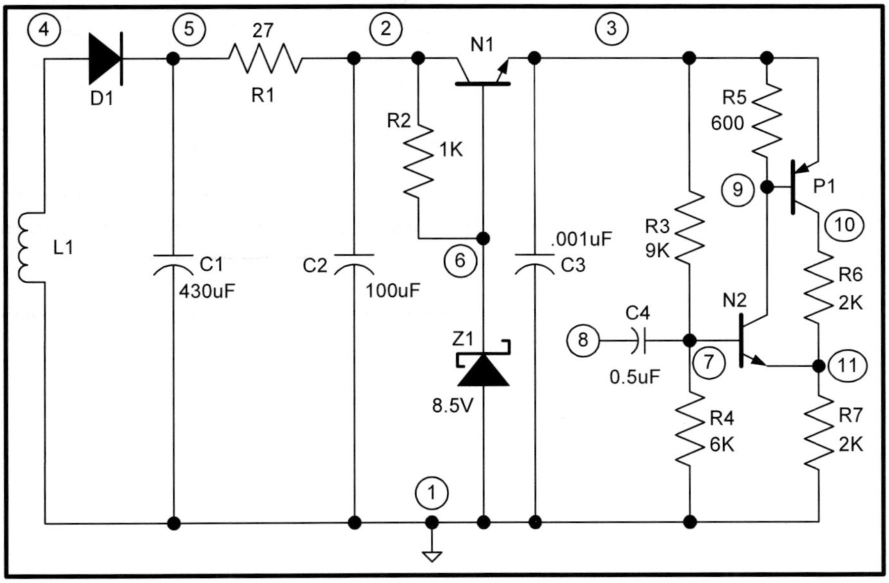

Figure 11-4: Example of a circuit tested for shorts.

With many of the devices effectively removed, we end up with an equivalent circuit as in Figure 11-6. We start by testing node 11 against all the others. If this shows no current flow, we can eliminate node 11 and move left to node 10. (Node 11 can float at this point). We continue testing each node and move to the left after each is done.

19 Since our tester relay network, fixture interface, fixture itself and probes contain some small series resistances, we cannot "see" values below this error term. Eight ohms assures we don't falsely pass a defective board.

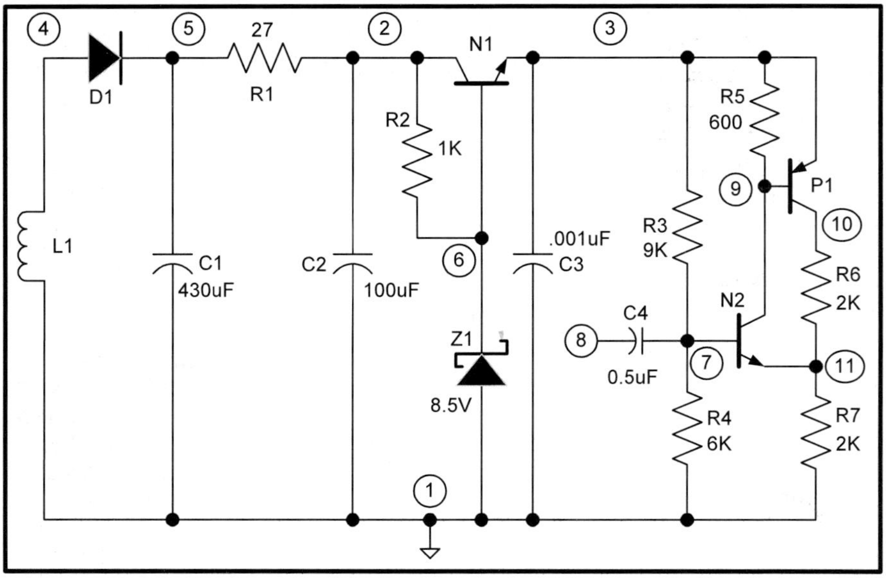

Figure 11-5: Circuit with junction devices and higher resistances omitted.

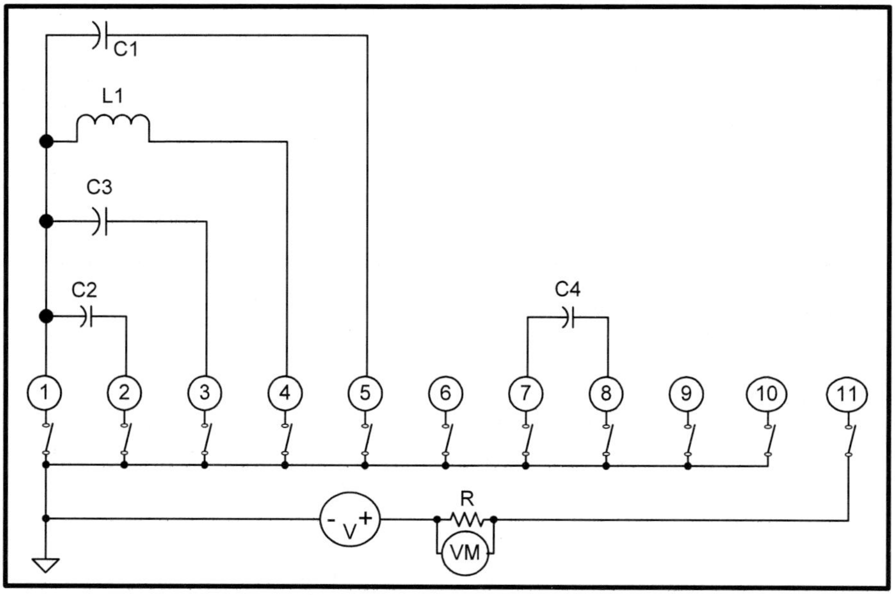

Figure 11-6: Equivalent node network for the shorts test example, set up to test node 11 versus all other nodes.

When we get to node 8 however, we now have a residual component between the stimulus and ground, the capacitor C4. When first stimulated, C4 will conduct a current that will begin to decay to zero as the capacitor charges. Knowing this will occur, we can set a small delay interval longer than the expected time constant of this circuit so as to not see the current flow as indicating a short. This similar event will occur when testing nodes 2, 3 and 5, though with differing time constants because of the different capacitances.

When we test node 4, we have a residual inductance between the stimulated node and ground. When first stimulated, no current will flow, but it will begin to ramp up. Depending on the time constant of this, we might or might not declare a short exists. We again use a delay to assure that current flow has been established before making a measurement. This means we are actually testing for an expected continuity, and thus a short from nodes 4 to (say) 1 will not be detected. If the time constant were reliably longer than the time it takes to close relays and make a measurement, then a short could be detected.

Let's say we test node 11 for current flow and we indeed see a short is indicated. What now? We need to provide better information than "Node 11 is shorted to some other node". When we detect a short, we must then *isolate* it so that a repair action becomes feasible.

When node 11 fails, we know that there is a current path to some subset of nodes in the 1 to 10 group that were grounded. We can determine this path by a process of *half-splitting* the grounded nodes. Here, we open one half of the grounded node relays, causing them to float. Current cannot pass through this half. If the short disappears, then the path must exist in the nodes that were opened. If the short does not disappear, then we haven't learned anything about either half.

Say the current flow did disappear. Then we open those nodes that were grounded because the current path does not go through those nodes. We close the originally opened half again to re-establish the current flow. Now, we have half as many nodes to consider for a new half-splitting process.

But if the current flow did not disappear, then we now have two groups of nodes that might contain a current path. We arbitrarily open one half and save it for later. We proceed with half splitting on the other group, recursively until we find the current path(s). Then we return to the first group (the second is now all open) and split it next until any paths are found. We can then report that node 11 is shorted to some other node(s) by name. Now the repair people can look for some defect that connects these nodes, typically a solder bridge between neighboring pins served by these nodes. An algorithm can search for these proximal neighbors given the nodes and positional data for each pin on those nodes to assist in the repair.

There is one hitch that can occur during isolation. Imagine in Figure 11-6 that there are two small resistors, one connecting node 10 to 9 and the other connecting node 10 to 8. Further, say the resistor values are 12 ohms. When node 10 is tested for current flow to combined nodes 1-9, there indeed is a flow equivalent to 6 ohms of connectivity (the two 12 ohm resistors in parallel). So, the half-splitting process begins. At some point, we are left with nodes 8 and 9 in a group that is then split. This causes the current flow to be halved, looking like 12 ohms, which is above the threshold. Alas, the short seems to disappear! This is called detecting a "*phantom short*". No harm is done, except we've wasted some time doing half-splitting when a real short did not exist[20]. Organizing the list of nodes by impedance can minimize this time waste. Put the nodes connected to larger resistances on the high (right) end of the list, and those connected to lower resistances near the left end. Then, when phantom searching begins, it will happen when we've gotten to the left hand side of the list, there are not many paths left to search and time is saved.

11.2.2 Unpowered Analog Tests

In 1979, Crook published a seminal work [5] on In-Circuit measurements. Little has changed since then in basic measurements involving accessible components. In the earlier 1990s, work was done to improve analog testing access via DFT that culminated in IEEE Std 1149.4 (analog Boundary-Scan) [6], [7]. In the later 1990s, McDermid published several works on how to do analog measurements when access becomes limited [8], [9], [10], see also [4]. We will review the basic workhorse technology here.

Measurement Accuracy

In-Circuit test is able to measure the value of analog components such as resistors, inductors and capacitors. These devices have nominal values specified for the design, and a tolerance on this value. For example, a resistor may have a nominal value of 4.7 Kohms, ±5%. Thus if we measure the resistor we expect it to have a value of 4.7 Kohms ±235 ohms. In a sample of these resistors, we might expect to see a truncated bell curve for the distribution of values as seen in Figure 11-7.

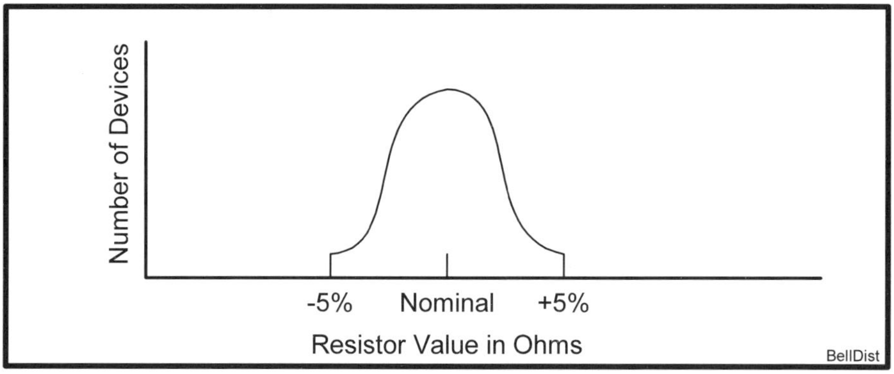

Figure 11-7: Distribution of resistance values for nominal resistors with a tolerance of ±5%.

What if we measure this resistor and see a value that is high or low by 240 ohms? Is this a failure? The answer is clouded by the fact that the measurement process itself may inject errors. It also happens that the circuit design itself could tolerate a

---

20 The act of opening or closing relays can take about 1 millisecond. The detection search involves opening one relay (the ground relay) and closing the stimulus relay. Thus we can search for shorts at roughly 500 nodes per second, or 10 seconds for a 5,000-node board. If many relay movements are needed during isolation, this time can add up, more so if phantom searches are processed.

10% deviation, but the designer only has 5% resistors available to save on inventory costs. In this case, testing for a 5% deviation could be failing perfectly functional boards. To avoid rejecting good boards, test engineers will add a guardband to the (true) tolerance on the device value. This might be an extra 1%, or it could be surprisingly large, for example, tripling the tolerance[21].

Measuring an Impedance

To measure the value of an impedance R, we make use of Ohm's law in one of the two configurations shown in Figure 11-8. Figure 11-8A shows an ideal current source forcing a known current through the impedance while a perfect voltmeter measures the voltage across the impedance. The value of R is computed by dividing the measured voltage by the known current:

$$R = V/i$$

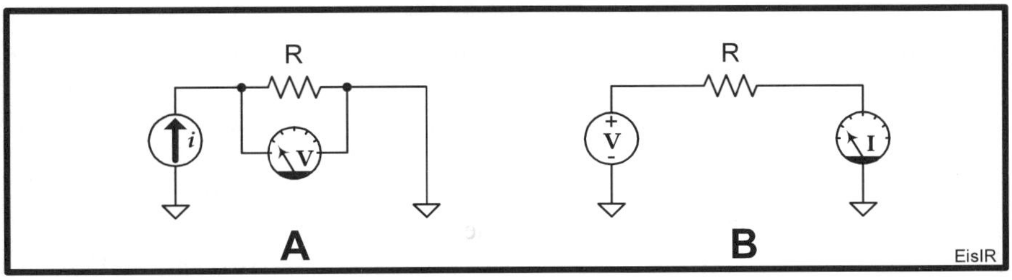

Figure 11-8: Measuring impedance (A) with current source stimulus and (B) with voltage source stimulus.

Next, we should take a moment to think about the current source. An ideal source will force a specified current, developing whatever voltage is required. However, if the device is a low-power device, it could conceivably be damaged by the power dissipation (V**i*) such a current and voltage would necessitate. Higher voltages could also damage diode junctions by causing voltage breakdown. This presents us with a problem; in order to keep the voltages in safe operating limits, we need to know an expected value (approximately) for the device being measured. But if it is truly an unknown value, then we need a compliance limit on the current source. A compliance limit is an upper bound on the voltage the source will develop and hence a bound on the both the voltage and the energy it will supply.

Whenever we use the setup in Figure 11-8A, it is assumed that the current source is not in compliance (not limited). If the device to be measured is a true unknown, the first selected current setting may produce a compliance limit signal and a different (lower) current should be tried. This process should eventually converge on a current setting that stays within the compliance limit and yet develops a measurable voltage across the resistor. However, if the current source is set too

[21] The test engineer must decide how to manipulate the tolerance based on several factors, including the "true" tolerance that the circuit requires, and how accurately the tester is able to measure the component's value. Measurement accuracy is governed by the range of the tester's detection circuitry vis-à-vis the device's value (is it near a measurement limit?) and by surrounding circuit topology that needs to be guarded (covered soon);

low, then the voltage across the resistor will be small, perhaps sacrificing some voltmeter accuracy as a result. Thus we look for a current setting that is high enough to utilize our voltmeter accuracy, but not too high to damage the device under test. (Other criteria will appear shortly.)

Another stimulus/measurement configuration is shown in Figure 11-8B. Here we use a voltage source to provide a known voltage across the resistor and a current meter to measure the resulting current. Note the ideal current meter has zero series impedance, so there is no voltage drop across it. (This means the right side of the resistor of Figure 11-8B is at zero volts.) The ideal voltage source will develop the desired voltage with whatever current is required by the circuit. This could result in a damaged resistor if the energy delivered is too large. Therefore we need a current compliance limit (an upper bound) on the voltage source current capability. Again, if the value of R is unknown, we may need to search for a voltage setting on the voltage source that does not cause a compliance condition, yet gives us good current measurement accuracy from the current meter.

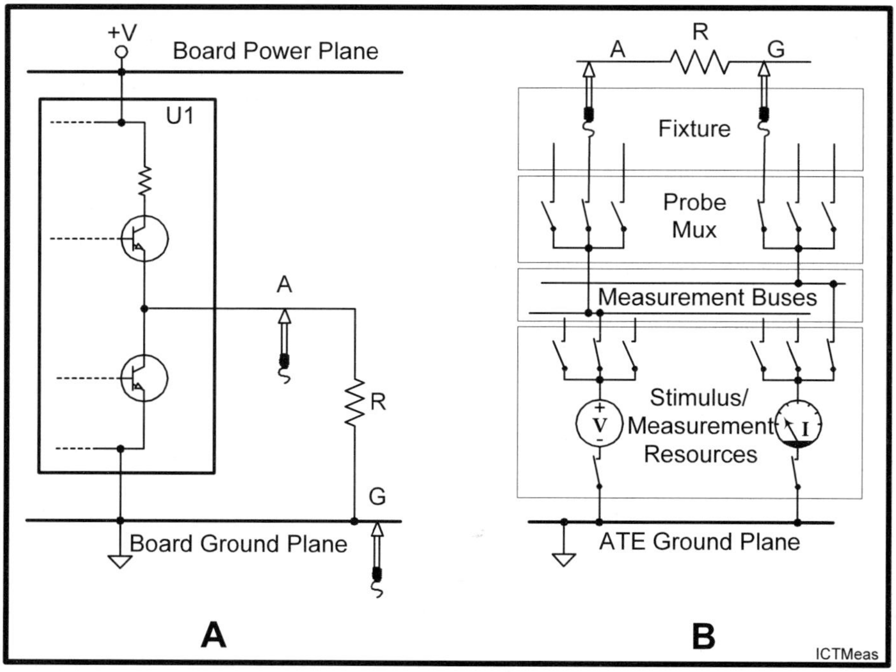

Figure 11-9: Measuring the impedance of a device on a board, (A) connected to a silicon device, and (B) as seen by an ICT system.

As we have seen it is not necessarily straightforward to measure the impedance of a simple freestanding resistor. If its value is unknown, or if there is an anomaly present such as a short or open circuit, this leads to special considerations. However, freestanding components will be the exception and not the rule when testing boards. The situation shown in Figure 11-9A will be quite common. Here, a resistor is connected between board ground and an integrated circuit output. How do we go about testing the value of the resistor?

In Figure 11-9A we have access to both sides of the resistor we want to test, but there are other components connected to the resistor as well. Inside the silicon device we see two transistors connected to the resistor. What effect will these have on our simple measurement process?

To answer this, we first remember that In-Circuit analog component testing is done with no power applied to the board. This means that the transistors inside the IC are not turned on. Further, if we limit the voltages used during testing to values that will not turn on a diode, perhaps less than 0.2 volts, then the silicon junctions within the IC will never conduct current. This makes the IC "disappear" from our problem. However, reality intrudes again when we consider the physical apparatus needed to measure the resistor. This is the In-Circuit tester itself.

Figure 11-9B shows the elements of a typical In-Circuit tester. Nails from the bed-of-nails fixture touch the nodes A and G on either side of resistor R. Within the fixture, wire-wrap wires connect the nails to the fixed array of tester channels that, in this diagram, are multiplexed to a measurement bus. The multiplexing is done with mechanical reed relays that have several desirable qualities. First, reed relays have very low "on" resistance, perhaps only 10^{-2} ohms. Second, when reed relays are open, they have very high "off" resistance, perhaps 10^{12} ohms. They come close to being "perfect" switches.

From the measurement buses (Figure 11-9B) another layer of reed relay multiplexing brings us to the stimulus and measurement resources of the In-Circuit tester. This is where we find the various forcing functions for voltage and current as well as measurement devices for current and voltage. Figure 11-9B shows how the appropriate relays are closed to set up the same voltage forcing measurement we saw in Figure 11-8B. The voltage source is set to less than 0.2 volts to prevent the silicon junctions in U1 from turning on. Next we examine complications in measurements due to parallel device topologies such as seen in Figure 11-10.

The parallel impedance problem is personified by the delta configuration of impedances shown in Figure 11-10A. Here we have three resistors and three In-Circuit nails, plus a connected IC. We can make the IC "disappear" by doing unpowered tests with low stimulus voltages as before. However, to measure the value of R when R_a and R_b are present, we need something more to deal with the parallel path problem. This is shown in Figure 11-10B.

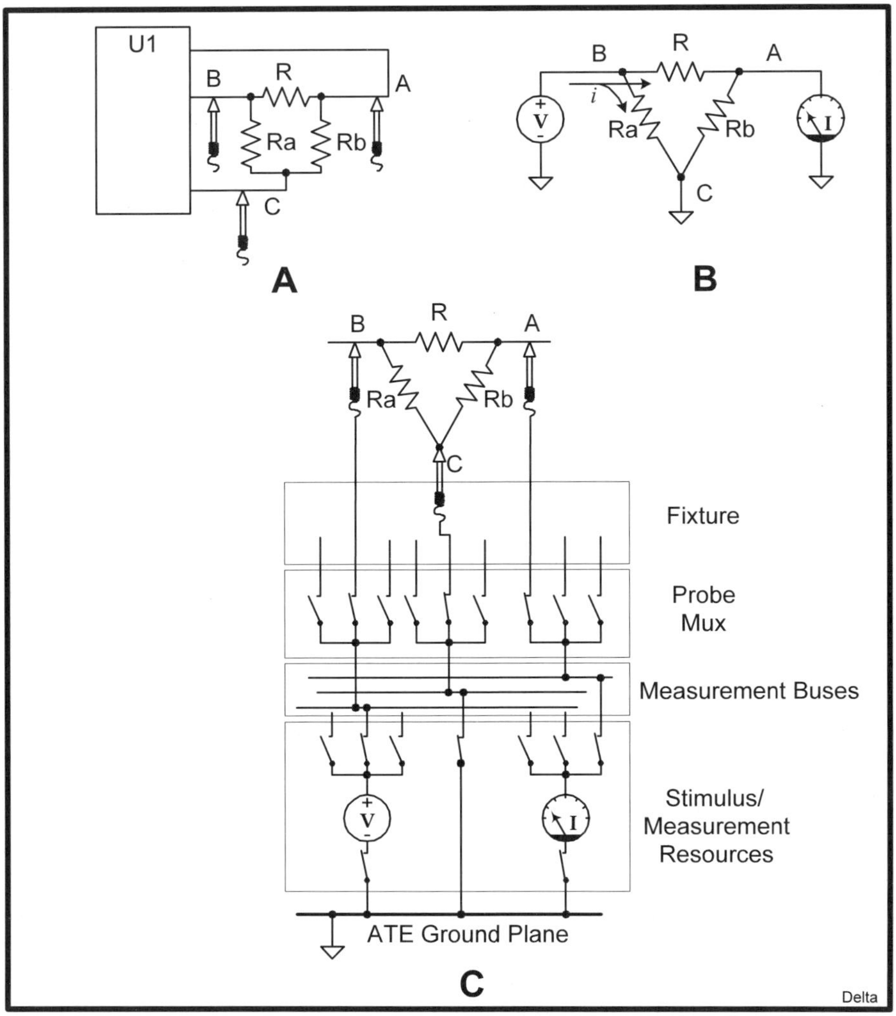

Figure 11-10: Devices may be connected into networks providing parallel pathways for currents.

In Figure 11-10B we see the familiar voltage forcing configuration on nodes B and A seen originally in Figure 11-8B, but with a new addition; a third node C is grounded using a third nail. This is called *guarding*. Guarding uses low impedance paths through reed relays to insert grounded points into the circuit. If you examine Figure 11-10B closely, you will see that current from the voltage source splits at node B and proceeds both to nodes A and C. However, because node C is grounded and because node A is also grounded (the current meter has zero impedance), the voltage across R_b is zero. No current can flow to node A from node C. This means the current meter measures only current through R. Thus we know the voltage across R (the voltage source value) and the current through it, which yields its impedance.

Figure 11-10C shows a typical ICT setup for measuring the resistor in a delta configuration[22].

When you look at the ICT setup you see a lot of path complexity and yes, this can introduce errors. Crook [5] discusses the sources of errors and how to account for them (see also [4], chapter 6).

Current meters are not usually supplied in ICT systems, but rather an operational amplifier is used as shown in Figure 11-11, where it is monitoring the same delta configuration we have used in other examples. In this configuration (often called a *Measuring Operational Amplifier* or MOA) the op-amp combined with the feedback resistor R_f will endeavor to keep node A at zero volts. The value of R is calculated by this formula:

$$R = -V * R_f / V_{out}$$

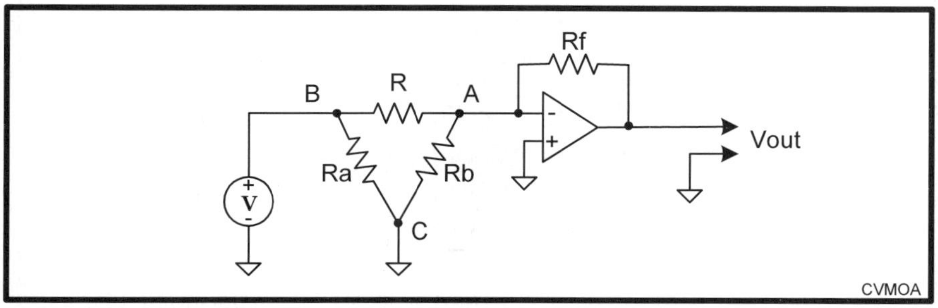

Figure 11-11: An operational amplifier with feedback resistor used as a current meter.

By measuring V_{out} and knowing the value of R_f, unknown impedance R can be determined.

All these examples have involved simple resistances. However, an ICT system can also measure the values of reactive components (inductors and capacitors) as well. The voltage stimulus source must be AC and both real and imaginary signals must be measured. (See chapter 6 in [4].) In the general case, an unknown impedance can be measured using the guarding and error correction techniques we use for simple resistances.

Finding Open Signal Pins on ICs

There is a way to test for open signal pins on ICs without applying power to the board. The most successful method within this genre (called "Unpowered Opens Test" by the test community) is Capacitive Leadframe Testing[23]. These methods have the advantage of finding open input pins without having to propagate their effects through an IC. (See the difficulty of this problem in Figure 11-1.) The

[22] The "delta" nomenclature arises from the similarity of the circuit to the uppercase Greek alphabetical character Δ (delta).

[23] Known at "TestJet", a trademark of Agilent Technologies.

powered solution to IC testing appears in the next section. The unpowered capacitive technique is examined here.

The Capacitive Lead frame technique exploits the fact that many ICs have a lead frame that forms the conductive path from the legs of the device to the die bond wire pads. Using the bed of nails, all but one node attached to the IC can be grounded and a small AC signal can be applied to the node that remains. An insulated metal plate pressed against the top of the IC forms the top plate of a capacitor and the stimulated IC leg and lead frame conductor form the bottom plate. See Figure 11-12.

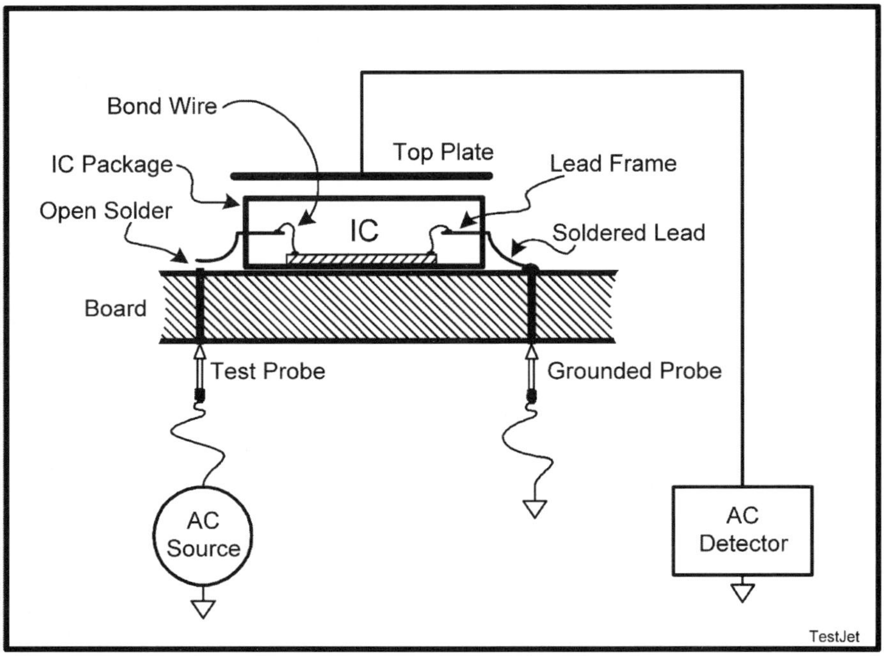

Figure 11-12: Capacitive lead frame testing used to find open solder joints on ICs (digital and/or analog).

The left and right parts of Figure 11-13 show an equivalent circuit for capacitive opens test, for a properly soldered IC lead and an open solder condition. The capacitance C_1 may be on the order of 20 to 100 femtoFarads (10^{-15} F), which is small enough to require sophisticated detection electronics to measure in the face of environmental noise. Now, if the IC leg is soldered to the stimulated board node, the correct capacitance will be measured. If the solder joint is open, then a second small capacitance C_2 now exists in series with the first. Thus a measured open capacitance is given by:

$$C_{\text{Measured Open}} = C_1 * C_2/(C_1+C_2)$$

This reduces the measured capacitance by a factor of from 2 to 10 in practice, a detectable difference. Capacitive lead frame test allows testing of complex ICs for solder opens on signal pins without knowing what the ICs actually do and without applying power. The technique requires no complicated programming, and gives accurate resolution of solder defects. The technique has been extended to allow testing of solder integrity for connectors and switches.

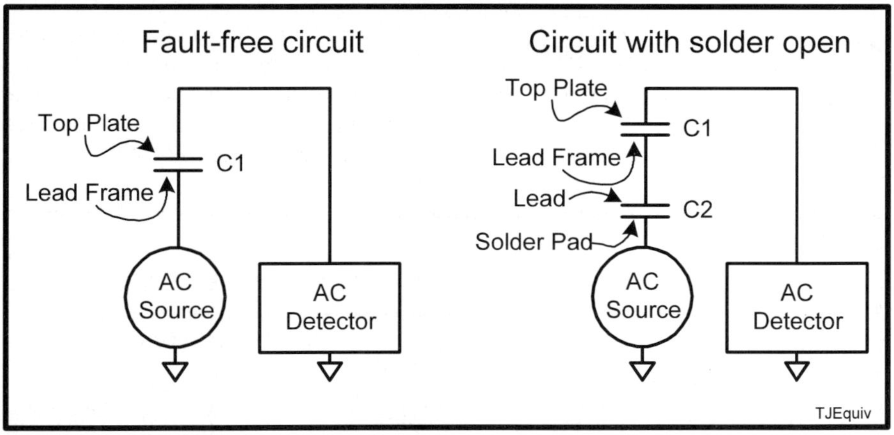

Figure 11-13: Equivalent circuits for fault-free and open solder conditions.

11.2.3 Powered In-Circuit Digital Tests

Digital devices, typically individual ICs, can also be tested with In-Circuit techniques[24]. While analog measurements implied an Analog Stimulus and Response capability, a Digital Test Sequencer is needed for applying digital data inputs and monitoring digital responses from ICs. Many such resources must be coordinated to run in parallel. Actual drivers and receiver/comparator circuits will need programmable analog parameters such as drive high/low voltages, slew rates, receiver high/low comparison windows, etc. The digital subsystem will require substantial amounts of memory to store digital stimulus and responses, and some sort of "vector sequencer" to apply all these in the proper order in time. ICT system can rarely achieve "natural" data application rates since the bed-of-nails fixture is basically a low-pass filter. Digital tests are applied and monitored with cycle rates typically less than 10 MHz, even though the devices being tested are usually far faster. Remember the PCOLA/SOQ model of defects that are tested; these defects are rarely dependent on "at-speed" testing. This is to say we do not expect to detect speed-related defects deep inside an IC[25] at In-Circuit board test. A model of an ICT test of a digital IC is shown in Figure 11-14.

---

[24] While the IC is on the board, connected to other components.

[25] Indeed, these are challenging to find during pins-free IC testing. If they escape both IC and board test, then one hopes to find them at functional and system test stages.

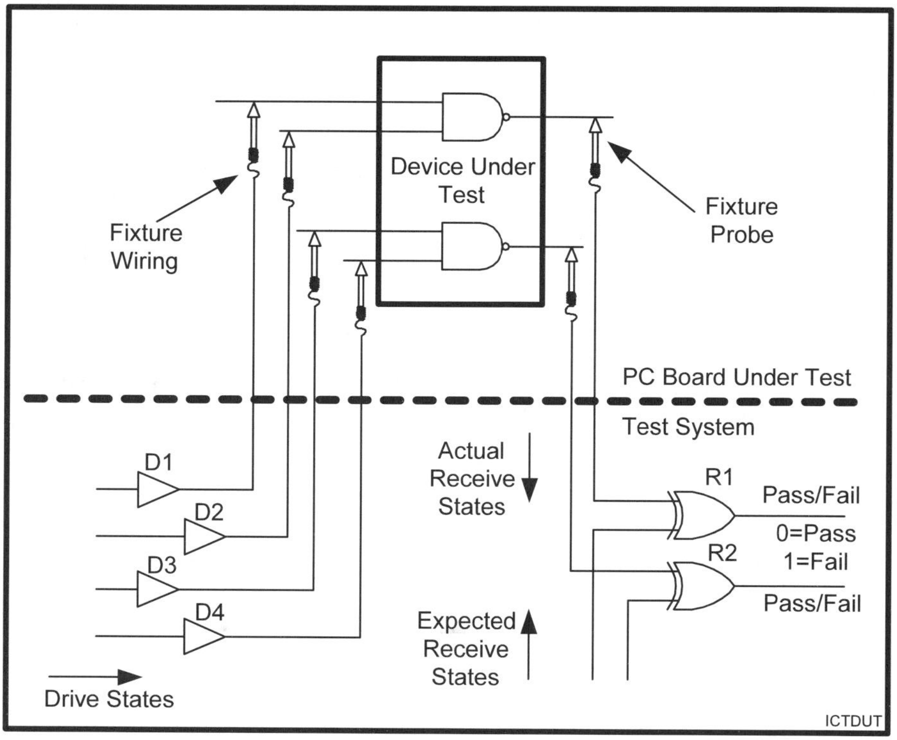

Figure 11-14: In-Circuit digital test setup with full nodal access.

Digital tests for individual devices are prepared beforehand and stored in libraries from which they can be called up as needed (see Figure 11-15). This alleviates the programming process, if the tests you need are indeed found there. This is quite different than Automatic Test Program Generation you often hear about for IC tests, where various DFT methodologies are used to reduce a complex IC to a collection of tractable combinational circuits that can be analyzed. Rather, a person must generate each library from a (sometimes incomplete) data sheet. The goal is to "wiggle all the pins" and not to prove each internal gate is working correctly. Thus, digital ATPG at board test is the process of matching devices found on a board to existing library tests. But there is more to consider.

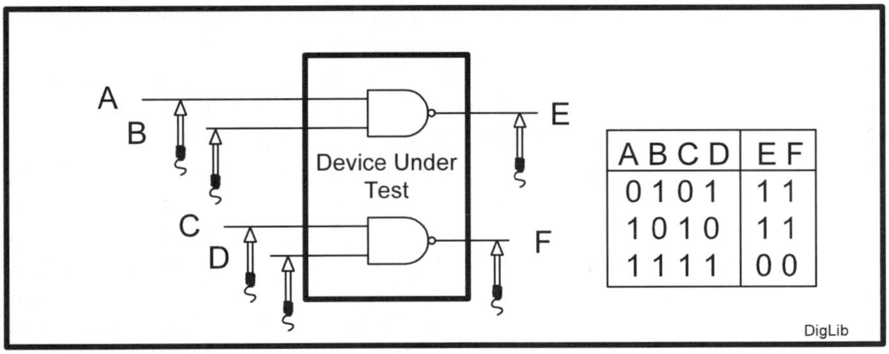

Figure 11-15: Test vectors for a given IC are called up from a manually generated library test.

One paradigm shift over digital testing that occurs in IC manufacturing is the idea that a digital device is tested on a board *while connected to other devices* on that board. During IC manufacturing, the IC stands in isolation, what ICT engineers call the "pins free" case. They rarely get to do such testing on boards. Thus, an existing test library may be defeated by board level topologies. See Figure 11-16.

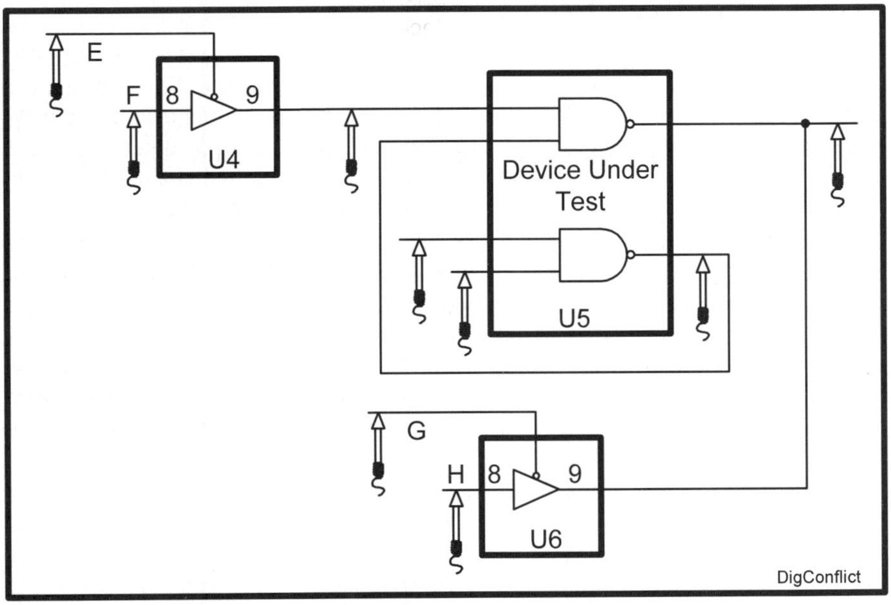

Figure 11-16: Surrounding devices and board topology can cause problems with digital test libraries.

Figure 11-16 illustrates three problems. First, when attempting to drive an input of the device we are testing (U5 in Figure 11-16), we may be *backdriving*[26] an

[26] A synonym for backdrive is "overdrive".

upstream device. In this case it is device U4, pin 9. Our tester should be equipped with relatively high-current drivers to achieve this. However, it begs the question of whether we are doing any damage to U4 in the process. This was studied intensively in the mid-1980s [11], [12] and it is known that backdrive time must be carefully controlled to eliminate heating effects in overdriven upstream drivers.

The second problem that arises is when multiple outputs are connected to the same node (a usual case in bus drivers, 3-state devices, etc). In this case, U6 pin 9 is connected to the output of the device we are testing. If U6 is driving out, then we have it conflicting with data we want to monitor during a test. Both the backdrive and output conflicts can be solved by *conditioning* drivers, usually by turning them off or putting them into an overdrivable state. In this example, both U4 and U6 have disable pins that can be added to the test such that they turn off conflicting drivers.

The third problem comes from board topology. In the example of Figure 11-16, one DUT output is fed back to a DUT input. In this case, the driver specified to provide data to this input will conflict with data we want to monitor coming out of the device. Some software tricks can be played to try to eliminate conflicting vectors, but this means we lose a bit of coverage, and it cannot be done reliably with sequential devices (and most DUTs are sequential). As you see, "ATPG" in the digital board test world is quite different than the IC test world.

11.2.4 Boundary-Scan Tests

Boundary-Scan, formally IEEE Std 1149.1, is a set of DFT rules that are applied to the design of a digital IC. When these rules are followed, board testing software can take advantage of the 1149.1 embedded feature set to facilitate board testing. Boundary-Scan nicely supports testing for the PCOLA/SOQ defect universe. Boundary-Scan devices can be used singly or in groups (called chains) to support test activities. Boundary-Scan tests require power to be applied to the board.

The most prominent contribution of Boundary-Scan is it allows boards to be better tested when probe access is compromised by board layout density, or, by the sheer size of the board overwhelming the resource set of the tester. In the theoretical limit, only 4 or 5 nodes (the board Test Access Port – TAP) would need to be accessed to test a board. This is never achieved in practice as boards almost always have many analog components and digital devices that do not contain Boundary-Scan. However, Boundary-Scan can make a significant contribution as a "force multiplier" for In-Circuit Testing.

Figure 11-17 shows a digital IC that has Boundary-Scan facilities included. A Test Access Port (4 or 5 signals; signal TRST is optional) drive a TAP Controller (a state machine) that in turn manipulates several added registers. There is a TAP Instruction register and some data registers (Bypass, ID Code and Boundary). The instruction register content defines a basic operation the device will perform in test mode. When Boundary-Scan is not in use, the System Circuitry will perform it normal function.

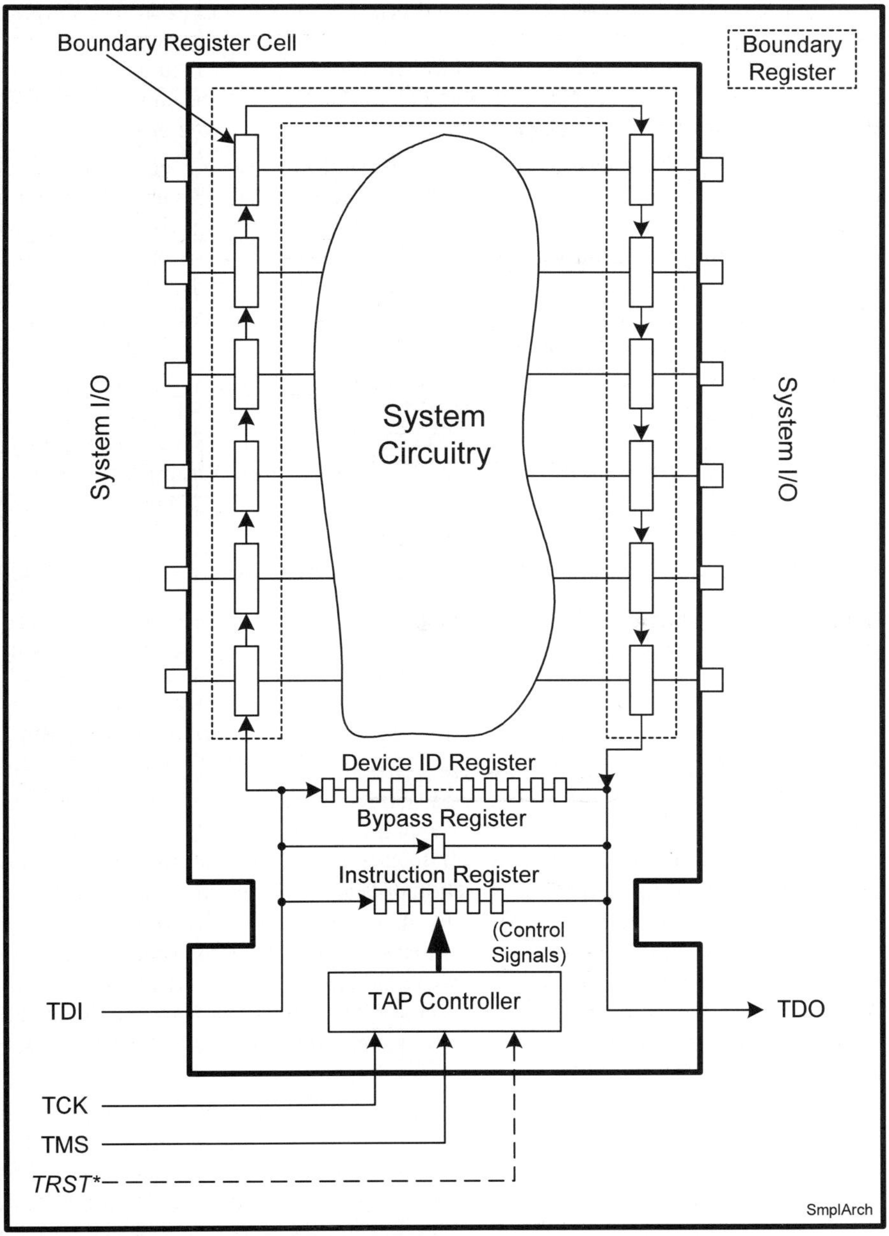

Figure 11-17: Boundary-Scan DFT additions to an IC.

In test mode when a test instruction (EXTEST) is operative, the Boundary Register takes over control of the I/O pins and the system circuitry is isolated. EXTEST can monitor input pins and produce states (0, 1 or Hi-Z) on outputs and bidirectional pins. Data can be shifted in/out of the device to read out captured input states, and produce new driven states on outputs. A chain of Boundary-Scan devices is shown in Figure 11-18.

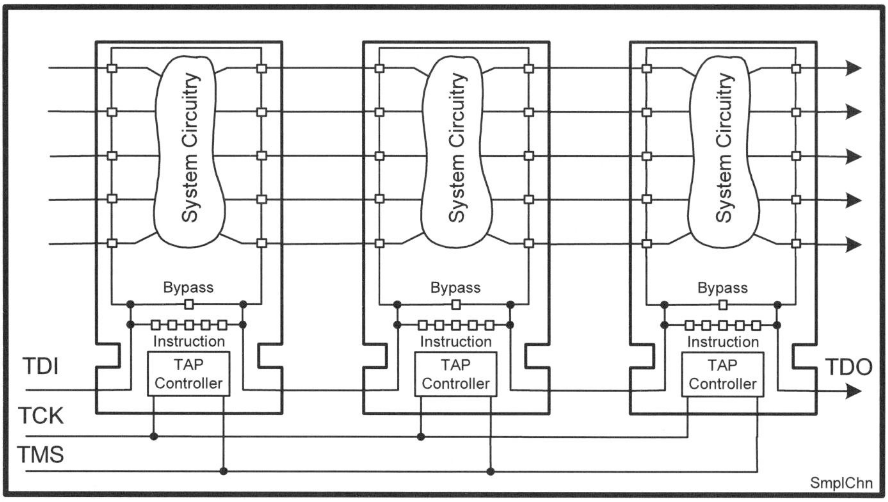

Figure 11-18: Multiple Boundary-Scan devices chained together are able to coordinate testing of their interconnections for shorts and opens.

As a quick example of how Boundary-Scan can be used to detect defects, consider the circuit of Figure 11-19, where a chain of four interconnected Boundary-Scan devices are being used to test their interconnections. Devices U1 and U3 control nets A, B, C and D. Devices U2 and U4 monitor the data values on these same nets. Each net is assigned a unique "ID code" of binary bits that are sequentially transmitted by U1 and U3. These bits are expected to make it across the interconnections to be serially received by devices U2 and U4. However, defects such as opens or shorts may prevent data from being properly received. In this example, a short across nets B and C damages the data (here it is wired-ORed) such that the received data for both nodes B and C are incorrect and also identical. Diagnostic routines can then determine the likelihood that these two nodes are shorted.

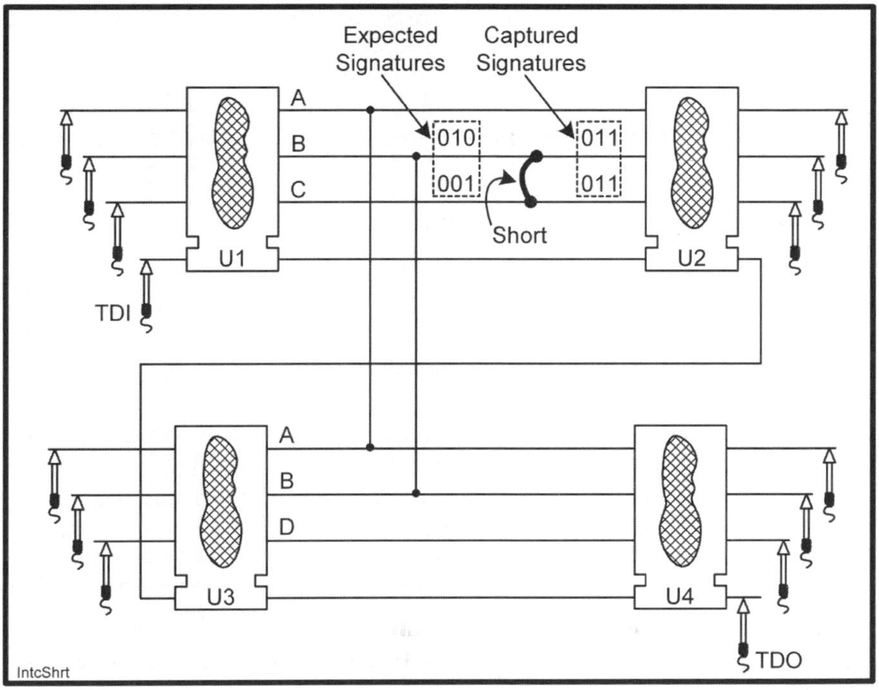

Figure 11-19: Example of Boundary-Scan signals generated by IC drivers and received by IC receivers, detecting a short.

Many papers have been written over the last 15 or so years on Boundary-Scan test techniques. See the proceedings of the International Test Conference during this period and also [4] for a discussion of how Boundary-Scan can be used to find shorts and opens. Reference [13] specifically discusses how Boundary-Scan provides board test coverage.

11.2.5 Powered Mixed-Signal Tests

An In-Circuit tester typically has both a digital test subsystem and a separate subsystem for analog tests. If these can be run in a coordinated fashion, then mixed-signal tests can be executed.

For example, imagine a board with a digital-to-analog converter (DAC) IC on it. The digital subsystem could stimulate the DAC with digital data that represents a sine wave. The analog subsystem could measure the frequency of this analog signal for the correct value, and maybe other parameters such as distortion. This would prove the DAC was basically alive and could provide other information about its functionality.

Mixed-signal tests applied at Board Testing are usually ad-hoc, limited in scope only by the capabilities of the tester and test programmer. They may be challenging to implement and maintain. Care should be taken to ask, "what is this test actually testing" with respect to the PCOLA/SOQ model. In some cases, the test may be of marginal value if the defects it uniquely covers are few or unlikely. For example, a mixed signal test might verify that a discrete transistor is working (live), present and properly connected, but the unpowered test for that same transistor would provide the same information. Thus the mixed signal test is redundant. It *would* give some functional feel for the transistor's performance, but this in general is not the goal of In-Circuit Test.

11.2.6 Pros and Cons of ICT

As with any other tool, In-Circuit testers have strong and weak points. With respect to the PCOLA/SOQ model, ICT tends to do well with Presence, Correctness, Orientation, Liveness, Shorts and Opens. ICT is essentially useless for device Alignment and joint Quality. These require some sort of inspection, covered in the next section.

There are weak spots in test coverage for ICT. Many boards will have hundreds of power supply bypass capacitors on them. One may see one very large capacitor (say, 10,000 μF) in parallel with several smaller (say, 100 μF) capacitors, also in parallel with hundreds of small, RF-quality capacitors. Since all of these capacitors are in parallel, the total capacitance is dominated by the large capacitor. Since most large capacitors have a large tolerance range (say, +/- 20%), a measurement will not be able to determine if one of the smaller capacitors is missing. Therefore in practice, only the very large capacitor can be "seen" and all the rest have poor coverage (essentially, only for shorted pins).

Another classical weak spot in coverage is in finding opens on redundant power and ground pins on ICs (see [2]). Most ICs today have these, but since they often are strapped together inside the IC, there are redundant paths for current to follow. Opens on these pins may not affect the results of a slow-speed board-level test, but the high-speed performance of the system may be compromised[27].

One of In-Circuit Testing's stronger points is that it *electrically* tests components (either with or without power applied). Inspection techniques can determine that a device *appears* to be present and possibly is correct, but it could be dead-as-a-stone and inspection cannot see that.

Aspects of In-Circuit testing can be challenging to accomplish. Getting a large digital device test to work may be difficult, especially if no library test exists and one has to be created from scratch. Board topologies may introduce constraints that are troublesome. Bed-of-nails fixturing for large, dense boards may also be a daunting task that consumes valuable time and energy. Test fixtures must be maintained, stored and shipped (they may weigh 100 pounds or more).

Preparing an ICT test takes some knowledge of electrical systems, and detailed knowledge of the board's design. Therefore, there may be worries about the security of a design and potential Intellectual Property (IP) problems when test development is done somewhere else.

---

[27] This may not be evident at system test, for example, if the defect-free power or ground pin is needed to assure performance margin over a range of environmental and power supply conditions.

In-Circuit testers have a cost roughly proportional to the size of the board being tested. Small boards may require a few pin electronics resources while a large board may require many resources. This can generate portability problems when you try to move testing to another site with similar ICT systems but with alternative configurations.

Sometimes, extra value can be added to a product while it sits on an In-Circuit tester. For example, some ICT systems can be used to download bit patterns into PLDs or flash RAM devices, once you are confident that no defects are present.

Access limitations have threatened the viability of ICT over the years. Improved DFT has been used to rectify this (e.g., [3], [7], [14]). The fact that people will strive to solve access problems to maintain ICT capabilities speaks to the value the ICT provides. Basically, ICT is well understood, trusted and valued. Test engineers fear the day that ICT becomes unusable.

11.3 LOADED BOARD INSPECTION SYSTEMS

Inspection is an "old" art, in that humans have inspected PC boards for many years (and still do). Human eyes combined with the unsurpassed recognition capability of the brain can find "obvious" defects that defeat sophisticated computer algorithms. However, humans tire and err, have low throughput rates and there are poor repeatability of results from person to person, or even from time to time with the same person. Thus, efforts to automate inspection continue today unabated.

All inspection systems work by illuminating an area of interest, focusing and capturing an image, and processing images to reduce them to qualitative and quantitative factors that can be judged against certain criteria. The major branches of automated inspection systems are delineated by the wavelength of operation: visible versus X-Ray illumination. Within the X-Ray branch, there are two subcategories: two-dimensional (2D) transmission systems and three-dimensional (3D) tomography systems.

One important difference about inspections systems, compared to ICT, is that they need far less information for programming. Essentially, they can be programmed with a list of component designators, their reference numbers, and XY positional data[28]. No netlist is required, nor any knowledge of a given component's electrical behavior. Indeed, a board can be programmed when all that is available is a mechanical sample, containing only dummy components. It is not uncommon to see an inspection program be developed in several hours, rather than in multiple days. And, no test fixture is needed. The cost of an inspection system is largely independent of the size of a board being tested.

---

[28] Do note that each side of a board must be treated as an independent problem. Thus a two-sided board will require two programming steps, and must be inspected "twice", flipping it over between passes.

11.3.1 Automatic Optical Inspection (AOI)

Automated Optical Inspection (AOI) systems utilize non-penetrating radiation to illuminate a board (Figure 11-20). Some utilize colors of light as part of their feature set, to adjust the contrast in an image. Non-penetrating radiation will reflect off exposed surfaces, so any feature not directly illuminated will not be visible (or testable). Devices such as Ball Grid Arrays, where some features of interest (the solder balls) are sandwiched between the device and board, are less testable than those such as in-line pinned packages.

Some AOI systems utilize multiple cameras and light sources mounted at varying angles to the board being tested. Flat surfaces will tend to reflect a lot of light while curved surfaces will tend to scatter light. Inspecting a surface from several angles can help gather additional information about a device or connection being examined.

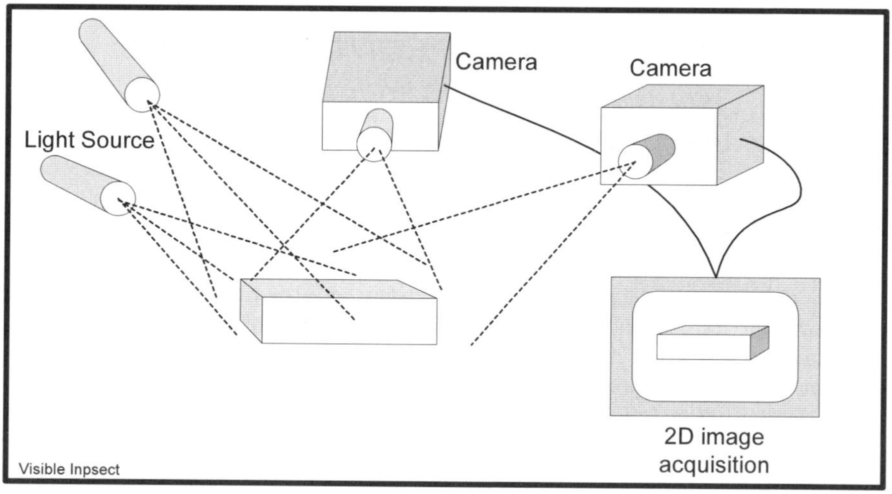

Figure 11-20: AOI system light source(s), camera(s) and image acquisition subsystems. Not shown are board handling/positioning mechanisms.

Device-related defects (PCOLA) such as those shown in Figure 11-21 are more testable. In the previous discussion about the untestability of bypass capacitors via In-Circuit test, we saw how potentially hundreds of devices were nearly "invisible" to ICT. If the device in Figure 11-21 is a bypass capacitor, then an optical system can easily see if it is present, correct (if the ID number can be imaged), the orientation, if important, is indicated (Figure 11-21) by a polarity stripe, and various types of alignment problems can be discerned. However, liveness cannot be tested by an optical inspection. If we inspected a board with bypass capacitors as well as using ICT, we could expect much higher total board test coverage as each technique's strength might augment another's weakness.

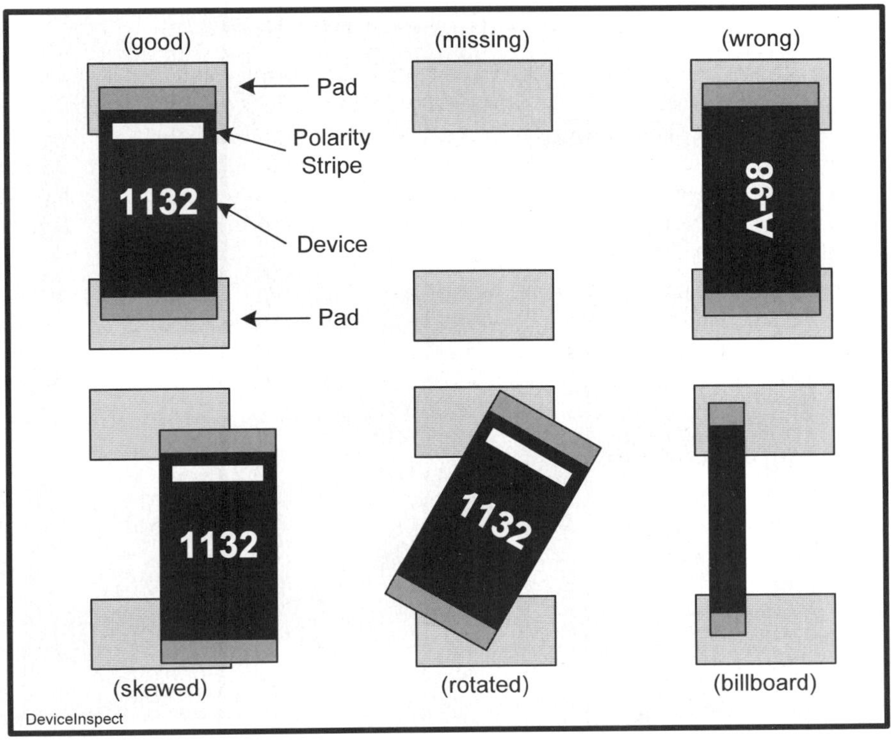

Figure 11-21: Device defects potentially tested by AOI.

Figure 11-22 shows a processed image of a region of a board where several defects are seen, a missing resistor, a broken resistor and one nicely placed resistor.

Inspection systems are also used in other places in the manufacturing line. For example, AOI systems are used to inspect solder paste on boards before devices are mounted. A missing or malformed paste brick can later become an open pin or a pin with insufficient or excess solder. AOI can also be used to verify parts placement before reflow to examine the placement process. This can be done between the placement of small, inexpensive components and before the placement of expensive, large components. This allows us to look for placement defects on the small components and solder paste defects that will be critical for the large components, before those components are committed to reflow.

Figure 11-22: Image of a missing resistor, a broken resistor and a well-placed resistor.

11.3.2 Automatic X-Ray Inspection (AXI)

X-Ray inspection systems use penetrating radiation to "see through" less dense objects, like copper traces, board materials, and integrated circuits in order to see denser material like solder. There are two forms of X-Ray inspection systems. The simpler type is based on two-dimensional transmission technology, where an X-Ray source on one side of a board illuminates an area. On the other side, an X-Ray detector converts X-radiation to visible light where a camera can capture an image as in Figure 11-23. All items on a board between the source and the detector contribute to the image.

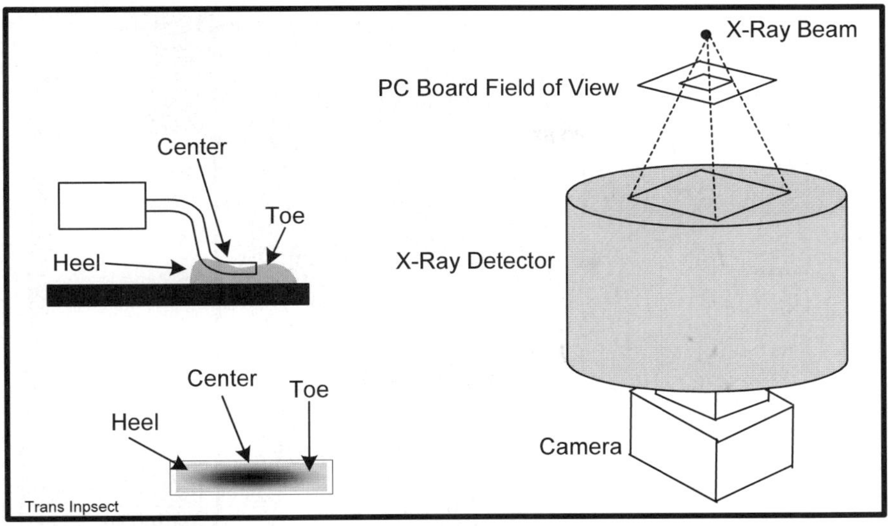

Figure 11-23: Transmission X-Ray system. X-Rays pass through a sector of a board and are converted into light by a detector. A camera records a visible image as a series of gray shades.

Two-dimensional images suffer degradation of resolution when components (and solder) are mounted on both sides of a board and there is overlap of their images. This can be addressed with three-dimensional X-Ray technology, at the cost of more mechanical complexity. Borrowing from the medical profession where three-dimensional imaging has been well adopted, a 3D imager can be constructed by moving both the X-Ray source and the detector such that items in a tight focal plane are imaged, while other objects outside the focal plane are de-emphasized. This is shown in Figure 11-24.

There, the X-Ray source is rotated, and the detector on the opposite side of the board is also rotated in synchrony. The board (with sides A and B) sits in the focal plane where objects on side A will image in the same place on the detector as the system rotates. Objects on side B are not in the focal plane and will tend to rotate such that their images are "smeared" across a large area rather than tightly focused. Their images merge into the background noise while objects on side A are resolved. Thus, a cross section in the Z-axis is possible. This can be taken to the logical conclusion of actually sectioning a solder joint and looking for the "signature" of a good joint. For a given type of pin connection, solder will wick up and flow as governed by surface tension, forming an expected shape. Typically there is a "heel" and "toe" that can be identified. Three-dimensional systems can make quantitative measurements of joints to determine the actual volume of solder and how it flowed. Thus, insufficient or excess solder can be found, and other qualitative problems such as solder voids and improper wetting.

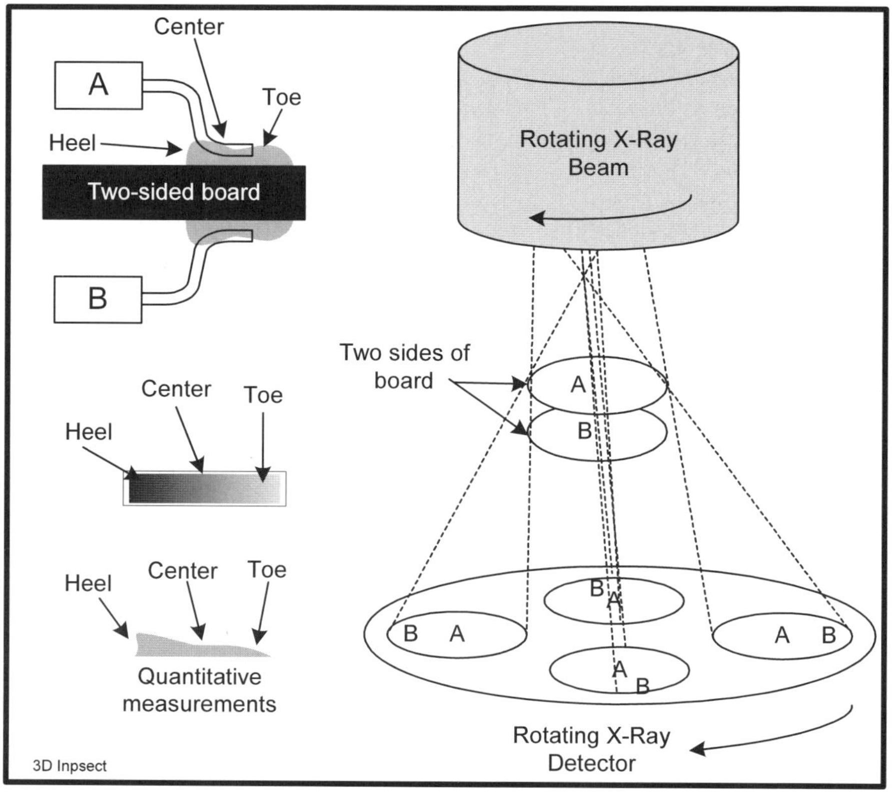

Figure 11-24: Three-dimensional X-Ray image is able to resolve a section perpendicular to the Z-axis.

Three-dimensional systems are mechanically complex in comparison with simple 2D transmission systems. Thus they are more expensive and also have somewhat lower throughput (inspections per second).

11.3.3 Pros and Cons of Inspection

The inspection process has the ability to find defects that are completely invisible to electrical tests done with ICT systems. These are "Alignment" problems with devices, and solder "Quality" issues of connections. Yet, AOI and AXI systems are completely unable to determine the "Liveness" of devices. Other areas of the PCOLA/SOQ model may have unique contributions made by each type of tester. For example, an ICT system might get a partial detection score for "Correct" when measuring a resistor. An AOI system could (in principle) actually identify the resistor by matching it to the expected image (literally, read the color code or serial number). Doing both tests would give high confidence that the resistor was really correct, live, and properly attached.

Imaging systems do take expertise in their programming and setup in order to correctly differentiate a real defect from an acceptable deviation. As such, their false call rate (false pass, false fail) may be higher than seen with ICT systems. Inspection of solder joints is necessarily a serial process, so that the rate in joints per second may be significantly lower than ICT, especially when ICT is using a DFT technology such as Boundary-Scan, where many thousands of joints are tested for opens and shorts in parallel. Again, the ICT will not be testing the quality of those joints. As such, ICT will not give as good process feedback on the soldering process.

Inspection systems take less time to program, require much less data, no electrical engineering knowledge and no bulky, expensive test fixture. AOI systems are capable of keeping up with the "beat rate" of a manufacturing line. AXI systems are somewhat slower and may be challenged in this area. Using an inspection system without first doing an intensive "test strategy" study is relatively risk-free, whereas, ICT will benefit by asking up front, "what do I need to test and how?" While all systems generate a DFT lore, DFT needed for inspections systems is considerably shorter and less complicated.

Inspection systems may run unattended, until a potential defect is detected. Then an operator must come over to the system and make a judgment on whether there really is a defect. ICT results are usually trusted on their face. Inspections systems cannot judge the electrical performance or basic liveness of a circuit. This only comes from ICT or downstream functional or system tests. Inspection systems can be used in several places in manufacturing, giving more immediate feedback on solder paste, placement and reflow processes. ICT is only performed at the end (post-reflow) of the manufacturing process and may not give as timely and clear process feedback.

11.4 THE FUTURE OF BOARD TEST

Board testing is continually challenged by industry trends towards increasing board density, lower costs, and process changes (such as the move to lead-free solders imposed by environmental concerns). The increases in density are of particular concern: it makes In-Circuit Test access, via fixtured probes, more difficult. It reduces the features sizes of solder joints viewed by X-Rays. And more connections are hidden from view by area-array connections such as ball-grid arrays technology. This tends to increase the cost of board testing, which moves in the wrong direction as products are expected to sell for less each year.

A recently introduced concept of "Bead Probes" [15] has promise to re-invigorate In-Circuit test. Briefly, the concept is to place a tiny solder bead on top of a trace, no wider than the trace and only 2 or 3 mils tall. This bead replaces the oversized test point target usually placed on a trace, so there is no need to re-route the board in that area to make room for it. Plus, it has negligible effect on the high-frequency performance of the board. This bead is still probed, by a flat-faced "target probe" in the fixture. This inverts the conventional fixture paradigm, where the old "hit a flat target on the board with a sharp pointed nail" is replaced by "hit a flat target in the fixture with a pointed solder structure (the bead) on the board". When the target probe hits the solder structure, a small, calculated amount of deformation of the bead creates good circuit contact needed for In-Circuit test.

Other people are striving to re-use testability structures inside ICs for the purpose of board test. One example is provided by Schuttert et al [16] where mixed signal test facilities inside an IC are co-opted for use at board test. This re-use of embedded IC test facilities may help "fill in the blanks" in test coverage at board test.

REFERENCES

[1] "Test Coverage: What does it mean when a board test passes", K. Hird, K. P. Parker and B. Follis, Proceedings, International Test Conference, 2002, pages 1066-1074.

[2] "Opens Board Test Coverage: When is 99% Really 40%", M. V. Tegethoff, K. P. Parker, K. Lee, Proceedings, International Test Conference, 1996, pages 333-339.

[3] "IEEE Standard Test Access Port and Boundary-Scan Architecture", IEEE Standard 1149.1 2001, IEEE Standards Board, 345 East 47th St. New York NY 10017, 2001.

[4] "The Boundary-Scan Handbook", 3rd Edition, Kenneth P. Parker, Kluwer Academic Publishers, Norwell Mass, 2003.

[5] "Analog In-Circuit Component Measurements: Problems and Solutions", D. T. Crook, *Hewlett-Packard Journal*, vol 30, No. 3, March 1979.

[6] "Structure and Metrology for an Analog Testability Bus", K. P. Parker, J. E. McDermid, S. Oresjo, *Proceedings, International Test Conference*, pp 309-322, Baltimore MD, Oct 1993.

[7] "IEEE Standard for a Mixed Signal Test Bus", *IEEE Standard 1149.4-1999, IEEE Standards Board*, 345 East 47th St. New York NY 10017, 1999.

[8] "Limited Access Testing: Ability and Requirements", J. McDermid, *Proceedings, NEPCON 1998*, Anaheim CA, Feb 1998.

[9] "Solving Limited Access Constraints in ICT", J. McDermid, *Electronics Engineer*, July 1998.

[10] "Limited Access Testing: 1149.4 Instrumentation and Methods", J. McDermid, *Proceedings, International Test Conference*, Washington DC, Oct 1998.

[11] "Measuring Thermal Rises Due to Digital Device Overdriving", G. S. Bushanam et al, *Proceedings, International Test Conference*, pp. 400-407, Philadelphia PA, Oct 1984.

[12] "SAFEGUARD In-Circuit, a Description of HP's Safe Implementation of Digital In-Circuit Test", *Product Note 3065-2*, Hewlett-Packard Manufacturing Test Division, Loveland CO, Sept 1985.

[13] "Defect Coverage of Boundary-Scan Tests: What does it mean when a Boundary-Scan test passes?", K. P. Parker, Proceedings, International Test Conference, pp.1268-1276, Charlotte NC, Sept 2003.

[14] "IEEE Standard for Boundary-Scan Testing of Advanced Digital Networks", *IEEE Standard 1149.6-2003, IEEE Standards Board*, 345 East 47th St. New York NY, 10017, 2003.

[15] "A New Probing Technique for High-Speed/High-Density Printed Circuit Boards", K. P. Parker, Proceedings, International Test Conference, pp. 365-374, Charlotte NC, Oct 2004.

[16] "On-chip Mixed-Signal Test Structures Re-used for Board Test", R. Schuttert, D. C. L. van Geest and A. Kumar, Proceedings, International Test Conference, pp. 375-383, Charlotte NC, Oct 2004.

Index